T0257674

Fundamentals of Wave Phenomena

The Mario Boella Series on Electromagnetism in Information & Communication

The Mario Boella series offers textbooks and monographs in all areas of radio science, with a special emphasis on the applications of electromagnetism to information and communication technologies. The series is scientifically and financially sponsored by the Istituto Superiore Mario Boella affiliated with the Politecnico di Torino, Italy, and is scientifically co-sponsored by the International Union of Radio Science (URSI). It is named to honor the memory of Professor Mario Boella of the Politecnico di Torino, who was a pioneer in the development of electronics and telecommunications in Italy for half a century, and a Vice President of URSI from 1966 to 1969.

Fundamentals of Wave Phenomena, Second Edition
Akira Hirose and Karl E. Lonngren

This volume is the first textbook produced in the series. It is a much expanded and revised edition of the original work that was published a quarter century ago. Written by two illustrious scientists with decades of teaching and research experience, it provides the ideal blueprint for a university course at the advanced undergraduate or first-year graduate levels. Since wave phenomena are of great importance in all areas of pure and applied natural sciences and engineering, this interdisciplinary textbook should appeal to students in such varied areas as physics, biomedical, electrical and mechanical engineering, and applied mathematics.

Piergiorgio L. E. Uslenghi – Series Editor
Chicago, March 2010

Forthcoming Series Books

The Wiener-Hopf Method in Electromagnetism
by Vito G. Daniele and Rodolfo S. Zich (2011)

Transmission Lines for Communication and Power
by Piergiorgio L. E. Uslenghi (2011)

Advanced Higher-order Methods for Computational Electromagnetics
by Roberto D. Graglia (2011)

Fundamentals of Wave Phenomena
Second Edition

Akira Hirose
University of Saskatchewan

Karl E. Lonngren
University of Iowa

SCITECH
PUBLISHING, INC.

Raleigh, NC
scitechpub.com

Published by SciTech Publishing, Inc.
911 Paverstone Drive, Suite B
Raleigh, NC 27615
(919) 847-2434, fax (919) 847-2568
scitechpublishing.com

Copyright © 2010 by SciTech Publishing, Raleigh, NC. All rights reserved.

No part of this publication may be reproduced, stored in a retrieval system or transmitted in any form or by any means, electronic, mechanical, photocopying, recording, scanning or otherwise, except as permitted under Sections 107 or 108 of the 1976 United Stated Copyright Act, without either the prior written permission of the Publisher, or authorization through payment of the appropriate per-copy fee to the Copyright Clearance Center, 222 Rosewood Drive, Danvers, MA 01923, (978) 750-8400, fax (978) 646-8600, or on the web at copyright.com. Requests to the Publisher for permission should be addressed to the Publisher, SciTech Publishing, Inc., 911 Paverstone Drive, Suite B, Raleigh, NC 27615, (919) 847-2434, fax (919) 847-2568, or email editor@scitechpub.com.

The publisher and the author make no representations or warranties with respect to the accuracy or completeness of the contents of this work and specifically disclaim all warranties, including without limitation warranties of fitness for a particular purpose.

Editor: Dudley R. Kay
Editorial Assistant: Katie Janelle
Production Manager: Robert Lawless
Typesetting: MPS Limited, A Macmillan Company
Cover Design: Brent Beckley
Printer: Sheridan Books, Inc., Chelsea, MI

Cover image, *Flame* by Clark Little and used by permission of Clark Little Photography.

This book is available at special quantity discounts to use as premiums and sales promotions, or for use in corporate training programs. For more information and quotes, please contact the publisher.

Printed in the United States of America
10 9 8 7 6 5 4 3 2

ISBN: 978-1-891121-92-0

Library of Congress Cataloging-in-Publication Data
Hirose, Akira, 1941-
  Fundamentals of wave phenomena / Akira Hirose, Karl E. Lonngren. – 2nd ed.
    p. cm. – (The Mario Boella series on electromagnetism in information & communication)
  Rev. ed. of: Introduction to wave phenomena. 1985.
  Includes bibliographical references and index.
    ISBN 978-1-891121-92-0 (hardcover : alk. paper) 1. Waves. 2. Wave-motion, Theory of. I. Lonngren, Karl E. (Karl Erik), 1938- II. Hirose, Akira, 1941- Introduction to wave phenomena. III. Title.
  QC157.H56 2010
    531'.1133–dc22   2010009089

In honor of our parents

Genji
Katsuyo
Bruno
Edith

Contents

Preface

Traditionally, the subjects dealing with wave phenomena have been taught in different disciplines. Indeed, sound waves and electromagnetic waves are entirely different physical phenomena. However, once one realizes that all waves, be they mechanical or electromagnetic, carry both energy and momentum as they propagate and that the observable quantities are necessarily associated with energy propagation, then one can formulate a unified understanding of wave phenomena. To this end, we have produced a textbook that elucidates the general properties of wave phenomena, both linear and nonlinear, and illustrates the physical contexts in which the various wave phenomena occur.

The first edition was published twenty-five years ago. Somewhat to our surprise, the book is still in demand, which is why we decided to revise and update it. This update includes a significant revision of the presentation and numerous new examples and problems.

This edition consists of fifteen chapters. We begin chapter 1 with a review of linear oscillations, both mechanical and electromagnetic. Oscillation systems can be the source of time harmonic waves since the mass in a mechanical oscillation system and the electrical charge in an electromagnetic oscillation system are all related to and can be used to excite various waves. The pendulum, or swing, is an oscillation system that readers may have encountered before.

In chapter 2 we study general properties of wave motion, that is, without specifying whether they are mechanical or electromagnetic waves. A mathematical expression for a sinusoidal wave and the wave equation are introduced. Phase and group velocities and beats are described.

Chapter 3 provides mathematical preparations that are needed to derive the wave equation from first principles,

$$\frac{\partial^2 \xi}{\partial t^2} = c_w^2 \frac{\partial^2 \xi}{\partial x^2}$$

for various kinds of waves. It is useful for the reader not already familiar with Taylor series expansion techniques.

In chapters 4 and 5, we study mechanical waves, including those on springs (longitudinal) and along a stretched string (transverse), and sound waves (longitudinal) in solids, liquids, and gases. We show how Newton's equation of motion can be converted into the wave equation for mechanical waves and can formulate the energy and momentum that is associated with these waves. A detailed analysis to find the relationship is presented: [wave momentum transfer rate = (wave energy transfer rate) / (wave velocity)].

In chapter 6, we examine the reflection of mechanical waves at various boundaries. The wave equation allows two distinct solutions that propagate in opposite directions. When these two waves coexist, standing waves are created. Standing mechanical waves play an important role in musical instruments. The concept of a mechanical impedance is introduced as an analogy to the characteristic impedance that one encounters in electromagnetic wave phenomena. The concept of wave reflection is explained in terms of the reflection of energy and momentum. In wave reflection at an impedance discontinuity, energy conservation is imposed to find the amplitudes of both the reflected and transmitted waves, assuming that the wave media are infinitely massive. Momentum conservation requires the consideration of momentum absorbed by the media. When applied to reflection of a light wave incident upon the surface of a semi-infinite glass slab, the radiation pressure is negative (force directed from the slab to air) as predicted by Poynting in 1905. Using this well-known phenomenon, the momentum of the wave transmitted into glass medium is uniquely determined, which is consistent with the classical formula.

When a loudspeaker creates sound waves in air, the wave amplitude becomes smaller as one moves away from the speaker. This effect is caused by a geometrical decrease of the amplitude of the wave as shown in chapter 7. The variation of the amplitude can also occur in one-dimensional waves if the medium is not uniform or the wave velocity varies from point to point. This interesting (but difficult to analyze) phenomenon is briefly discussed.

In chapter 8, the Doppler effect of sound waves is studied. Whenever a sound source or an observer are moving relative to the ambient air, the observer hears a frequency that is different from the true frequency emitted from the source. If the source is moving faster than the speed of sound, a shock wave can occur. A Doppler shift in electromagnetic waves is similar to that in sound waves, but there is a fundamental difference between the two. In electromagnetic waves, only the relative velocity between the wave source and the observer enters into the description of the Doppler shift. This is a consequence of one of Einstein's postulates in relativity theory that the velocity of electromagnetic waves in a vacuum is independent of the source's and the observer's velocity.

In chapter 9, we study the propagation and radiation of electromagnetic waves. We start with an LC transmission line that is an analogue to the mass-spring mechanical transmission line described in chapter 3. The reflection of electromagnetic waves is discussed in terms of the possible mismatch of the electrical characteristic impedances, as was done in chapter 6 for mechanical waves. Electromagnetic waves in conducting materials such as metals or plasmas require a special treatment. In such media, the wave equation is drastically modified and electromagnetic waves become strongly dispersive, implying that the propagation properties depend upon the frequency of the waves.

In chapter 10, we learn that the radiation of electromagnetic waves requires the acceleration or deceleration of electric charges. The electric field lines associated with a charge are continuous. In the presence of charge acceleration or deceleration, the electric field consists of a nonradiation Coulomb field and a transverse radiation field. The fields are linearly related and the radiation electric field can thus be obtained from the known Coulomb electric field.

In chapter 11, interference and diffraction, which are caused by the constructive and destructive interference between more than one wave propagating in the same direction, are described. Depending on the phase of each wave, the resulting wave amplitude is either strengthened or weakened. Light does not always travel along a straight line because of the diffractive nature of the wave. Both Fraunhofer and Fresnel diffractions are analyzed. Diffraction imposes a limit on the resolving power of all optical instruments.

Geometrical optics is one branch of optics in which we are able to neglect the wave nature of light, and this is discussed in chapter 12. Light can be assumed to propagate in a straight line in a uniform medium. Of course, when light hits a boundary between two media, light changes its propagation direction. A familiar example is the reflection and refraction at the boundary of air and glass. Optical devices such as mirrors and lenses are discussed, followed by a look at various optical instruments. The matrix method in geometrical optics is a powerful tool when analyzing multilens systems and thick lenses.

Chapter 13 is an introduction to modern quantum physics. It is shown that under certain circumstances, light behaves as a collection of particles that are called photons. Briefly discussed here is that energetic particles such as accelerated electrons also have a wave nature. For example, the (deBloglie) wavelength of energetic electrons in an electron microscope is orders of magnitude shorter than an optical wavelength and this is why an electron microscope can have a higher resolving power than an optical microscope.

Chapter 14 contains an introduction to Fourier series and Laplace transforms. The concept of the frequency spectrum is explained. Also, it is shown how Laplace transforms can convert a differential equation into an algebraic equation.

The material covered in the previous chapters made the tacit assumption that the equations describing the physical phenomena could be linearized in that all perturbations of any physical quantity were small. In chapter 15, we introduce some mathematical techniques that are used to obtain analytical solutions to certain relevant nonlinear equations. The subjects of solitons and chaos are also described.

The MKS (meter, kilogram, second) unit system is used throughout the book. Some traditional and some conventional units are employed such as the angstrom ($1\text{Å} = 1 \times 10^{-10}$ m) in optics and the electron volt ($1 \text{ eV} = 1.6 \times 10^{-19}$ J) for the energy of elementary particles. In physics, a traveling wave is conventionally written as $A\cos(kx - \omega t)$ or ($Ae^{i(kx-\omega t)}$), and in engineering, it is written as $A\cos(\omega t - kx)$ or ($Ae^{j(\omega t - kx)}$) where $i = j = \sqrt{-1}$. We use the former representation but there are no fundamental differences between the two.

The website for an undergraduate wave course at the University of Saskatchewan can be accessed at http://physics.usask.ca/~hirose/ep225/, which contains animation of various wave motions.

The authors wish to acknowledge the helpful and professional assistance that we received from our Editor Dudley Kay, his assistant Katie Janelle, and Production Manager Robert Lawless and the entire staff at SciTech in the preparation of this book.

We also thank our colleagues and students for their comments and more importantly, our wives, Kimiko and Vicki, for their tireless support.

Publisher's Acknowledgements

SciTech acknowledges the unsung heroes who help shape a manuscript into the final book form by reviewing chapter drafts with an objective eye and offering numerous helpful suggestions. The following reviewers gave unselfishly of their time and experience:

Swamy Jagannatham
University of California Los Angeles

Dr. Jerome Helffrich
University of Texas, San Antonio

Dr. Ronald Riechers
Wright State University

Dr. W. Doyle St. John
University of Wisconsin, Platteville

Dr. Mustapha C.E. Yagoub
University of Ottawa

Dr. Dimitris E. Anagnostou
*South Dakota School of Mines
    and Technology*

Jack Adams PhDEE
Merrimack College

Dr. Andrew F. Peterson
Georgia Institute of Technology

Dr. Edward Wheeler
Rose-Hulman Institute of Technology

Dr. Matthew Sadiku
Prairie View A&M University

Dr. Erdem Topsakal
Mississippi State University

Dr. Banmali Rawat
University of Nevada, Reno

Review of Oscillations

1.1 Introduction

Most waves we encounter, either mechanical or electromagnetic, are created by something vibrating or oscillating. In the classroom, the instructor's voice reaches your ears as sound waves in air. To create the waves, the instructor uses vocal chords which are forced to vibrate by the airflow through the throat. Similarly, radio waves emitted from a radio station also originate from something that is oscillating. In this case, free or conduction electrons in a vertically erected antenna execute up and down oscillatory motion with a certain frequency, which is determined by an electrical oscillator connected to the antenna. Whenever physical objects oscillate or vibrate, there is a possibility that waves are created in the medium surrounding those objects.

In this chapter, we review oscillation phenomena, both mechanical and electromagnetic, since oscillations and waves have many common properties, hence understanding oscillations can greatly help us understand wave phenomena. More importantly, harmonic (or sinusoidal) waves that we frequently encounter in daily life are created by physical objects undergoing oscillatory motions. It is recommended that you refresh your knowledge (and skills) of properties of trigonometric functions, such as

$$\frac{d}{dx}\sin ax = a\cos ax, \qquad \frac{d}{dx}\cos ax = -a\sin ax$$

and so on.

Mass Spring System

Consider a mass M (kg) on a frictionless plane that is connected to a spring with a spring constant k_s (N/m), and an initial (natural) length l (m) (Figure 1–1). Without any external disturbance, the mass would stay at the equilibrium position, $x = 0$. Suppose one now pulls the mass a certain distance and then releases it. The mass would start oscillating with a certain frequency. If one pushes the mass and then releases it, the mass would also start oscillating with the *same* frequency. Otherwise, one could hit the mass with a hammer to cause it to start oscillating about its equilibrium position. No matter how the oscillation is started, the frequency will be the same.

One of the major objectives in studying oscillations is to find oscillation frequencies that are determined by physical quantities. As we will see later, the mass M and the spring constant k_s determine the oscillation frequency in the preceding example.

What makes the mass–spring system oscillate? When one pulls the mass, the spring must be elongated, and it tends to pull the mass back to its equilibrium position, $x = 0$. Therefore on being released, the mass starts moving to the left being pulled by the spring. This pulling force is given by *Hooke's law*,

$$F = \begin{cases} k_s x & \text{(directed to the \textit{left})} \\ -k_s x & \text{(directed to the \textit{right})} \end{cases} \tag{1.1}$$

where x is the deviation of the spring length from the initial length, l. The minus sign appears when the mass is at a location $x < 0$. The pulling force provided by the spring disappears at the instant when the mass reaches the equilibrium position, $x = 0$. By this time, however, the mass has acquired a kinetic energy (which will be shown to be equal to the potential energy initially stored in the spring). Because of its inertia, it cannot stop at the equilibrium position, but keeps moving, overshooting or passing the equilibrium position. Think of a swing in your backyard. You do not abruptly stop at the lowest point of the trajectory of the swing but keep moving beyond this point. In Figure 1–1,

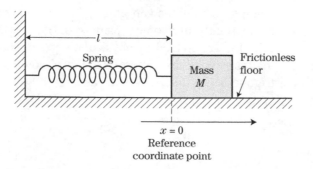

FIGURE 1–1

Mass–spring system in equilibrium. The spring has a spring constant k_s, and l is the initial natural length of the spring.

the spring will be squeezed and will push the mass back to the equilibrium position. This time the force is given by

$$F = -k_s x \quad (\text{directed to the } right) \tag{1.2}$$

since x is now a negative quantity. The mass keeps moving to the left until the kinetic energy, which the mass had when it passed through the equilibrium position, $x = 0$, is all converted into potential energy that is stored in the spring. After this instant, the mass again starts moving to the right toward the equilibrium position. This process continues and appears as an oscillation.

The key agent in the oscillatory motion in the mass-spring system is the force provided by the spring. This force always acts on the mass so as to make it seek its equilibrium position, $x = 0$. Such a force is called *a restoring force*. In any mechanical oscillating system, there is always a restoring force (or torque). In the case of a grandfather clock, gravity provides the restoring force, and for a wheel balance in a watch, a spiral hair-spring does the job by providing a restoring torque. If the spring constant k_s is greater (stronger spring), the spring can pull or push the mass more quickly and we expect that the oscillation frequency will be larger. On the other hand, if the mass is larger, the mass should move more slowly and we expect that the oscillation frequency will be smaller. Indeed, as we will see, the oscillation frequency ν for the mass–spring is given by

$$\nu = \frac{1}{2\pi} \sqrt{\frac{k_s}{M}} \quad (\text{cycles/sec} = \text{Hertz}) \tag{1.3}$$

Let us now find out what kind of mathematical expression can describe the oscillatory motion of the mass–spring system. We assume that the mass is gradually pulled a distance x_0 to the right from the equilibrium position, $x = 0$, and then released at time $t = 0$. Suppose that at a certain time, the mass is a distance x away from the equilibrium position, $x = 0$ (Figures 1–2 and 1–3). The instantaneous velocity of the mass is given by

$$v = \frac{dx}{dt} \quad (\text{m/s}) \tag{1.4}$$

and the acceleration is

$$a = \frac{dv}{dt} = \frac{d^2 x}{dt^2} \quad (\text{m/s}^2) \tag{1.5}$$

But the force acting on the mass—from Eqs. (1.1) and (1.2)—is

$$F = -k_s x \quad (\text{N}) \tag{1.6}$$

(Note that in Eq. (1.1), the force was given by $k_s x$ directed to the *left*, which is equivalent to a force $-k_s x$ directed to the *right*. Therefore the restoring force can be generalized as $-k_s x$ regardless of the sign of x.) Applying Newton's

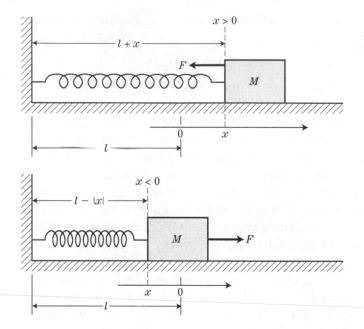

FIGURE 1–2

Displacement $x > 0$. The spring is stretched and pulls the mass toward the equilibrium position, $x = 0$.

FIGURE 1–3

Displacement $x < 0$. The spring is squeezed and pushes the mass toward the equilibrium position.

second law (mass $\times$ acceleration = force), we find

$$M\frac{d^2x}{dt^2} = -k_s x \tag{1.7}$$

This is the equation of motion for the mass to follow. The position x of the mass is to be found from this *differential equation* as a function of time t. Remember that the mass was located at $x = x_0$ at $t = 0$, when the oscillation starts, or

$$x(0) = x_0 \tag{1.8}$$

where $x(0)$ means the value of x at $t = 0$. We know that the second order derivatives of sinusoidal functions, $\sin a\theta$ and $\cos a\theta$, are

$$\frac{d^2}{d\theta^2} \sin a\theta = -a^2 \sin a\theta$$

$$\frac{d^2}{d\theta^2} \cos a\theta = -a^2 \cos a\theta$$

Therefore it is very likely that Eq. (1.7) has a sinusoidal solution. Let the solution for $x(t)$ be

$$x(t) = A \cos \omega t \tag{1.9}$$

where A and ω are constants that are to be determined. Since $\cos 0 = 1$, we find

$$x(0) = A \tag{1.10}$$

Comparing Eq. (1.8) with Eq. (1.10), we must have $A = x_0$. This quantity is called the amplitude of the oscillation.

To find ω, we calculate the second derivative of $x(t) = x_0 \cos \omega t$:

$$\frac{dx}{dt} = x_0 \frac{d}{dt} \cos \omega t = -x_0 \omega \sin \omega t$$

$$\frac{d^2 x}{dt^2} = x_0 \frac{d^2}{dt^2} \cos \omega t = -x_0 \omega \frac{d}{dt} \sin \omega t = -x_0 \omega^2 \cos \omega t \tag{1.11}$$

After substituting Eq. (1.11) into Eq. (1.7), we find

$$-M\omega^2 x_0 \cos \omega t = -k_s x_0 \cos \omega t$$

which yields

$$\omega = \sqrt{\frac{k_s}{M}} \tag{1.12}$$

This quantity ω is called the angular frequency and has the dimensions of radians/second. Since the function $\cos \omega t$ has a period of 2π radians, the temporal period T is given by

$$T = \frac{2\pi}{\omega} \quad \text{(s)} \tag{1.13}$$

In 1 second, the oscillation repeats itself $1/T$ times (Figure 1–4). This number is defined as the frequency, ν (cycles/s = Hertz). It is obvious that

$$\nu = \frac{1}{T} = \frac{\omega}{2\pi} \quad \text{(Hz)} \tag{1.14}$$

FIGURE 1–4

The displacement $x(t)$ as a function of time t.

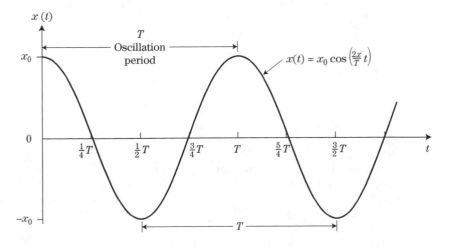

EXAMPLE 1.1

Show that the quantity $\sqrt{k_s/M}$ indeed has the dimensions of $1/s$.

Solution

Since the spring constant k_s has the dimensions of $N/m = (kg \cdot m/s^2)/m = kg/s^2$, and the mass has the dimensions of kg, we find $\sqrt{k_s/M}$ has the dimensions of

$$\sqrt{\frac{kg/s^2}{kg}} = \frac{1}{s}$$

Note that the angle (radians) is a dimensionless quantity.

EXAMPLE 1.2

A spring 2 m long hangs from the ceiling as shown in Figure 1–5. When a mass of 1.5 kg is suspended from the spring, the spring is elongated by 30 cm in equilibrium. The mass is then pulled down an additional 5 cm and released. Neglecting the mass of the spring or any air resistance, find an equation to describe the oscillatory motion of the mass.

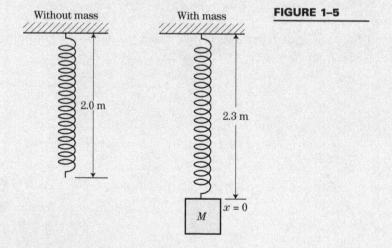

FIGURE 1–5

Solution

The mass is expected to oscillate about the equilibrium position, $x = 0 \, (x > 0$ upward). The spring constant k_s is found from

$$Mg = k_s \Delta l$$

where $M = 1.5$ kg, $\Delta l = 0.3$ m. Then

$$k_s = \frac{1.5 \text{ kg} \times 9.8 \text{ N/kg}}{0.3 \text{ m}} = 49.0 \text{ N/m}$$

and

$$\omega = \sqrt{\frac{k_s}{M}} = \sqrt{\frac{49 \text{ N/m}}{1.5 \text{ kg}}} = 5.7 \text{ rad/s}$$

Since the initial position is $x_0 = -0.05$ m, the equation to describe the motion of the mass is given by

$$x(t) = -0.05 \cos (5.7t) \text{ m}$$

The period of the oscillation T is

$$T = \frac{2\pi}{\omega} = 1.1 \text{ s}$$

and the frequency ν is $\nu = 1/T = 0.91$ Hz.

We have seen that a cosine function can appropriately describe the case in which a mass is released from a position that is removed from its equilibrium position. A solution that is proportional to a sine function $A \sin \omega t$ cannot describe the case since it cannot yield the initial position x_0. (Remember that $\sin 0 = 0$.) However, the function $A \sin \omega t$ can describe other cases in which the mass is given an impulse at the equilibrium position. For example, if one hits the mass with a hammer, the mass executes an oscillation *starting from* $x = 0$ and this case is correctly described by a function of the form $A \sin \omega t$. However, once the oscillations start, it is rather immaterial which form (cos or sin) to use, since the oscillation frequency (or ν) is the same in either case. No matter how we let oscillations start, the mass–spring system oscillates with the frequency $\omega = \sqrt{k_s/M}$ that is totally determined by the physical nature of the system. Such a frequency is called the *natural* (or *resonance*) *frequency* and appears when an oscillation system is *isolated* from the external driving force. Oscillation systems, however, can be forced to oscillate at a frequency other than their natural frequencies. We will study forced oscillation phenomena later in this chapter.

 Energy Tossing in Mechanical Oscillations

We again consider the mass–spring system oscillating according to

$$x(t) = x_0 \cos \omega t \qquad (1.15)$$

As we have seen, this equation describes the case in which the mass is pulled a distance x_0 and then released at $t = 0$. The spring is also elongated by the distance x_0, and before releasing the mass, the spring had stored a potential energy given by

$$\frac{1}{2}k_s x_0^2 \quad (J)$$

(1.16)

(Recall that the energy required to elongate the spring gradually by a length x_0 is $\int_0^{x_0} k_s x \, dx = \frac{1}{2}k_s x_0^2$.) This potential energy must be supplied by an external agent, such as our hands in Example 1.2, and this provides the energy source for the oscillations.

After being released, the mass starts moving toward the negative x direction and acquires a kinetic energy. At the same time, the spring loses its potential energy, since x is now smaller than x_0. We expect, however, that the sum of the potential energy and kinetic energy remains equal to the initial energy given by Eq. (1.16) since the system is isolated from external agents and hence the total energy must be conserved provided the friction loss is ignorable.

In order to see this, we calculate each form of energy at an arbitrary instant of time. The potential energy is

$$\text{potential energy} = \frac{1}{2}k_s x^2 = \frac{1}{2}k_s x_0^2 \cos^2 \omega t$$

(1.17)

The kinetic energy is

$$\text{kinetic energy} = \frac{1}{2}M v^2$$

Since

$$v = \frac{dx}{dt} = -\omega x_0 \sin \omega t$$

we find

$$\text{kinetic energy} = \frac{1}{2}M \omega^2 x_0^2 \sin^2 \omega t$$

(1.18)

Then the total energy is

$$\text{total energy} = \frac{1}{2}k_s x_0^2 \cos^2 \omega t + \frac{1}{2}M \omega^2 x_0^2 \sin^2 \omega t$$

However, the frequency of oscillation ω is

$$\omega = \sqrt{\frac{k_s}{M}}$$

and thus the total energy becomes

$$\text{total energy } = \frac{1}{2}k_s x_0^2(\cos^2 \omega t + \sin^2 \omega t) = \frac{1}{2}k_s x_0^2$$

This holds true at any instant and our guess is indeed justified.

What oscillates in the mass spring system is energy. The mass and the spring periodically exchange energy. Figure 1–6 qualitatively illustrates this energy tossing mechanism. In Figure 1–7, the position of the mass, $x(t)$; its velocity, $v(t)$; and potential and kinetic energies are shown as functions of time.

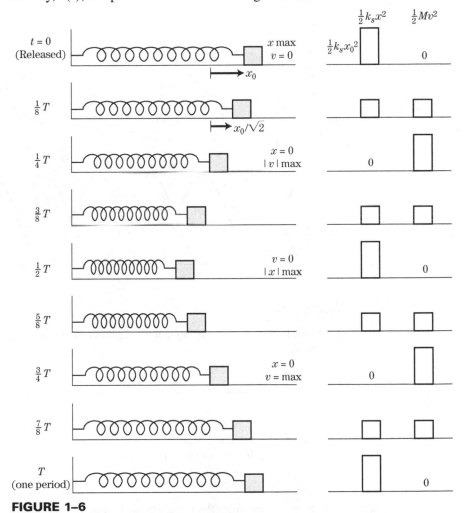

FIGURE 1–6

Location of the mass at various times. The corresponding potential and kinetic energies are also schematically shown. The total energy, which is the sum of the potential and kinetic energies, is a constant.

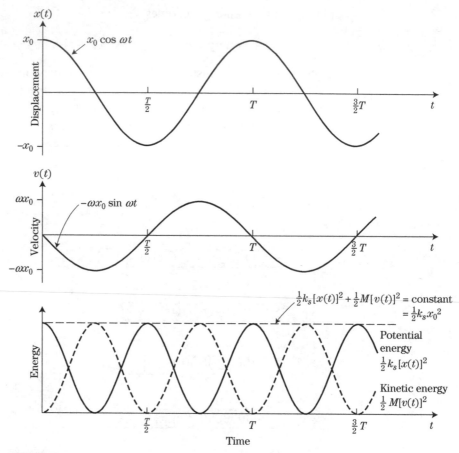

FIGURE 1–7

Displacement $x(t)$, velocity $v(t)$, and the potential and kinetic energies as functions of time t.

EXAMPLE 1.3

In the configuration shown in Figure 1–1, the mass is hit by a hammer and instantly acquires a kinetic energy of 0.1 J. Assuming $k_s = 30$ N/m and $M = 0.5$ kg, find an expression to describe the oscillatory motion of the mass.

Solution

Since the mass starts oscillating from the equilibrium position in this case, we have to choose a sine solution,

$$x(t) = A \sin \omega t$$

where

$$\omega = \sqrt{\frac{k_s}{M}} = \sqrt{\frac{30\ \text{N/m}}{0.5\ \text{kg}}} = 7.75\ \text{rad/s}$$

The amplitude A can be determined from

$$\text{initial kinetic energy} = \frac{1}{2}k_s A^2 = 0.1\ \text{J}$$

Then

$$A = \pm 0.08\ \text{m} = \pm 8\ \text{cm} \quad \text{and} \quad x(t) = \pm 8 \sin \omega t\ \text{cm}$$

The plus sign describes the case in which the impulse is directed to the positive x (right) direction and the minus sign, the negative x direction.

That the total energy of the oscillating mass–spring system is conserved or constant can alternatively be shown directly from the equation of motion

$$M\frac{dv}{dt} = -k_s x \tag{1.19}$$

Let us multiply this equation by the velocity v.

$$Mv\frac{dv}{dt} = -k_s vx \tag{1.20}$$

Since

$$\frac{dv^2}{dt} = \frac{dv^2}{dv}\frac{dv}{dt} - 2v\frac{dv}{dt}$$

$$xv = x\frac{dx}{dt} = \frac{1}{2}\frac{dx^2}{dt}$$

Eq. (1.20) becomes

$$\frac{d}{dt}\left(\frac{1}{2}Mv^2 + \frac{1}{2}k_s x^2\right) = 0 \tag{1.21}$$

which indeed states that the total energy is conserved,

$$\frac{1}{2}Mv^2 + \frac{1}{2}k_s x^2 = \text{constant}$$

Other Mechanical Oscillation Systems

Whenever there is a restoring force to act on a mass, oscillations are likely to occur. As we have seen, there must be two agents for mechanical oscillations to take place, one capable of storing potential energy (such as the spring) and the other capable of storing kinetic energy (such as the mass). In rotational devices (such as the wheel balance in watches), the restoring torque and rotational inertia replace the restoring force and mass, respectively, but the energy relations still hold.

Pendulum

A grandfather clock's accuracy is largely determined by the regularity of the pendulum oscillation frequency. You may already know that the pendulum frequency is totally determined by the length of the support l (Figure 1–8) and it does not depend on the mass M.

The restoring force to act on the mass M in Figure 1–8 is provided by the earth's gravity, which tends to make the mass stay at the equilibrium position, P, or the lowest position. The restoring force F is given by

$$F = Mg \sin \theta \quad \text{(toward } P\text{)} \tag{1.22}$$

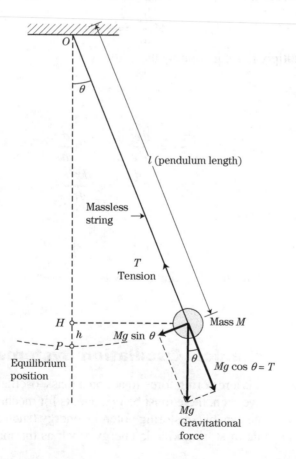

FIGURE 1–8

Pendulum having a mass M and a length l. The acceleration due to gravity is $g = 9.8 \text{ m/s}^2$.

and the equation of motion for the mass becomes

$$M\frac{dv}{dt} = -Mg\sin\theta \tag{1.23}$$

(Do you see why the minus sign appears in this equation? See the discussion given in the previous section.) Since the velocity v is given by

$$v = l\frac{d\theta}{dt} \tag{1.24}$$

Eq. (1.23) becomes

$$\frac{d^2\theta}{dt^2} = -\frac{g}{l}\sin\theta \tag{1.25}$$

Although this equation looks simple, it is a *nonlinear differential equation* and its solution cannot be expressed in terms of sine or cosine functions unless $|\theta|$ is much smaller than 1 rad. If $|\theta| \ll 1$ rad, $\sin\theta$ can be well approximated simply by θ (see Chapter 3), and Eq. (1.25) reduces to a linear form,

$$\frac{d^2\theta}{dt^2} \simeq -\frac{g}{l}\theta \tag{1.26}$$

(This procedure is called linearization of the differential equation. The solution of Eq. (1.25) can be obtained using numerical techniques as will be shown in Chapter 15.) Eq. (1.26) is mathematically identical to our previous equation, Eq. (1.7). We can immediately find the oscillation frequency as

$$\omega = \sqrt{\frac{g}{l}} \quad \text{(rad/s)} \tag{1.27}$$

You should check that $\sqrt{g/l}$ indeed has the dimensions of $1/s$.

EXAMPLE 1.4

Find the length of the pendulum rod of a grandfather clock having an oscillation period of 2.0 s. Assume that the mass of the rod is negligible compared with the mass to be attached.

Solution

From $T = 2\pi/\omega = 2\pi\sqrt{l/g}$, we find $l = (T/2\pi)^2 g$. Substituting $T = 2$ s and $g = 9.8$ m/s^2, we find $l = 0.99$ m.

EXAMPLE 1.5

Assuming a solution of the form

$$\theta(t) = \theta_0 \cos \omega t$$

for Eq. (1.26), show that the total energy of the mass (potential and kinetic) is constant.

Solution

The kinetic energy is

$$\text{kinetic energy } = \frac{1}{2}Mv^2 = \frac{1}{2}M\left(l\frac{d\theta}{dt}\right)^2 = \frac{1}{2}Ml^2\omega^2\theta_0^2 \sin^2 \omega t$$

The potential energy is Mgh, where h is the height measured from the equilibrium position and is given by

$$h = OP - OH = l - l\cos\theta$$

as seen in Fig. 1–8. If θ is small, $\cos\theta$ can be approximated by (see Chapter 3)

$$\cos\theta \simeq 1 - \frac{1}{2}\theta^2$$

Then

$$h \simeq \frac{1}{2}l\theta^2$$

and the potential energy becomes

$$\text{potential energy } = Mgh = Mg\frac{1}{2}l\theta^2 = \frac{1}{2}Mgl\,\theta_0^2 \cos^2 \omega t$$

Recalling $\omega^2 = g/l$, we find

$$\text{kinetic energy} + \text{potential energy} = \frac{1}{2}Mgl\,\theta_0^2 \quad \text{(constant)}$$

Rotational Inertial Systems

The balance wheel (Figure 1–9) in watches oscillates about its center. A spiral spring connected to the wheel balance provides a restoring torque rather than restoring force, and the rotational inertia of the wheel balance plays the role of mass inertia (translational inertia) in the mass-spring system.

Let the moment of inertia of the wheel balance be I (kg $\cdot$ m^2) and the restoring torque provided by the spring be

$$\text{torque } = -k_\tau\theta \quad (\text{N} \cdot \text{m}) \tag{1.28}$$

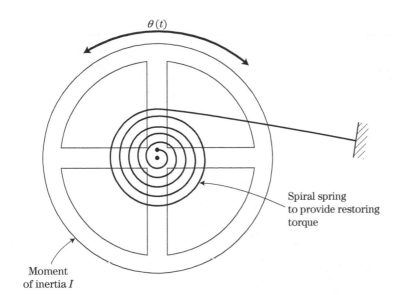

$\theta(t)$

Moment
of inertia I

Spiral spring
to provide restoring
torque

FIGURE 1–9

Wheel balance of watches
connected to a spiral spring, an
example of a linear oscillator.

where k_τ is a constant (torsional constant, $\mathrm{N \cdot m}$) that plays the same role as
the spring constant in the mass spring system, and θ is the rotational angle
of the wheel balance measured from the equilibrium (zero torsion) angular
position.

Since the equation of motion for the rotational system is given by

$$I\frac{d^2\theta}{dt^2} = \text{torque} \tag{1.29}$$

where I is the moment of inertia, we find

$$I\frac{d^2\theta}{dt^2} = -k_\tau\theta \tag{1.30}$$

This is again identical in mathematical form to Eq. (1.7). (Compare Eq. (1.30)
with Eq. (1.25). A wheel balance is a linear oscillator irrespective of how large
the angle θ is, as long as it stays within elastic limit.) The oscillation frequency
is then given by

$$\omega = \sqrt{\frac{k_\tau}{I}} \tag{1.31}$$

You should check that $\sqrt{k_\tau/I}$ indeed has the dimensions of frequency.

EXAMPLE 1.6

A straight uniform stick having a length l (m) and a mass M (kg) is freely pivoted at one end as shown in Figure 1–10. Find the frequency of oscillation about the pivot assuming that the angle θ is small.

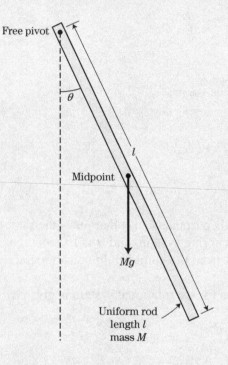

Free pivot

θ

l

Midpoint

Mg

Uniform rod
length l
mass M

FIGURE 1–10

Pivoted rod as a pendulum.

Solution

The gravity force Mg acts on the center of mass. The restoring torque to act on the stick is thus

$$Mg \sin \theta \times \frac{l}{2} \simeq \frac{1}{2} Mgl\theta$$

Therefore, $Mgl/2$ plays the role of restoring torque constant k_τ provided $|\theta| \ll 1$. (Recall $\sin \theta \simeq \theta$ if $|\theta| \ll 1$ rad.) The moment of inertia about the end of the stick is

$$I = \int_0^l \frac{M}{l} x^2 dx = \frac{1}{3} Ml^2$$

Then

$$\omega = \sqrt{\frac{k_\tau}{I}} = \sqrt{\frac{3g}{2l}}$$

Electromagnetic Oscillation

One of the first electrical circuits that one encounters is an isolated LC (inductance and capacitance) circuit that oscillates with an angular frequency

$$\omega = \frac{1}{\sqrt{LC}} \quad \text{(rad/s)} \tag{1.32}$$

Although physical quantities we treat in the electromagnetic oscillation are quite different from those in the mechanical oscillation, the fundamental concept of the oscillation mechanism—namely, the energy tossing mechanism—remains the same. Instead of kinetic and potential energies found in the mass-spring system, we now have electric and magnetic energies stored in the capacitor and inductor, respectively.

Consider a capacitor charged to a charge q_0 (Coulombs) suddenly connected to an inductor L with the closing of the switch (Figure 1–11). The charge initially stored in the capacitor tends to flow toward the inductor and creates a current along the circuit. The voltage across the capacitor is

$$V_C(t) = \frac{q(t)}{C}$$

and that across the inductor is

$$V_L(t) = L\frac{dI(t)}{dt}$$

Then *Kirchhoff's voltage law* requires

$$\frac{q(t)}{C} = L\frac{dI(t)}{dt} \tag{1.33}$$

Since we have chosen the direction of the current corresponding to a discharging capacitor, we write

$$I(t) = -\frac{dq(t)}{dt} \tag{1.34}$$

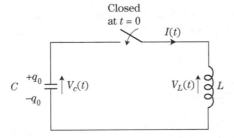

FIGURE 1–11

LC resonant circuit. The initial charge on the capacitor is q_0.

Substituting Eq. (1.34) into Eq. (1.33), we find the following differential equation for the temporal variation of the charge $q(t)$,

$$\frac{d^2 q(t)}{dt^2} = -\frac{1}{LC} q(t) \tag{1.35}$$

This is again mathematically identical to Eq. (1.7) for the mass-spring system and we immediately find that the LC circuit would oscillate with the frequency

$$\omega = \frac{1}{\sqrt{LC}}$$

Since the capacitor had an initial charge q_0, the equation to describe the charge at an arbitrary instant should be chosen as

$$q(t) = q_0 \cos \omega t \tag{1.36}$$

Using Eq. (1.34), the current $I(t)$ becomes

$$I(t) = \omega q_0 \sin \omega t \tag{1.37}$$

The electric energy stored in the capacitor is

$$\text{electric energy} = \frac{1}{2} \frac{q^2(t)}{C} = \frac{1}{2C} q_0^2 \cos^2 \omega t \tag{1.38}$$

and the magnetic energy stored in the inductor is

$$\text{magnetic energy} = \frac{1}{2} L I^2(t) = \frac{1}{2} L \omega^2 q_0^2 \sin^2 \omega t \tag{1.39}$$

Recalling that $\omega^2 = 1/LC$, we find that the sum of the two energies is a constant,

$$\text{electric energy} + \text{magnetic energy} = \frac{1}{2} \frac{q_0^2}{C}$$

and the total energy is equal to the initial electric energy stored in the capacitor.

The capacitor and inductor exchange energy periodically as the mass and spring do in mechanical oscillations, and we see that this energy-tossing mechanism is common to any kind of oscillation, mechanical or electromagnetic.

EXAMPLE 1.7

In the LC circuit shown in Figure 1–12, the switch S is closed for a long time. Then the switch is opened at $t = 0$. Find the expressions for the current to flow in the LC circuit and the charge on the capacitor.

Solution

The initial current flowing through the inductor is

$$I_0 = \frac{12\,\text{V}}{2\,\Omega} = 6\,\text{A}$$

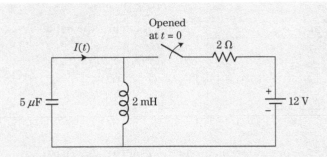

FIGURE 1–12

An example in which the initial current is not zero.

Then the current chosen clockwise is described by

$$I(t) = I_0 \cos \omega t = 6 \cos \omega t \text{ A}$$

where

$$\omega = \frac{1}{\sqrt{LC}} = \frac{1}{\sqrt{2 \times 10^{-3} \times 5 \times 10^{-6}}} = 10^4 \text{ rad/s}$$

The charge on the lower plate of the capacitor is given by

$$q(t) = \int_0^t I(t)\, dt = \int_0^t I_0 \cos \omega t\, dt = \frac{I_0}{\omega} \sin \omega t = 6 \times 10^{-4} \sin \omega t \quad \text{(Coulombs)}$$

Note that the initial condition in this example is different from that in Figure 1–11.

Damped Oscillation

So far we have considered ideal cases in which energy dissipation can be completely neglected. For example, in the mass-spring system, we assumed that the floor on which the mass is placed is frictionless. Also, in the LC circuit, we neglected the resistance in the circuit. Both mechanical friction and electric resistance give rise to energy dissipation, and the oscillation cannot continue forever, but should eventually be damped and approach 0. The oscillation energy is converted into heat in an irreversible manner or radiated into space to be lost forever.

Consider now a capacitor C with a charge q_0 suddenly connected to an inductor L through a finite resistance R (Figure 1–13). Using Kirchhoff's voltage theorem, we find

$$\frac{q(t)}{C} = RI(t) + L\frac{dI(t)}{dt} \tag{1.40}$$

Recalling

$$I(t) = -\frac{dq(t)}{dt}$$

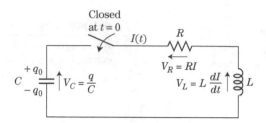

FIGURE 1-13

LCR circuit. The initial charge is q_0. An example of a system that yields a damped oscillation.

we now have the following differential equation for the charge $q(t)$:

$$\frac{d^2q(t)}{dt^2} + \frac{R}{L}\frac{dq(t)}{dt} + \frac{1}{LC}q(t) = 0 \tag{1.41}$$

In the limit of $R \to 0$ (zero resistance), we recover Eq. (1.35).

Solving Eq. (1.41) is not straightforward because of the presence of the first-order derivative. However, in the absence of the inductance, we expect that the charge on the capacitor will exponentially decrease in time,

$$q(t) = q_0 e^{-t/RC} \tag{1.42}$$

where RC is the time constant. A time constant is defined as the time that it takes for the initial charge to decrease to $q_0/e \simeq 0.37q_0$. Therefore, we may expect the solution to Eq. (1.41) to be a combination of an oscillatory function and an exponential function, and we write

$$q(t) = q_0 e^{-\gamma t}\cos\omega t \tag{1.43}$$

where γ is the damping constant that is to be determined. The preceding form of solution, however, is valid only for the case of weak damping such that $\gamma \ll \omega$. The general case will be given as a problem of this chapter. Also in Chapter 13, the same problem will be solved using the method of Laplace transformation.

Note that the solution for $q(t)$ given by Eq. (1.43) satisfies the initial condition

$$q(0) = q_0$$

We now calculate $dq(t)/dt$ and $d^2q(t)/dt^2$:

$$\frac{dq(t)}{dt} = q_0(-\gamma\cos\omega t - \omega\sin\omega t)e^{-\gamma t} \tag{1.44}$$

$$\frac{d^2q(t)}{dt^2} = q_0\left[(\gamma^2 - \omega^2)\cos\omega t + 2\gamma\omega\sin\omega t\right]e^{-\gamma t} \tag{1.45}$$

Substituting Eqs. (1.44) and (1.45) into Eq. (1.41) and eliminating the common factors q_0 and $e^{-\gamma t}$, we find

$$\left(\gamma^2 - \omega^2 - \frac{R}{L}\gamma + \frac{1}{LC}\right)\cos \omega t + \left(2\gamma\omega - \frac{R}{L}\omega\right)\sin \omega t = 0 \qquad (1.46)$$

which must hold at any time. Then the coefficients of $\cos \omega t$ and $\sin \omega t$ must identically be zero,

$$\gamma^2 - \omega^2 - \frac{R}{L}\gamma + \frac{1}{LC} = 0 \qquad (1.47)$$

$$2\gamma - \frac{R}{L} = 0 \qquad (1.48)$$

From these we find

$$\gamma = \frac{R}{2L}, \quad \omega \simeq \frac{1}{\sqrt{LC}} \qquad (1.49)$$

where in Eq. (1.47) we have neglected terms containing γ since we have assumed $\gamma \ll \omega$.

The function $q(t) = q_0 e^{-\gamma t} \cos \omega t$ is qualitatively shown in Figure 1–14. The damped oscillation is confined between the two curves $\pm q_0 e^{-\gamma t}$, which are called envelopes. It should be emphasized again that the solution we have found is correct only for the case of small damping, $\gamma \ll \omega$, or equivalently,

$$R \ll \sqrt{\frac{2L}{C}} \qquad (1.50)$$

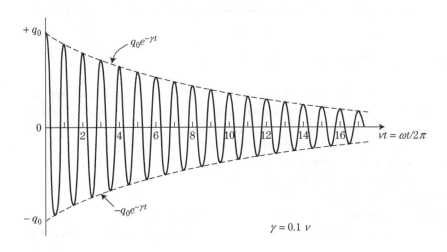

$\gamma = 0.1\, \nu$

FIGURE 1–14

Behavior of the charge on the capacitor in Figure 1–13.

EXAMPLE 1.8

In the mass-spring oscillation system, assume there exists small but finite friction force between the mass and floor, which is proportional to the velocity of the mass,

$$F_{\text{friction}} = -fv = -f\frac{dx}{dt} \tag{1.51}$$

where f is a constant. Show that the differential equation for the displacement $x(t)$ is mathematically identical to Eq. (1.41) and find the condition for weakly damped oscillation.

Solution

The equation of motion for the mass is now given by

$$M\frac{d^2x}{dt^2} = -k_s x - f\frac{dx}{dt}$$

or

$$\frac{d^2x}{dt^2} + \frac{f}{M}\frac{dx}{dt} + \frac{k_s}{M}x = 0 \tag{1.52}$$

Comparing this with Eq. (1.41), we see that if the following substitution is made

$$x \to q, \quad M \to L, \quad f \to R, \quad k_s \to \frac{1}{C}$$

both equations are identical.

The condition for weakly damped oscillation

$$R \ll \sqrt{\frac{2L}{C}}$$

can thus be translated as

$$f \ll \sqrt{2k_s M} \tag{1.53}$$

Forced Oscillation

In previous sections we found several natural oscillation frequencies appearing in both mechanical and electromagnetic systems. Those oscillation frequencies ($\omega = \sqrt{k_s/M}$, $1/\sqrt{LC}$, etc.) are also specifically called natural (or resonance) frequencies, since they appear when the oscillation systems are left alone, or isolated from external driving forces. Both mechanical and electromagnetic systems, however, can be forced to oscillate with frequencies other than the natural frequency.

A typical example of a forced oscillation is an ac (alternating current) circuit, in which an oscillating voltage generator with an angular frequency ω

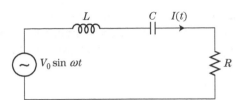

FIGURE 1–15

LCR ac circuit. An example of a forced oscillation.

is driving a current through L, C, R elements (Figure 1–15). Even though there is a resistor R in the circuit, the current $I(t)$ does not damp, in contrast to the case we studied in Section 1.6, since the generator can continuously feed energy to compensate the amount of energy dissipated in the resistor.

The current flowing in the circuit will attain steady oscillation with the same frequency ω after the transient stage is over. This time is typically several multiples of $1/\gamma$. The amplitude of the current in the circuit is given by

$$I_0 = \frac{V_0}{\sqrt{R^2 + (\omega L - 1/\omega C)^2}} \tag{1.54}$$

This takes a maximum when

$$\omega L = \frac{1}{\omega C} \quad \text{or} \quad \omega = \frac{1}{\sqrt{LC}}$$

as shown in Figure 1–16. The frequency determined from $\omega = 1/\sqrt{LC}$ is thus appropriately called a resonance frequency.

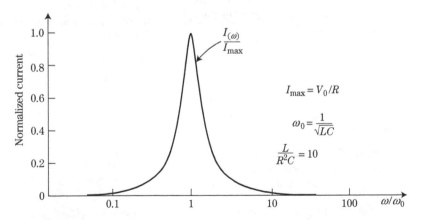

FIGURE 1–16

Plot of Eq. (1.54) as a function of frequency ω. If the frequency is plotted on a logarithmic scale, the graph becomes symmetric about the resonance frequency, $\omega_0 = 1/\sqrt{LC}$. The graph shown corresponds to the case $L/(R^2C) = 10$.

In mechanical oscillation systems, similar resonance phenomena can be found. When one pushes a swing, one naturally matches the pushing frequency with the natural or resonance frequency of the swing. We also know that a person riding on a swing can increase the swing amplitude by periodically shifting the center of mass. In the process, the person gains a kinetic energy by doing a work against the gravity and centrifugal force. A simple swing is actually a self-exciting oscillation system (parametric amplifier).

1.8 Problems

1. Calculate

$$\frac{d}{dx}\sin 5x, \quad \frac{d^2}{dx^2}\sin 5x, \quad \frac{d}{dx}\cos 3x, \quad \frac{d^2}{dx^2}\cos 3x$$

2. If $|x| \ll 1$, a function $(1+x)^n$ can be approximated by

$$(1+x)^n \simeq 1 + nx \quad \text{(binomial expansion)}.$$

   (a) Find the percent error caused by the approximation

   $$\sqrt{1+x} \simeq 1 + \frac{1}{2}x \quad \text{for } x = 0.1, 0.01$$

   (b) Repeat (a) for

   $$(1+x)^{-1/2} \simeq 1 - \frac{1}{2}x$$

3. If $|\theta| \ll 1$ rad, $\sin\theta$ may be approximated by $\sin\theta \simeq \theta$. Calculate the percent error of this approximation for $\theta = 0.1$ rad, 0.01 rad.

4. Show that functions
   (a) $x = A\sin\omega t$,
   (b) $x = A\sin\omega t + B\cos\omega t$,
   (c) $x = A\cos(\omega t + \phi)$
   all satisfy Eq. (1.7), provided $\omega = \sqrt{k_s/M}$. A, B, and ϕ are constants.

5. In the configuration of Figure 1–1, the mass (1.5 kg) is displaced 10 cm to the left and then released. Twenty oscillations are observed in 1 minute. Find
   (a) The spring constant.
   (b) The equation describing the oscillation.
   (c) The energy associated with the oscillation.

6. A meter stick is freely pivoted about a horizontal axis at (a) the end of the stick or 100-cm mark and (b) the 75-cm mark. Find the oscillation frequencies

in each case, assuming that the oscillation angle is small.

7. A thin circular hoop of radius a is hung over a sharp horizontal knife edge. Show that the hoop oscillates with an oscillation frequency $\omega = \sqrt{g/2a}$.

8. A marble thrown into a bowl executes oscillatory motion. Assuming that the inner surface of the bowl is parabolic $(y = ax^2)$ and the marble has a mass m, find the oscillation frequency. Neglect friction and assume a small oscillation amplitude.

9. Place an object on a turntable of a record player. Observe the motion of the object from the side. The motion is harmonic or sinusoidal of the form $x = x_0 \cos\omega t$. Prove this. If the turntable is revolving at $33\frac{1}{2}$ rpm, what is ω? What is v? What is the length of a pendulum to oscillate with the preceding frequency?

10. In the configuration of Figure 1–11,
    (a) Show that the current $I(t)$ is described by

    $$I(t) = \frac{V_0}{\sqrt{L/C}}\sin\omega t$$

    where $V_0 = q_0/C$ is the initial voltage on the capacitor.
    (b) The preceding expression indicates that the quantity $\sqrt{L/C}$ has the dimensions of ohms.

    Prove this. $\sqrt{L/C}$ is called the *characteristic impedance* of the LC circuit.

11. A capacitor of 5 μF charged to 1 kV is discharged through an inductor of 2 μH. The total resistance in the circuit is 5 mΩ.

(a) Is this a weakly damped *LCR* circuit?

(b) Find the time by which one-half the initial energy stored in the capacitor has been dissipated. The time is measured from the instant when discharge is started.

12. To solve Eq. (1.41) without the restriction of weak damping, assume

$$q(t) = e^{-\gamma t}(A \cos \omega t + B \sin \omega t)$$

where $\omega, \gamma, A,$ and B are to be determined. From the initial condition, $q(0) = q_0$, we must have

$$A = q_0 \qquad \text{(i)}$$

(a) Calculate dq/dt and d^2q/dt^2. Then substitute these into Eq. (1.41). You will obtain a relation like

$$f \cos \omega t + g \sin \omega t = 0 \qquad \text{(ii)}$$

where f and g contain A, B, ω, γ. For Eq. (ii) to hold at any time, $f = g = 0$ must hold,

$$f - 0 \qquad \text{(iii)}$$

$$g = 0 \qquad \text{(iv)}$$

(b) Another initial condition is that at $t = 0$, the current is zero, since the inductor behaves as if it were an infinitely large resistor right after the switch is closed. (Recall that inductors tend to resist any current variation.) Thus

$$I(0) = \left. \frac{dq}{dt} \right|_{t=0} = 0 \qquad \text{(v)}$$

(c) Equations (i), (iii), (iv), and (v) constitute four simultaneous equations for four unknowns A, B, ω, γ. (We already have found A.) Solve these.

(d) Find the condition for ω to be real.

13. Prepare two eggs of approximately the same weight, one boiled and another raw. Make two pendulums

using the eggs. (You can use Scotch tape or paper cups. Be careful with the raw egg.) Let the pendulums start oscillating. The pendulum with the raw egg would damp faster. Explain why. (Make sure that the pendulums both have the same length.)

14. Explain the function of shock absorbers installed on automobiles. What would happen without them?

15. Consider two cascaded mass-spring systems.

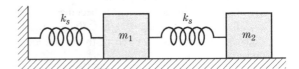

FIGURE 1–17

Problem 15.

(a) Write down the equation of motion for *each* mass, assigning displacements $x_1(t)$ and $x_2(t)$ for the masses m_1 and m_2, respectively. You may assume that the springs are identical.

(b) Then eliminate $x_2(t)$ between the two equations to show that the differential equation for $x_1(t)$ is given by

$$m_1 m_2 \frac{d^4 x_1}{dt^4} + k(m_1 + 2m_2) \frac{d^2 x_1}{dt^2} + k_s^2 x_1 = 0$$

(c) Show that the oscillation frequency ω is given as solutions to

$$m_1 m_2 \omega^4 - k(m_1 + 2m_2) \omega^2 + k_s^2 = 0$$

which allows two possible solutions for $|\omega|$.

16. Repeat Problem 15 for a two-mass, three-spring system whose both ends are clamped. You may assume that the springs are identical.

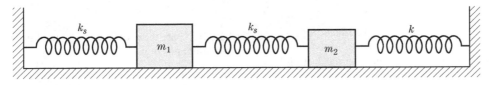

FIGURE 1–18

Problem 16.

17. If $|\theta|$ is much smaller than 1 rad, $\sin\theta$ can be approximated by

$$\sin\theta \simeq \theta - \frac{1}{6}\theta^3$$

and the pendulum equation Eq. (1.25) becomes

$$\frac{d^2\theta}{dt^2} = -\frac{g}{l}\theta\left(1 - \frac{1}{6}\theta^2\right)$$

This is still a nonlinear equation. However, it can tell us that as the oscillation amplitude of a pendulum increases, the oscillation frequency becomes smaller than the linear value, $\omega_0 = \sqrt{g/l}$. Explain (qualitatively) why this is so, referring to the preceding equation.

Hint: The new frequency $\omega(\theta)$ is found approximately from

$$\omega^2(\theta) \simeq -\frac{1}{\theta}\frac{d^2\theta}{dt^2}$$

Calculate the average of $1 - \frac{1}{6}\theta^2$, assuming $\theta = \theta_0 \sin\omega t$ to find

$$\omega^2(\theta_0) \simeq \frac{g}{l}\left(1 - \frac{1}{12}\theta_0^2\right)$$

A more exact analysis to be presented in Chapter 15 yields the following correction to the frequency,

$$\omega^2(\theta_0) \simeq \frac{g}{l}\left(1 - \frac{1}{8}\theta_0^2\right)$$

or

$$\omega(\theta_0) \simeq \sqrt{\frac{g}{l}}\left(1 - \frac{1}{16}\theta_0^2\right)$$

Numerical method is presented in Problem 18.

18. Multiplying the nonlinear pendulum equation

$$\frac{d^2\theta}{dt^2} + \frac{g}{l}\sin\theta = 0$$

by $d\theta/dt$, we obtain the following form of energy conservation,

$$\frac{1}{2}\left(\frac{d\theta}{dt}\right)^2 - \frac{g}{l}\cos\theta = \text{constant}$$

Show that the nonlinear oscillation period $T(\theta_0)$ is given by the following integral,

$$T(\theta_0) = 4\sqrt{\frac{l}{2g}}\int_0^{\theta_0}\frac{1}{\sqrt{\cos\theta - \cos\theta_0}}d\theta$$

where $\theta_{\max}$ is the amplitude of the oscillation. In the limit of small amplitude $\theta_0 \ll 1$ rad, the integral yields

$$T(\theta_0) = 4\sqrt{\frac{l}{g}}\int_0^{\theta_0}\frac{d\theta}{\sqrt{\theta_0^2 - \theta^2}}$$

$$= 4\sqrt{\frac{l}{g}}\int_0^1\frac{1}{\sqrt{1 - x^2}}dx = 2\pi\sqrt{\frac{l}{g}}$$

as expected. Note that

$$\int_0^1\frac{1}{\sqrt{1 - x^2}}dx = \arcsin(1) = \frac{\pi}{2}$$

Find T by numerical integration when $\theta_{\max} = 0.1$ and 1 rad.

19. Water in a rectangular pan (or lake) undergoes sloshing oscillation at a frequency

$$\omega = \sqrt{\frac{12hg}{L^2}}$$

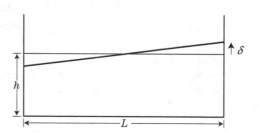

FIGURE 1–19

Problem 19.

where L is the larger side of the rectangle and h is the depth of water. Derive the frequency by showing that the center of mass of the water follows a parabolic trajectory

$$y = 6\frac{h}{L^2}x^2$$

where (x, y) are the coordinates measured from the equilibrium position.

20. The moment of inertia of a body having a mass M about an arbitrary pivot point is given by

$$I = I_{CM} + Md^2$$

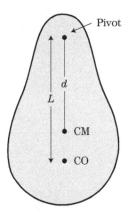

FIGURE 1–20

Problem 20.

where I_{CM} is the moment of inertia about the center of mass and d is the distance between the pivot and center of mass.

(a) Show that the oscillation frequency of the physical pendulum is

$$\omega = \sqrt{\frac{Mdg}{I}}$$

(b) The length of a simple pendulum having the same frequency can be found from

$$\sqrt{\frac{g}{L}} = \sqrt{\frac{Mdg}{I}} \quad L = \frac{I_{CM} + Md^2}{Md} = \frac{I_{CM}}{Md} + d$$

Explain that the sweet point of the body is located at distance L from the pivot. (If a baseball is hit at the sweet point of a bat, the grip experiences minimum impact.)

21. Show that an incomplete arc of radius a pivoted at the midpoint oscillates at the frequency

$$\omega = \sqrt{\frac{g}{2a}}$$

independent of the arc angle α. (The case in Problem 7 is for a complete hoop $\alpha = 2\pi$.)

Wave Motion

2.1   Introduction

The world is full of all kinds of waves. Sound waves, waves on the water surface, electromagnetic waves (radio and TV waves, microwaves, visible light, ultraviolet that are invisible, and X rays), earthquake waves, and brain waves are only a few examples. An overview of wave properties is presented in this chapter.

Waves are different from oscillations in the sense that waves propagate through a certain medium or they move in space, while oscillations are localized at one point in space. Consider sound waves in air. Our vocal cord is essentially an oscillator that is localized in our throat. But sound waves created by a vocal cord propagate through air, which is the medium for sound waves. Thus whenever we talk about waves, we have to have a media in which the waves propagate and the media must have spatial spread, whether large or small. For oscillations, time is only the *independent variable*. If we choose time t, we can automatically find the instantaneous values of any oscillating physical quantities. For waves, however, *we have another independent variable, the spatial coordinate x.* The variable x should not be confused with that used in Chapter 1 to describe the displacement of the oscillating mass. There $x(t)$ was a *dependent variable* that depended only on the independent variable of time t. In this chapter, the general properties of wave motion are outlined.

Creation of Waves on a String

Consider a mass M hanging from the ceiling with a spring (Figure 2–1). This mass-spring system would start oscillating if we gave it an initial pull or push. The oscillation would continue forever if all friction losses are negligibly small. Now connect a light string to the oscillating mass, at a certain instant. The end connected to the mass would start oscillating with the mass. At the same time a wavy structure starts propagating along the string, with a well-defined *spatial period* λ. We denote propagation velocity by c_w. The spatial period becomes shorter if the oscillation frequency of the mass increases (Figure 2–2).

Figure 2–1 depicts snapshots taken at increasing values of time. A point on the string moves only up and down although the wavy structure itself is moving from left to right. These are mechanical waves. The oscillating mass is the driver for the wave in this example. What is transferred from the oscillating

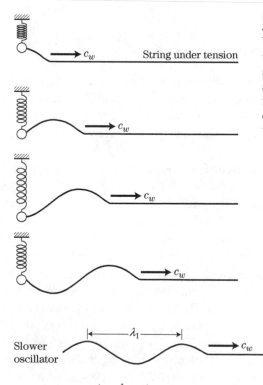

FIGURE 2–1

When a string is suddenly connected to an oscillating mass-spring system, sinusoidal waves start propagating along the string. One can view these figures as snapshots taken by a camera at sequentially increasing values of time.

FIGURE 2–2

The spatial period λ (m) decreases if the frequency is increased.

mass to the wave is nothing more than energy (and momentum, as we will see later). Thus it is expected that the oscillation amplitude of the mass will become smaller and smaller since the mass-spring system is transferring its energy to the wave on the rope. In other words, the oscillating mass-spring system is doing work. Of course, the mass-spring system can be replaced with other driving mechanisms such as our hands. Then the energy reserve can be large and we can create waves as long as we want.

The spatial period λ (m) is called the *wavelength*. The wavelength is related to the frequency ν through

$$\nu\lambda = c_w \tag{2.1}$$

since in one second the oscillating mass creates ν waves, which propagate with the velocity c_w. Thus an observer looking at a certain point on the string will see ν oscillations passing by in one second, or the oscillation period is $1/\nu = T$ (Figure 2–3). If we know the propagation velocity c_w we can immediately find the wavelength that the wave with a given frequency ν should have.

One of the major objectives of wave studies is to find this propagation velocity c_w, for various kinds of waves, mechanical or electromagnetic. The

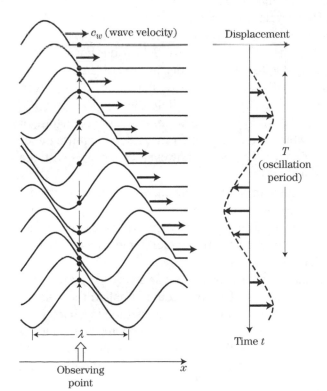

FIGURE 2–3

Sinusoidal waves as seen by a stationary observer. Note that the observer detects a sinusoidal variation in time.

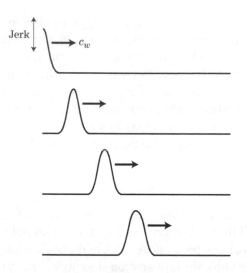

FIGURE 2–4

Waves do not have to be sinusoidal. Here a single pulse is created by a sudden jerk at one end of the line.

propagation velocity is the fundamental quantity to characterize waves and is determined by the physical constants of the wave medium. You may know that the sound velocity in air is approximately 340 m/s and the velocity of light in free space is 3.0×10^8 m/s. From where do these numbers come? How can we predict the sound velocity in water and the velocity of light in free space? These questions will be answered in subsequent chapters.

The waveform created on the string was sinusoidal (or harmonic) in the example. But this was simply due to the oscillatory motion of the mass. If a wave source oscillates sinusoidally, sinusoidal waves are created. However, waves do not have to be sinusoidal. Suppose we hold the rope and give it a sudden jerk at one end (Figure 2–4). This would be similar to the launching of a wave that propagates on a whip. Then the waveform created is a pulse rather than a continuous sinusoidal wave. The pulse, however, propagates with the same speed as the sinusoidal wave.

Actually, waves we encounter in daily life are hardly simple sinusoidal waves. They all have complicated wave structures. The human voice can easily be observed by detecting it with a microphone and displaying the signal on an oscilloscope. As you will find, the waveform is rather complicated, it hardly resembles a clear sinusoidal wave.

No matter how complicated the waveform, there is a way to approximate it in terms of many sinusoidal waves. This procedure is called the *Fourier analysis* and we will study this powerful mathematical method in Chapter 14. Here all you have to remember is that any function can be approximated by adding many sinusoidal functions. For this reason, we initially can discuss waves in terms of simple sinusoidal functions, cos or sin, without much of a loss in generality.

Sinusoidal (Harmonic) Waves

Consider a sinusoidal wave with an amplitude A, frequency of ν cycles per second, and wavelength λ (m) propagating in the positive x direction with a velocity c_w (m/s). It should be noted at this point that the wave could also propagate in the negative x direction and its velocity would be $-c_w$ where c_w is a positive number and the minus sign indicates the direction of propagation. We want to find a mathematical expression that contains all the preceding information. We again consider snapshots taken at equal time intervals, as shown in Figure 2–5(a) for the wave that is propagating in the positive x direction. We choose the reference time $t = 0$ so that the snapshot taken at this time is described by the function

$$A \sin\left(\frac{2\pi}{\lambda}x\right), \quad t = 0 \tag{2.2}$$

which is indeed periodic with a spatial period λ (m). The next snapshot is taken at $t = \tau$, by which time the whole wave pattern has moved in the positive x direction by a distance $c_w \tau$ (m). Since any function $f(x)$ shifted in the positive x direction by a distance a is given by $f(x - a)$, the equation to describe the wave pattern at $t = \tau$ is given by

$$A \sin\left(\frac{2\pi}{\lambda}(x - c_w \tau)\right) \tag{2.3}$$

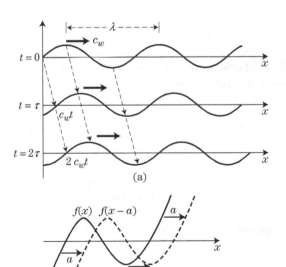

(a)

(b)

FIGURE 2–5

(a) Snapshots taken at successive times. The sinusoidal wavetrain at $t = 0$ is shifted or translated by a distance $c_w t$. (b) If a function $f(x)$ is shifted in the positive x direction by a distance a, the function is described by $f(x - a)$. If the shift is in the negative x direction by a distance a, $f(x + a)$ should be used.

At $t = 2\tau$, a third snapshot is taken, and the equation for this wave pattern is given by

$$A \sin\left(\frac{2\pi}{\lambda}(x - 2c_w\tau)\right) \tag{2.4}$$

and so on. We can easily generalize this argument to the case of an arbitrary time t and the general equation to describe the wave propagation is given by

$$A \sin\left(\frac{2\pi}{\lambda}(x - c_w t)\right) \tag{2.5}$$

Here we introduce a function $f(x, t)$ and equate this to the previous expression:

$$f(x, t) = A \sin\left(\frac{2\pi}{\lambda}(x - c_w t)\right) \tag{2.6}$$

The wave that is propagating in the $-x$ direction with the same exciting signal will be written as

$$g(x, t) = A \sin\left(\frac{2\pi}{\lambda}(x + c_w t)\right) \tag{2.7}$$

Although we have assumed that the excitation of the wave propagating in the $+x$ direction was due to a sinusoidal signal, the excitation could have been due to a cosinusoidal signal and the resulting wave would be described with the expression

$$\hat{f}(x, t) = A \cos\left(\frac{2\pi}{\lambda}(x - c_w t)\right) \tag{2.8}$$

Finally, it is also possible to make use of the the Euler identity, which relates the exponential function to the trigonometric functions by separating the real and imaginary parts

$$e^{i\theta} = \cos\theta + i\sin\theta = \text{Re}(e^{i\theta}) + i\text{Im}(e^{i\theta}) \tag{2.9}$$

and write that

$$f_{\sin}(x, t) = A \, \text{Im} \, \exp\left[i\frac{2\pi}{\lambda}(x - c_w t)\right] \tag{2.10}$$

where the amplitude A is assumed to be a real quantity. Similarly, Eq. (2.8) can be written as

$$\hat{f}_{\cos}(x, t) = A \, \text{Re} \, \exp\left[i\frac{2\pi}{\lambda}(x - c_w t)\right] \tag{2.11}$$

We note that there are several methods of expressing the same quantity and they will be used throughout this book.

The quantity A can represent the amplitude associated with any wave. For the case of the wave on the string, A can be chosen as the transverse displacement associated with the wave, and thus has the dimensions of distance (m). For sound waves in air, A can be the displacement of air molecules about their equilibrium positions; for electromagnetic waves A can be the amplitudes of the electric and magnetic fields associated with the waves; and so on. All kinds of sinusoidal waves can be cast into the preceding general expression.

Let us fix the point $x = x_0$ (constant). This corresponds to observing the waves passing by as a function of time only. The observer must observe the point on the string at $x = x_0$ moving up and down in a sinusoidal manner. The frequency for this oscillation is equal to v. Mathematically, we have

$$f(x_0, t) = A \sin\left(\frac{2\pi}{\lambda} - \omega t\right) \tag{2.12}$$

which must also have a sinusoidal time dependence with a frequency v. Thus

$$\frac{2\pi}{\lambda} c_w = 2\pi v$$

or

$$c_w = v\lambda$$

This result is consistent with our previous qualitative argument.

Since $\omega = 2\pi v$, we may write $f(x, t)$ as

$$f(x, t) = A \sin\left(\frac{2\pi x}{\lambda} - \omega t\right) \tag{2.13}$$

Furthermore, we introduce a quantity defined by

$$k = \frac{2\pi}{\lambda} \tag{2.14}$$

which has the dimensions of rad/m. The quantity k is called the *wavenumber*, although the actual number of waves in 1 m is given by $1/\lambda$. More appropriately, k should be called the angular wavenumber just as we identified ω as the angular frequency. However, we follow tradition and use only the word wavenumber. You must be careful about the factor 2π whenever you use ω and k. Also, in electrical engineering, β is frequently used instead of k and is called the *phase constant*.

Using the angular frequency ω and the wavenumber k, Eq. (2.1) can be rewritten as

$$c_w = \frac{\omega}{k} \qquad (2.15)$$

You may wonder why we use ω and k instead of ν and λ, particularly since ν and λ alone can fully describe wave motion. The reason is more than just eliminating the factor 2π in Eq. (2.14). There is a lot more to it. The difference between ω and ν is not significant. They differ only by a factor 2π. The use of k instead of λ (or $1/\lambda$) is more fundamental. We will see later that k is actually a vector directed in the direction of the wave propagation.

EXAMPLE 2.1

The velocity of *displacement waves* on a string is 40 m/s. A driver (i.e., source) oscillating with a frequency 15 Hz is connected to one end of the string. Find the wavelength of the displacement wave created on the string.

Solution

Since the wavelength is given by

$$\lambda = \frac{c_w}{\nu}$$

we find

$$\lambda = \frac{40 \, \text{m/sec}}{15 / \text{sec}} = 2.67 \, \text{m}$$

EXAMPLE 2.2

In the preceding example, determine A, k, and ω if the wave is written as

$$\xi(x, t) = A \sin (kx - \omega t)$$

Assume the driver oscillates with an amplitude of 1.0 cm. The amplitude A is 1.0 cm.

Solution

From $k = 2\pi/\lambda$, we find

$$k = \frac{2\pi}{2.67 \, \text{m}} = 2.36 \, \text{rad/m}$$

and from $\omega = 2\pi\nu$,

$$\omega = 2\pi \times 15 = 94.25 \, \text{rad/s}$$

Wave Differential Equation, Partial Differentiation

In Chapter 1 we found the solution for oscillation satisfies a differential equation of the form

$$\frac{d^2 f}{dt^2} + \omega^2 f = 0 \qquad (2.16)$$

The solutions of this differential equation are indeed sinusoidal as we have already seen. Here we ask the following question: What differential equation can yield the wave function such as that we have previously described, namely $A \sin(kx - \omega t)$?

Since $\sin(kx - \omega t)$ is periodic for both spatial coordinate x and time t, we expect that the differential equation contains both

$$\frac{d^2 f}{dx^2} \quad \text{and} \quad \frac{d^2 f}{dt^2} \qquad (2.17)$$

Let us then try to calculate these second order derivatives for the particular function $f(x, t) = A \sin(kx - \omega t)$. Since

$$\frac{df}{dt} = \frac{df}{d(kx - \omega t)} \frac{d(kx - \omega t)}{dt} \qquad (2.18)$$

using the chain rule, we find

$$\frac{df}{dt} = -\omega A \cos(kx - \omega t) \qquad (2.19)$$

Further differentiation yields

$$\frac{d^2 f}{dt^2} = -\omega^2 A \sin(kx - \omega t) \qquad (2.20)$$

Similarly, for $\dfrac{d^2 f}{dx^2}$ we find

$$\frac{d^2 f}{dx^2} = -k^2 A \sin(kx - \omega t) \qquad (2.21)$$

But we know that

$$\frac{\omega}{k} = c_w$$

Then the function $f(x, t) = A \sin(kx - \omega t)$ satisfies the following differential equation:

$$\frac{d^2 f}{dt^2} = c_w^2 \frac{d^2 f}{dx^2} \qquad (2.22)$$

This may not be the only differential equation satisfied by the function $f(x, t)$, but it is at least one candidate.

Here we need some mathematics regarding the extension of differentiation. In the preceding calculation, we took for granted the following variable transformation:

$$\frac{df}{dt} = \frac{df}{d(kx - \omega t)}\frac{d}{dt}(kx - \omega t) \tag{2.23}$$

$$\frac{df}{dx} = \frac{df}{d(kx - \omega t)}\frac{d}{dx}(kx - \omega t) \tag{2.24}$$

with

$$\frac{d}{dt}(kx - \omega t) = -\omega \tag{2.25}$$

$$\frac{d}{dx}(kx - \omega t) = k \tag{2.26}$$

However

$$\frac{d}{dt}(kx - \omega t) = -\omega$$

is valid (or meaningful) only if x does not depend on t. Since x and t are independent variables, we can choose in fact any values of x and t to calculate the preceding differentiation. However, it is meaningful only if we fix the value of x, and as long as the differentiation is concerned, the spatial coordinate x should be momentarily frozen, x = constant. Physically, this corresponds to the situation in which we fix the observation point on the string and look at only the time variation of the wave motion at that particular point. Thus, more rigorously, we should have written Eq. (2.25) as

$$\frac{d}{dt}(kx - \omega t)|_{x=\text{const.}} = -\omega \tag{2.27}$$

Such a differentiation is called *partial differentiation*, and instead of the usual d/dt, we write

$$\frac{\partial}{\partial t}(kx - \omega t) = -\omega \tag{2.28}$$

The notation $\partial/\partial t$ automatically means that the spatial coordinate x is frozen and we are to carry out the differentiation with respect to time in the conventional manner.

Partial comes from the fact that the differentiation $\partial/\partial t$ can give us the time derivative only and says nothing about the spatial derivative. The spatial partial derivative can be written as

$$\frac{\partial}{\partial x}(kx - \omega t) = k \qquad (2.29)$$

in which we now freeze the time t and vary only the spatial coordinate x. Physically, this corresponds to taking a snapshot picture at a given time.

The partial derivative of the function $f(x, t)$, which depends on both x and t. can be defined in exactly the same manner. For example, $\partial/\partial t$ indicates that we carry out the usual differentiation with respect to t by freezing x or assuming that x is a constant. Then for $f(x, t) = A \sin(kx - \omega t)$, we have

$$\frac{\partial f}{\partial t} = \frac{d[A \sin(kx - \omega t)]}{d(kx - \omega t)}\frac{\partial}{\partial t}(kx - \omega t) = -\omega A \cos(kx - \omega t)$$

Note that we did not write

$$\frac{\partial[A \sin(kx - \omega t)]}{\partial(kx - \omega t)}$$

since $A \sin(kx - \omega t)$ contains both variables x and t in the form $kx - \omega t$, which is now a single variable with respect to which we differentiate the function $A \sin(kx - \omega t)$. In this case we can define a new variable X where $X = kx - \omega t$ and the derivative becomes

$$\frac{d}{dX}A \sin X = A \cos X = A \cos(kx - \omega t)$$

This is simply an ordinary derivative with respect to X. Differentiation in this form is called *total differentiation.*

Using the symbols $\partial/\partial t$ and $\partial/\partial x$, we can rewrite Eq. (2.17) as

$$\frac{\partial^2 f}{\partial t^2} = c_w^2 \frac{\partial^2 f}{\partial x^2} \qquad (2.30)$$

where

$$\frac{\partial}{\partial t}\left(\frac{\partial f}{\partial t}\right) = \frac{\partial^2 f}{\partial t^2} \quad \text{and} \quad \frac{\partial}{\partial x}\left(\frac{\partial f}{\partial x}\right) = \frac{\partial^2 f}{\partial x^2}$$

as in ordinary differentiation.

The partial differential equation we just found is called the wave (differential) equation, and we will encounter this many times in the chapters to follow. Most waves, mechanical and electromagnetic, can be cast into this form of a partial differential equation. Whenever we end up with a differential equation

$$\frac{\partial^2 f}{\partial t^2} = \text{constant}\frac{\partial^2 f}{\partial x^2} \qquad (2.31)$$

where the constant is *positive,* we can immediately find a propagation velocity from

$$c_w = \sqrt{\text{constant}} \qquad (2.32)$$

For example, for electromagnetic waves in free space, the constant will become

$$\text{constant} = \frac{1}{\varepsilon_0 \mu_0} \qquad (2.33)$$

where $\varepsilon_0 = 8.85 \times 10^{-12} \simeq \dfrac{1}{36\pi} \times 10^{-9}$ F/m is the vacuum permittivity and $\mu_0 = 4\pi \times 10^{-7}$ H/m is the vacuum permeability. The numerical value of $c = \dfrac{1}{\sqrt{\varepsilon_0 \mu_0}}$ is 3.0×10^8 m/s, the velocity of light in free space.

EXAMPLE 2.3

A two-variable function is given by $f(x, y) = x^2 + y^2$.

(a) Make a three-dimensional plot of $f(x, y)$.
(b) Calculate $\partial f/\partial x$ and $\partial f/\partial y$.
(c) Indicate how $\partial f/\partial x$ and $\partial f/\partial y$ can be graphically shown in the plot for $f(x, y)$.

Solution

(a) See Figure 2–6.

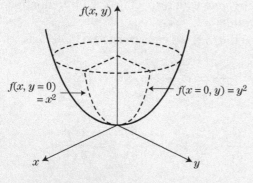

FIGURE 2–6

Plot of $f(x, y) = x^2 + y^2$.

(b) $\dfrac{\partial f}{\partial x} = 2x, \dfrac{\partial f}{\partial y} = 2y.$

(c) $\partial f/\partial x$ indicates the slope of a straight line tangent to the parabola $f(x, y_0) = x^2 + y_0^2$ which is contained in the plane determined by $y = y_0$ (const.) as shown in Figure 2–7. $\partial f/\partial y$ has a similar meaning. It indicates the slope of a straight line tangent to the parabola $f(x_0, y^2) = x_0^2 + y^2$ in the plane $x = x_0$.

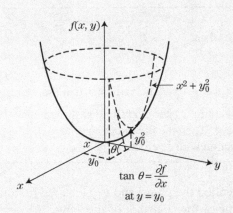

FIGURE 2–7

The partial derivative $\partial f/\partial x$ indicates the slope of the tangentline to a parabola in the plane $y = $ constant.

We can make a similar three-dimensional plot of the sinusoidal wave (see Figure 2–8),

$$f(x, y) = A \sin (kx - \omega t)$$

on the plane of x and t. The spatial wave structure at $t = 0$ progressively moves in the positive direction of x as t increases. The trajectory of points (x, t) satisfying

$$\phi = kx - \omega t = \text{constant} \tag{2.34}$$

indicates the propagation velocity c_w.

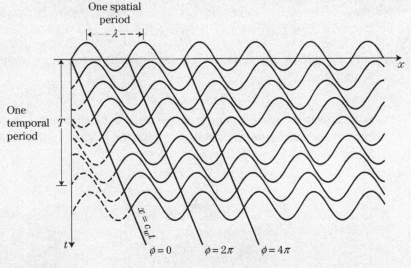

FIGURE 2–8

Propagation of sinusoidal waves shown in $x - t$ plane.

EXAMPLE 2.4

Interpret the expression $f(x, t) = A \sin(kx + \omega t)$, where $k, \omega > 0$.

Solution

We rewrite $f(x, t) = A \sin(kx + \omega t) = A \sin\left[k\left(x + \frac{\omega}{k}t\right)\right]$. This indicates that the spatially sinusoidal wave $f(x, t = 0) = A \sin(kx)$ at times later than $t = 0$ is shifted in the *negative* x direction by the distance $(\omega/k)t = c_w t$. Then the function expresses a sinusoidal wave propagating in the negative x direction with a velocity ω/k.

The quantity $\phi = kx - \omega t$ is called the phase and is measured in units of radians or degrees. For a fixed time or the snapshot of the wave, the phase ϕ varies linearly with x,

$$\phi = kx - \omega t_0$$

However, since sinusoidal functions are periodic with the period of 2π, the two phases ϕ_1 and ϕ_2 that differ by a multiple of 2π (or $2m\pi$ where m is an integer) yield exactly the same value for the wave quantity.

Nonsinusoidal Waves

The wave equation

$$\frac{\partial^2 f}{\partial t^2} = c_w^2 \frac{\partial^2 f}{\partial x^2}$$

actually allows any function $f(x, t)$ to be its solution as long as f can be written as

$$f(x, t) = f(x - c_w t) \text{ or } f(x, t) = f(x + c_w t) \tag{2.35}$$

Solutions to the wave equation *do not have to be sinusoidal*, in contrast to the case of the oscillation we studied in Chapter 1. The proof is straightforward. Since

$$X = x \pm c_w t$$

$$\frac{\partial f}{\partial t} = \frac{df}{dX}\frac{\partial X}{\partial t} = \pm c_w \frac{df}{dX}$$

$$\frac{\partial^2 f}{\partial t^2} = c_w^2 \frac{d^2 f}{dX^2}$$

$$\frac{\partial f}{\partial x} = \frac{df}{dX}\frac{\partial X}{\partial x} = \frac{df}{dX}$$

$$\frac{\partial^2 f}{\partial x^2} = \frac{d^2 f}{dX^2}$$

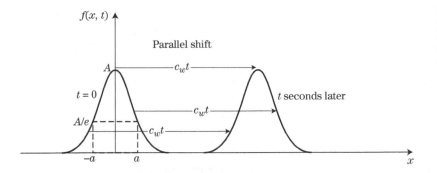

FIGURE 2–9

Exponential pulse wave
shown at two different
times.

we immediately find that

$$\frac{\partial^2 f}{\partial t^2} = c_w^2 \frac{\partial^2 f}{\partial x^2}$$

This is a surprising feature of the wave equation. The function $f(x - c_w t)$
can take any form, sinusoidal, pulse, triangle, square, and so on. Let $f(x - c_w t)$
be a Gaussian function,

$$f(x - c_w t) = A e^{-(x - c_w t)^2 / a^2} \tag{2.36}$$

where A is the amplitude and a (m) determines the spatial width of the pulse
(Figure 2–9). At the initial time $t = 0$, we have an exponential function

$$f(x, 0) = A e^{-x^2 / a^2} \tag{2.37}$$

This function is parallel-shifted or translated by a distance $c_w t$ after time t.
That is, the exponential profile propagates with the velocity c_w in the positive
x direction. This exponential waveform can approximate the wave created on
a string when we suddenly give a jerk to the end of the rope (Figure 2–4). This
is also reminiscent of the "cracking of a whip," although the propagation of
the perturbation is on an inhomogeneous line.

No matter what waves are created, they all propagate with the same speed
c_w determined by the medium. As briefly discussed earlier, the waveforms
of sounds we hear are extremely complicated, nothing like a clear sinusoidal
wave. However, as long as the propagation velocity is concerned, any sound
waveforms propagate at the same speed, that is, approximately 340 m/s at
room temperature. Sound waves of a violin and those of a flute propagate
with the same velocity. Imagine what would happen to a symphony played by
an orchestra if this were not the case. Sounds from various instruments would
be all mixed up, since they would reach your ears at different times!

You should be cautioned here that waves are not always described by
the wave equation that predicts the constancy of the propagation velocity

independent of the waveform, or wave frequency. Sound waves and electro-magnetic waves in free space can be well described by the preceding wave equation, but the waves on water surface, which is probably the most visible example of wave motion, cannot be described with this simple wave equation. The constancy of propagation velocity completely breaks down for water waves, and the propagation velocity depends on the waveform, or the wave frequency. The differential equation to describe the waves on water surface cannot be given by Eq. (2.30), but is more complicated. Another example is electromagnetic waves in matter, such as light waves in glass and water. As we will see, the velocity of light in glass is smaller than $c = 3 \times 10^8$ m/s and is approximately $0.67c$ at the wavelengths region of visible light, $\lambda = 4 \times 10^{-7} - 7 \times 10^{-7}$ m. More important, the velocity depends on the wavelength (or the wave frequency), even in that narrow wavelength region, and this in fact explains how a prism and rainbow work (Chapter 7).

2.6 Phase and Group Velocities, Dispersion

At the end of Section 2.5 it was pointed out that certain waves cannot be described by the simple wave equation

$$\frac{\partial^2 f}{\partial t^2} = c_w^2 \frac{\partial^2 f}{\partial x^2} \tag{2.38}$$

Waves described by this differential equation all have constant propagation velocities irrespective of wave frequencies or wavelengths,

$$c_w = \frac{\omega}{k} = \text{constant} \tag{2.39}$$

If we plot ω as a function of k, the graph is simply a straight line and its slope gives the propagation velocity (Figure 2–10).

Such waves with a constant propagation velocity are called nondispersive or dispersionless. If the propagation velocity depends on wave frequency, such

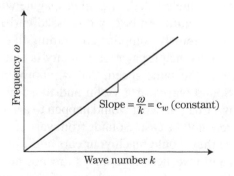

FIGURE 2–10

A graph of the frequency versus the wave number that is called an $\omega - k$ diagram. For nondispersive waves, the ratio $\omega/k = c_w = $ constant.

Slope $= \frac{\omega}{k} = c_w$ (constant)

Frequency ω

Wave number k

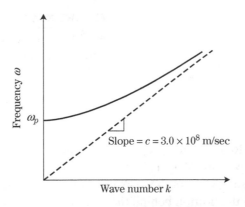

FIGURE 2–11

The $\omega - k$ *diagram* for electromagnetic waves in a plasma. The frequency ω_p is called the plasma frequency and is determined by the density of free electrons. A similar $\omega - k$ relationship holds for electromagnetic waves confined in waveguides (conducting hollow tubes).

waves are called dispersive. Dispersive waves do not have linear proportionality between ω and k. An example of dispersive waves is shown in Figure 2–11. It describes how the wave frequency ω depends on the wavenumber k for the electromagnetic waves in a highly conductive medium, such as ionospheric plasma surrounding the earth. (This we will study in Chapter 9.) Note that the slope ω / k varies depending on the frequency ω or the wavelength $\lambda = 2\pi / k$.

Nondispersive waves are described by the differential equation in Eq. (2.38). As we have seen in previous sections, a wave pattern initially created is simply translated as time goes on, keeping its waveform unchanged. If a pulse is created, the pulse propagates without deforming its shape (Figure 2–12).

For dispersive waves this simple translation does not hold. The pulse initially created is severely deformed as it propagates. The pulse width Δx becomes larger and larger, and the pulse becomes widely spread or dispersed (Figure 2–13). Now you see why we call such waves dispersive. The pulse, initially well-defined and narrow, deforms its shape. After a sufficiently long time, it becomes difficult to tell where the wave is located. Also, you can see

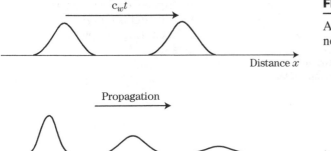

FIGURE 2–12

A pulse propagates undeformed in a nondispersive medium.

FIGURE 2–13

In a dispersive medium, the pulse is strongly deformed or dispersed.

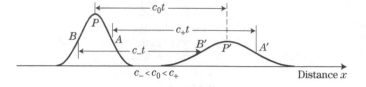

FIGURE 2–14

In a dispersive medium, the pulse is deformed as it propagates. The portion (A) in front of the peak propagates faster than the peak and the portion (B) behind the peak propagates more slowly than the peak.

that it is not easy to define the propagation velocity. By locating the peak at successive times, we can calculate how fast the peak is moving. However, the portion of the pulse in front of the peak must propagate with a faster velocity than the peak and the portion behind the peak must propagate with a slower velocity (Figure 2–14). This spreading of the pulse can be fully described once we find differential equations for dispersive waves.

Let us again consider a sinusoidal wave $f(x, t) = A \sin (kx - \omega t)$. As we have seen in Figures 2–5 and 2–8, the collection of points all having the same phase,

$$kx - \omega t = \text{constant}$$

forms a straight line on the $x - t$ plane. The slope of the line gives the propagation velocity

$$\frac{\omega}{k} = c_w$$

The velocity defined by ω / k is called the *phase velocity* since the points with the same phase propagate with this velocity. In the $\omega - k$ diagram, the phase velocity is simply given by the slope of the straight line connecting the origin 0 and a certain point on the curve describing ω as a function of k, $\omega = \omega(k)$ (Figure 2–15).

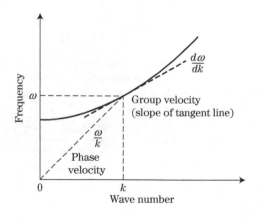

FIGURE 2–15

In dispersive waves, the group velocity defined as $d\omega/dk$ is different from the phase velocity ω / k.

The velocity determined by the slope of a tangent line at a point on the curve, $d\omega/dk$, is called the *group velocity* and can be different from the phase velocity ω/k. For nondispersive waves,

$$\omega = c_w k, \quad c_w = \text{constant}$$

and the phase and group velocities are the same. In the example shown in Figure 2–15, the group velocity is slower than the phase velocity as you can easily see. The peak of the pulse will propagate with the group velocity.

Thus we may redefine dispersive and nondispersive waves as follows:

$$\text{Dispersive wave: } \frac{d\omega}{dk} \neq \frac{\omega}{k} \tag{2.40}$$

$$\text{Nondispersive wave: } \frac{d\omega}{dk} = \frac{\omega}{k} = \text{constant} \tag{2.41}$$

Superposition of Two Waves, Beats

From where does the terminology *group velocity* come? What do we mean by a group? To answer these questions, let us consider two sinusoidal waves of equal amplitude A but slightly different frequencies ω_1 and ω_2 that are both simultaneously propagating in the positive x direction. The two waves thus have wavenumbers k_1 and k_2, respectively. The wavenumbers can be found if the relationship between ω and k is known. The sum of the two waves then becomes

$$f(x,t) = A[\sin(k_1 x - \omega_1 t) + \sin(k_2 x - \omega_2 t)] \tag{2.42}$$

We make use of the trigonometric identity

$$\sin \alpha + \sin \beta = 2\sin\frac{\alpha + \beta}{2}\cos\frac{\alpha - \beta}{2} \tag{2.43}$$

Then

$$f(x,t) = 2A\sin\left[\frac{(k_1 + k_2)x - (\omega_1 + \omega_2)t}{2}\right]$$
$$\times \cos\left[\frac{(k_1 - k_2)x - (\omega_1 - \omega_2)t}{2}\right] \tag{2.44}$$

If ω_1 and ω_2 are exactly equal, then k_1 and k_2 are too, and we simply have

$$f(x,t) = 2A\sin(kx - \omega t) \tag{2.45}$$

or the amplitude is merely doubled, as it should be. Now let us consider a case in which ω_1 and ω_2 are slightly different, so that

$$\omega_1 = \omega_2 + \Delta\omega, \quad \Delta\omega \text{ small} \tag{2.46}$$

Similarly, we let

$$k_1 = k_2 + \Delta k, \quad \Delta k \text{ small} \tag{2.47}$$

Then

$$f(x, t) = 2A \sin(k_1 x - \omega_1 t) \cos\left(\frac{\Delta k}{2} x - \frac{\Delta \omega}{2} t\right) \tag{2.48}$$

since

$$\frac{\omega_1 + \omega_2}{2} = \frac{2\omega_1 + \Delta \omega}{2} \simeq \omega_1$$

and

$$\frac{k_1 + k_2}{2} = \frac{2k_1 + \Delta k}{2} \simeq k_1$$

Let us see what this new waveform looks like. For this we take a snapshot at $t = 0$

$$f(x, t = 0) = 2A \sin(k_1 x) \cos\left(\frac{\Delta k}{2} x\right) \tag{2.49}$$

Since $\Delta k \ll k_1$, the wavelength associated with $\Delta k / 2$,

$$\lambda = \frac{2\pi}{\Delta k / 2} \tag{2.50}$$

is much longer than that corresponding to k_1,

$$\lambda_1 = \frac{2\pi}{k_1} \tag{2.51}$$

Thus the function that is a product of two sinusoidal functions as shown in Figure 2–16.

The fine ripples of the short wavelength propagate with the phase velocity

$$c_{ph} = \frac{\omega_1}{k_1}$$

Distance x

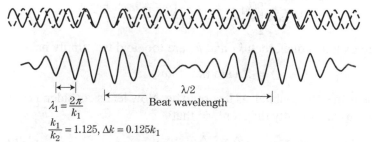

$$\lambda_1 = \frac{2\pi}{k_1}$$
$$\frac{k_1}{k_2} = 1.125, \Delta k = 0.125 k_1$$

$\lambda/2$

Beat wavelength

FIGURE 2–16

When two waves (solid and broken lines in the top figure) with slightly different wavelengths (and thus frequencies) are added, the amplitude is modulated and clumps are formed.

But the envelope determined by the function

$$\cos\left(\frac{\Delta k}{2}x - \frac{\Delta\omega}{2}t\right)$$

propagates with the velocity

$$\frac{\Delta\omega/2}{\Delta k/2} = \frac{\Delta\omega}{\Delta k}$$

By making $\Delta\omega$ and Δk sufficiently small, $\Delta\omega/\Delta k$ approaches the group velocity

$$c_g = \frac{d\omega}{dk} \qquad (2.52)$$

The clumps formed by several short waves may appropriately be called groups of waves, and these clumps propagate with the group velocity, which can be different from the phase velocity for dispersive waves. In certain instances, the two velocities may actually be in the opposite direction.

We will see in later chapters that any waves must carry energy and momentum. It will be shown that *energy is transferred with the group velocity, rather than the phase velocity*, and in this respect, the group velocity is a more fundamental quantity.

EXAMPLE 2.5

The dispersion relation of electromagnetic waves in the ionospheric plasma is given by

$$\omega^2 = \omega_p^2 + c^2 k^2$$

where ω_p is a constant (called the plasma frequency) and c is the speed of light in free space, 3×10^8 m/s. Assuming $\omega_p = 1 \times 10^8$ rad/s, calculate the phase and group velocities at a frequency of $\nu = 20$ MHz (see Figure 2–17).

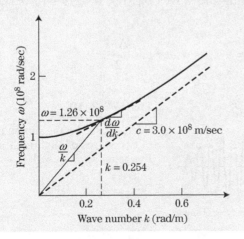

FIGURE 2–17

Dispersion relation of electromagnetic waves in the atmospheric plasma.

Solution

Since $\omega = 2\pi\nu = 2\pi \times 20 \times 10^6 = 1.26 \times 10^8$ rad/s, we find

$$k = \frac{1}{c}\sqrt{\omega^2 - \omega_p^2} = 0.254 \text{ rad/m}$$

Then the phase velocity is

$$c_{ph} = \frac{\omega}{k} = \frac{1.26 \times 10^8}{0.254} = 4.97 \times 10^8 \text{ m/s}$$

The group velocity is given by

$$c_g = \frac{d\omega}{dk} = \frac{d}{dk}\sqrt{\omega_p^2 + c^2k^2} = \frac{c^2k}{\sqrt{\omega_p^2 + c^2k^2}} = \frac{c^2}{\omega/k}$$

Then

$$c_g = \frac{(3 \times 10^8)^2}{4.97 \times 10^8} = 1.81 \times 10^8 \text{ m/s}$$

In the preceding example the phase velocity is actually larger than the speed of light c. This, however, is not in contradiction to Einstein's relativity theory, which postulates that nothing can travel faster than the speed of light. As briefly mentioned earlier, the energy of waves travels with the group velocity rather than the phase velocity, and as long as the group velocity does not exceed c, no contradiction to the relativity theory should arise.

The superposition of two waves of slightly different frequencies yields an important phenomenon called beats (Figure 2–18). Let us fix the spatial coordinate x in Eq. (2.48), say at $x = 0$. This corresponds to an observer standing at $x = 0$ and watching the waves passing by. He will observe a waveform given by

$$f(0, t) = -2A \sin \omega_1 t \cos \frac{\Delta\omega}{2}t \tag{2.53}$$

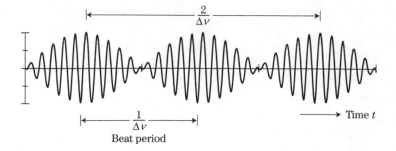

FIGURE 2–18

Beats caused by the superposition of two waves of slightly different frequencies.

Time t

Beat period

which indicates that the amplitudes of high-frequency ω_1 oscillations are modulated by the slowly varying ($\Delta\omega \ll \omega_1$) sinusoidal function, cos ($\Delta\omega t/2$). The clumps appear every $2\pi/\Delta\omega = 1/\Delta\nu$ sec. Thus in the case of sound waves, for example, one hears the sound intensity going up and down with a frequency

$$\Delta\nu = \frac{\Delta\omega}{2\pi} = |\nu_2 - \nu_1| \qquad (2.54)$$

This intensity modulation is called the beats. A piano tuner uses this beat phenomenon when tuning a piano with tuning forks.

EXAMPLE 2.6

When a certain note of a piano is sounded with a tuning fork of a frequency 580 Hz, 5 beats are heard every second. Find the frequency of the note.

Solution

Let the frequency be ν. Then

$$|580 - \nu| = 5, \quad \nu = 575 \text{ Hz or } 585 \text{ Hz}$$

The beat frequency alone cannot determine which is the frequency of the note. However, in practice the piano tuner can tell which (piano or fork) has a higher frequency by sounding one immediately after another. Beats are frequently used in tuning musical instruments. Zero beat indicates that two sound sources have the same frequency.

2.8   **Problems**

1. A displacement wave on a string is described by $0.02 \sin [2\pi(0.5x - 10t)]$ (m), where x is in meters and t in seconds. Find
   (a) The propagation velocity.
   (b) Wavelength λ and wavenumber k.
   (c) Frequency ν and angular frequency ω.
   (d) Direction of propagation.
   (e) Amplitude of the wave.

2. Find the expression for the wave in Problem 1 propagating in the negative x direction.

3. Repeat Problem 1 for a wave given by $f(x, t) = 0.03 \sin (5x - 150t)$ (m)

4. Plot $f(x, t) = 0.1 \sin [2\pi(0.5x - 20t)]$ as a function of x at $t =$ (a) 0 s, (b) 0.0125 s, (c) 0.025 s, (d) 0.0375 s, and (e) 0.05 s. Convince yourself that the wave pattern progresses in the positive x direction as time goes on.

5. The speed of electromagnetic waves in free space is 3×10^8 m/s. An AM radio station uses a frequency of 540 kHz (Hz, hertz = cycles/second). Find the wavelength. (AM is for amplitude modulation.)

6. Find the wavelength of wavy groove structure on an LP record ($33\frac{1}{3}$ rev/min) for a sound wave with a frequency 1 kHz. Consider two radial positions on the record, $r =$ (a) 15 cm, and (b) 10 cm.

7. Calculate

$$\frac{\partial f}{\partial x}, \frac{\partial f}{\partial y}, \frac{\partial^2 f}{\partial x^2}, \frac{\partial^2 f}{\partial y^2}, \frac{\partial^2 f}{\partial x \partial y}$$

   for the functions
   (a) $f(x, y) = \sin (x - 2y)$,
   (b) $f(x, y) = x^2 + xy + y^2$.

8. **(a)** Show that $f(x, t) = 0.1 \sin [2\pi (0.5x - 20t)]$ satisfies the wave equation

$$\frac{\partial^2 f}{\partial t^2} = 1600 \frac{\partial^2 f}{\partial x^2}$$

   (b) What is the propagation velocity?

9. **(a)** Plot $f(x, t) = 0.5e^{-(x-5t)^2}$ as functions of x (m) at $t = 0, 0.5$ s, 1.0 s. Convince yourself that the preceding expression describes a pulse propagating in the positive x direction.

   (b) Find the differential equation that the preceding expression should satisfy.

10. **(a)** A certain wave has the $\omega - k$ relationship (or dispersion relation) given by $\omega = 10^3 k - 3 \times 10^{-5} k^3$ Plot the frequency ω versus the wave number k for $0 \leq k \leq 3 \times 10^3$ rad/m.

   (b) Is the wave dispersive or nondispersive? Find the phase and group velocities at $k = 1 \times 10^3$ rad/m

11. **(a)** Add two sinusoidal waves

   (b) $2 \sin (5x - 1500t)$ and $2 \sin (5.1x - 1530t)$. What is the beat frequency?

   (c) What is the beat wavelength?

12. The phase velocity of surface waves in deep water is approximately given by

$$\frac{\omega}{k} = \sqrt{\frac{g}{k} + \frac{T_s}{\rho_m} k}$$

where T_s is the surface tension (7.3×10^{-2} N/m) and $\rho_m = 10^3$ kg/m^3 is the water mass density.

   (a) Find the expression for the group velocity.

   (b) When a stone is thrown into a pond, waves of many wavelength components start propagating. What would be the slowest wavelength? Note: A stone thrown into water gives an impulse-like disturbance, which contains many wavelength components.

13. A certain wave in a magnetized plasma obeys the dispersion relation

$$\omega = \text{constant } k^2$$

where k is the wavenumber along the magnetic field. (The wave of concern is the whistler wave propagating along the earth magnetic field.) Argue that short wavelength (thus high-frequency) components propagate faster than long wavelength components.

Some Mathematics

Taylor Series Expansion

The Taylor expansion of mathematical functions is frequently used in the following chapters. To familiarize you with this powerful mathematical technique, we devote one chapter to it and related mathematical formulas. Taylor's theorem states that: A function $f(x)$ can be expanded in terms of a power series of x as

$$f(x) = f(a) + f'(a)(x - a) + \frac{f''(a)}{2!}(x - a)^2 + \frac{f'''(a)}{3!}(x - a)^3$$
$$+ \cdots + \frac{f^{(n)}(a)}{n!}(x - a)^n \cdots \tag{3.1}$$

where a is an arbitrary value of x and $f^{(n)}(a)$ is the nth order derivative of $f(x)$ evaluated *at* $x = a$,

$$f^{(n)}(a) = \frac{d^n}{dx^n} f(x)|_{x=a}$$

The proof is straightforward. Let us assume that $f(x)$ can be expanded in a power series as

$$f(x) = A_0 + A_1(x - a) + A_2(x - a)^2 + \cdots = \sum_{m=0}^{\infty} A_m(x - a)^m$$

where A_m's are constants that are to be determined. A_0 can be immediately specified as $f(a)$ by letting $x = a$. To determine A_m, we differentiate $f(x)$ n times,

$$\frac{d^n}{dx^n} f(x) = \sum_{m \geq n} A_m m(m - 1)(m - 2) \cdots (m - (n - 1))(x - a)^{m-n}$$

and let $x = a$. Then only the term with $m = n$ remains nonzero and

$$\frac{d^n}{dx^n} f(x)|_{x=a} = n(n - 1) \cdots 3 \cdot 2 \cdot 1 \, A_n = n! A_n$$

or

$$A_n = \frac{1}{n!} f^{(n)}(a)$$

which gives Eq. (3.1).

As an example, let $f(x) = \sin x$. Since

$$\frac{d}{dx} \sin x = \cos x$$

$$\frac{d^2}{dx^2} \sin x = - \sin x$$

$$\frac{d^3}{dx^3} \sin x = - \cos x$$

$$\vdots$$

we find

$$\sin x = \sin a + (\cos a)(x - a)$$

$$- \frac{\sin a}{2!}(x - a)^2 - \frac{\cos a}{3!}(x - a)^3$$

$$+ \frac{\sin a}{4!}(x - a)^4 + \frac{\cos a}{5!}(x - a)^5 \cdots \tag{3.2}$$

If we choose $a = 0$, then $\sin 0 = 0$ and $\cos 0 = 1$ and we find

$$\sin x = x - \frac{1}{3!}x^3 + \frac{1}{5!}x^5 - \frac{1}{7!}x^7 + \cdots \tag{3.3}$$

Eq. (3.3) is the series expansion of $\sin x$ about the origin $x = 0$. If x is small (x is measured in radians here; small x means $|x| \ll 1$ rad), $\sin x$ can be approximated by $\sin x \simeq x$. (Recall that this approximation was used for analyzing pendulum oscillation in Chapter 1.) The evaluation of a Taylor series about the origin is sometimes called a MacLaurin series.

A similar power series expansion can be found for other widely used functions that you should verify for yourself.

$$\cos x = 1 - \frac{1}{2!}x^2 + \frac{1}{4!}x^4 - \frac{1}{6}x^6 + \cdots \tag{3.4}$$

$$\tan x = x + \frac{1}{3}x^3 + \frac{2}{15}x^5 + \cdots \tag{3.5}$$

$$e^x = 1 + x + \frac{1}{2!}x^2 + \frac{1}{3!}x^3 + \frac{1}{4!}x^4 + \cdots \tag{3.6}$$

A particularly important case that we will frequently encounter in this book is the binomial expansion

$$(1 + x)^n = 1 + nx + \frac{n(n-1)}{2!}x^2 + \frac{n(n-1)(n-2)}{3!}x^3 + \cdots \quad (3.7)$$

where n is an arbitrary number (not necessarily an integer). For this we let $f(x) = (1 + x)^n$ in Eq. (3.1), and $a = 0$. Then

$$\left.\frac{df(x)}{dx}\right|_{x=0} = n$$

$$\left.\frac{d^2 f(x)}{dx^2}\right|_{x=0} = n(n-1)$$

$$\left.\frac{d^3 f(x)}{dx^3}\right|_{x=0} = n(n-1)(n-2)$$

$$\vdots$$

and we readily obtain Eq. (3.7). We have already used this binomial expansion in Chapter 1; n can be any number, positive, negative, or even complex. For example, let $n = \frac{1}{2}$. Then

$$\sqrt{1+x} = (1+x)^{1/2} = 1 + \frac{1}{2}x - \frac{1}{8}x^2 + \frac{3}{48}x^3 - \cdots \quad (3.8)$$

For $n = -1$, we find

$$\frac{1}{1+x} = 1 - x + x^2 - x^3 + x^4 \cdots \quad (3.9)$$

If x is small, we may approximate $(1 + x)^n$ with

$$(1 + x)^n \simeq 1 + nx \quad (3.10)$$

Let x be replaced by $x + a$ in Eq. (3.1). Then we find another form of the Taylor series expansion

$$f(x + a) = f(x)|_{x=a} + \left.\frac{df(x)}{dx}\right|_{x=a} x + \frac{1}{2!}\left.\frac{d^2 f(x)}{dx^2}\right|_{x=a} x^2 + \cdots \quad (3.11)$$

Thus if x is small, we can approximate $f(x + a)$ by

$$f(x + a) = f(x)|_{x=a} + \left.\frac{df(x)}{dx}\right|_{x=a} x \quad (3.12)$$

retaining only the first-order term in the power series. This approximation will be frequently used in the following chapters. Only terms with a linear variation are retained in the expansion of a nonlinear function. In other words, we have *linearized* the nonlinear function by assuming that the variations of the function about its linear value will be small.

Complex Numbers and Variables

Using Eqs. (3.3), (3.4), and (3.6), we can prove Euler's formula which we have used in the previous chapter:

$$e^{i\theta} = \cos\theta + i\sin\theta \tag{3.13}$$

where $i = \sqrt{-1}$. To prove this, let $x = i\theta$ in Eq. (3.6) where θ is a real number (positive or negative). Then

$$e^{i\theta} = 1 + i\theta + \frac{1}{2!}(i\theta)^2 + \frac{1}{3!}(i\theta)^3 + \frac{1}{4!}(i\theta)^4 + \cdots$$

$$= 1 + i\theta - \frac{1}{2!}\theta^2 - i\frac{1}{3!}\theta^3 + \frac{1}{4!}\theta^4 + i\frac{1}{5!}\theta^5 - \frac{1}{6!}\theta^6 + \cdots$$

$$= \left[1 - \frac{1}{2!}\theta^2 + \frac{1}{4!}\theta^4 + \frac{1}{6!}\theta^6 \cdots\right] + i\left[\theta - \frac{1}{3!}\theta^3 + \frac{1}{5!}\theta^5 - \cdots\right]$$

$$= \cos\theta + i\sin\theta \tag{3.14}$$

where the terms in the expansion have been regrouped to yield the trigonometric functions. Using this result, we can also depict a complex number $a + ib$ (a, b real) in Cartesian coordinates where the horizontal axis is the real part of the complex number and the vertical axis is the imaginary part of the complex number, as shown in Figure 3–1. This leads to the polar representation of the complex number. This is sometimes called the *phasor* representation of a complex number. It is frequently used in the analysis of electrical circuits where the excitation source is a sinusoidal source with a well-defined frequency of oscillation. The polar or phasor representation is

$$re^{i\theta} \tag{3.15}$$

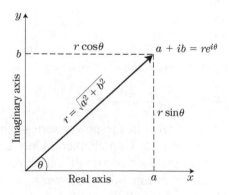

FIGURE 3–1

Polar representation of the complex number $a + ib = re^{i\theta}$.

The magnitude and the phase angle in the polar or phasor representation are given by

$$r = \sqrt{a^2 + b^2} \tag{3.16}$$

$$\theta = \tan^{-1}\left(\frac{b}{a}\right) \tag{3.17}$$

From Figure 3–1, we note

$$re^{i\theta} = \sqrt{a^2 + b^2}\,(\cos\theta + i\sin\theta)$$

and

$$\cos\theta = \frac{a}{\sqrt{a^2 + b^2}}$$

$$\sin\theta = \frac{b}{\sqrt{a^2 + b^2}}$$

EXAMPLE 3.1

Write $1 \pm i\sqrt{3}$ in the polar representation form (Figure 3–2).

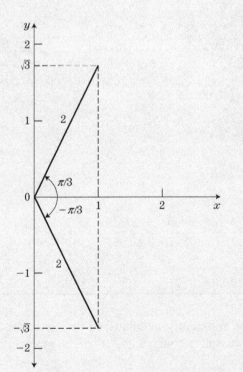

FIGURE 3–2

The complex number $1 \pm i\sqrt{3}$ is represented on the complex plane.

Solution

Since $a = 1, b = \pm\sqrt{3}$, we find $1 \pm i\sqrt{3} = \sqrt{1+3}e^{\pm i\theta}$, where

$$\theta = \tan^{-1}\left(\frac{\sqrt{3}}{1}\right) = \frac{\pi}{3}\text{rad}$$

Then

$$1 \pm i\sqrt{3} = 2e^{\pm i\frac{\pi}{3}}$$

EXAMPLE 3.2

Assume that a function of the form $f(x, t) = Ae^{i(kx-\omega t)}$ is a solution of the wave equation

$$\frac{\partial^2 f}{\partial t^2} = c_w^2 \frac{\partial^2 f}{\partial x^2}$$

Show that f satisfies the wave equation provided that $\omega^2 = c_w^2 k^2$.

Solution

Since

$$\frac{\partial}{\partial x} Ae^{i(kx-\omega t)} = Aike^{i(kx-\omega t)}$$

$$\frac{\partial^2}{\partial x^2} Ae^{i(kx-\omega t)} = A(ik)^2 e^{i(kx-\omega t)}$$

$$\frac{\partial}{\partial t} Ae^{i(kx-\omega t)} = A(-i\omega)e^{i(kx-\omega t)}$$

$$\frac{\partial^2}{\partial t^2} Ae^{i(kx-\omega t)} = A(-i\omega)^2 e^{i(kx-\omega t)}$$

we find

$$-A\omega^2 e^{i(kx-\omega t)} = -c_w^2 Ak^2 e^{i(kx-\omega t)}$$

or

$$\omega^2 = c_w^2 k^2$$

This particular function will be frequently encountered in this book. This example also indicates that the partial derivative $\partial f/\partial t$ can be replaced with $-i\omega f$ if $f(x, t)$ is assumed to be

$$Ae^{i(kx-\omega t)}$$

Similarly, $\partial f / \partial x$ can be replaced with ikf. When we need a solution of the form $A \cos(kx - \omega t)$, we can take the real part of $Ae^{i(kx-\omega t)}$. For $A \sin(kx - \omega t)$, we can take the imaginary part of $Ae^{i(kx-\omega t)}$. This so-called *operator method* greatly simplifies mathematical analyses of oscillations and wave phenomena. We will study this subject in more detail in Chapter 13. This technique is used to obtain *dispersion relations* in wave propagation systems and is very important in understanding wave phenomena.

3.3 Problems

1. Prove Eqs. (3.4), (3.5), and (3.6).

2. What is the power series expansion of e^{-x}?

3. Show that

$$\cos x = \frac{e^{ix} + e^{-ix}}{2}, \quad \sin x = \frac{e^{ix} - e^{-ix}}{2i}$$

4. Electromagnetic waves in the ionosphere are described with the following differential equation,

$$\frac{\partial^2 E}{\partial t^2} + \omega_p^2 E = c^2 \frac{\partial^2 E}{\partial x^2}$$

where ω_p (called the plasma frequency) is a constant, $c = 3.0 \times 10^8$ m/s is the speed of light in a vacuum, and $E(x, t)$ is the electric field associated with the waves. Using the operator method, show that the dispersion relation for the wave is given by

$$\omega^2 = \omega_p^2 + c^2 k^2$$

5. Show that the wave differential equation to be satisfied by the dispersion relation given in Problem 10 (Chapter 2) is

$$\frac{\partial f}{\partial t} = -10^3 \frac{\partial f}{\partial x} - 3 \times 10^{-5} \frac{\partial^3 f}{\partial x^3}$$

6. Show that if Δx is small, $f(x + \Delta x)$ can be approximated by

$$f(x) + \frac{\partial f(x)}{\partial x} \Delta x$$

7. Show that the differential equation to yield the dispersion relation of water wave $\omega^2 = gk$ can be given by either

$$\frac{\partial^2 \xi}{\partial t^2} = ig \frac{\partial \xi}{\partial x} (i = \sqrt{-1})$$

or

$$\frac{\partial^4 \xi}{\partial t^4} + g^2 \frac{\partial^2 \xi}{\partial x^2} = 0$$

8. Calculate $\sqrt{i}$.

Fundamentals of Mechanical Waves

Introduction

Waves can be classified into two major categories, mechanical waves and electromagnetic waves. Mechanical waves are those that can be created and propagated in elastic material media. Sound waves in gases, liquids, and solids, and waves on an elastic string are typical examples. All mechanical waves can be described fully by Newton's equation of motion once we can find the appropriate forces that act on a small volume (or segment) of the media. Electromagnetic waves, on the other hand, do not require any material media. Vacuum is an excellent medium that supports the propagation of electromagnetic waves (light, radio and TV waves, etc.). However, electromagnetic waves can also propagate in material media, depending on the frequency of the wave. Examples include light waves that propagate in glass and water and X-rays that penetrate tissues and even metals. Maxwell's equations are needed to describe electromagnetic waves. Electromagnetic waves in material media need both Maxwell's equations and Newton's equations of motion.

In this chapter, general aspects of mechanical waves are discussed. One important requirement for a medium to accommodate mechanical waves is that the medium be elastic. If it is compressed or expanded by a force, it should be able to restore its original shape when the force is removed. Sound waves can propagate in hard soil but not in soft clay. Waves can be created on a rope under a tension but not on a rope without tension.

Another essential fact is that any material media (gas, liquid, or solid) have a mass density. Air is light but it has a finite mass density of 1.29 kg/m^3 at $0°\text{C}$ and 1 atmospheric pressure, which determines, together with the pressure, the propagation velocity of sound waves in air.

Elasticity and inertia (mass) are two major physical properties that determine the propagation velocity of mechanical waves. You recall that we have already encountered similar requirements for mechanical *oscillations*. For a mass-spring system, the spring provides the elasticity or restoring force and determines the potential energy. The mass provides the inertia of the system and determines the kinetic energy. Thus, the elasticity and inertia are physical qualities common to oscillating systems and the appropriate medium that supports the propagation of mechanical waves. The main difference is that in a propagating wave without reflection, the kinetic energy and the potential energy are the same, while in oscillations they are mutually exclusive, and the sum of the two energies is a constant.

4.2 Mass-Spring Transmission Line

We model mechanical wave propagation in a media using a large number of identical units of discrete mass-spring combinations as shown in Figure 4–1. A single unit consisting of one spring and one mass cannot accommodate waves since it has no spatial spread. For waves, we need a medium that is spread spatially and thus many mass-spring units must be connected in series.

Suppose that each unit contains a spring with a spring constant k_s and a mass M and has a length Δx that is also assumed to be the natural length of each spring. Thus if we do not apply any force to disturb the system, the masses are located at equilibrium positions: $x = \Delta x, 2\Delta x, 3\Delta x$, etc. Let us give the left end a sudden push displacing it by a distance ξ_0. The first spring is then compressed and tends to push the first mass. Because of its inertia it takes the mass a finite time before it moves to the right and then pushes the second spring. If the mass finally moves the distance ξ_0, the first spring relaxes back to its natural length. Since the first mass now moves the distance

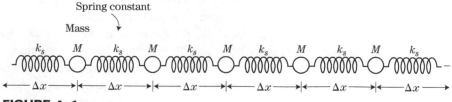

FIGURE 4–1

Mechanical transmission line composed of mass-spring units.

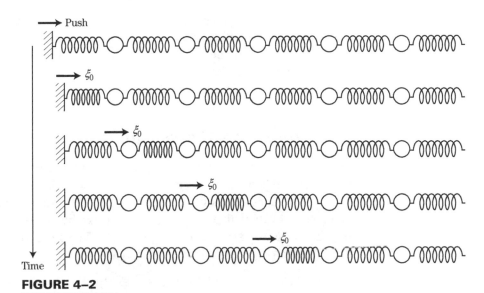

FIGURE 4-2

When a sudden push is given to one end of the transmission line, the disturbance starts propagating. The displacement of the masses from their original position is ξ_0.

ξ_0, the second spring repeats what the first spring did but after some delay. The sequence repeats on and on along the system (Figure 4–2).

The time delay depends on how strong the spring is and how heavy the mass is. If the spring is strong (larger k_s), we expect the spring can restore its original length quicker and the time delay will be shorter. If the mass (M) is larger, the time delay will be larger. Since the delay time is a measure of the wave velocity, we expect that the expression for the wave velocity should contain a factor of k_s/M.

In the preceding example, we can already see some of the fundamental properties of mechanical waves. First of all, the model transmission line has both elasticity (provided by the spring) and a mass density $M/\Delta x$ (kg/m). Therefore it is expected to be a medium for mechanical waves. Second, we can see that what propagates with the disturbance (or wave) is energy. By squeezing the first spring, we have provided it with an initial potential energy $\frac{1}{2}k_s\xi_0^2$. The spring then transfers this energy to the first mass, which now acquires a kinetic energy that is equal to the potential energy $\frac{1}{2}k_s\xi_0^2$. Since the mass then tends to squeeze the second spring, the energy is transferred to the second spring, and so on. Notice that neither springs nor masses move *along* with the disturbance. They stay more or less at their original positions that have been slightly displaced by a distance ξ_0 after the disturbance has passed. The wave velocity thus has nothing to do with the material velocity.

In fact the two velocities are independent from each other as long as the displacement ξ is sufficiently small (linear wave medium). This is about as far as we can go without using mathematics. But we wish to find out exactly how the propagation velocity is related to the factor k_s/M. We can do this with our newly acquired knowledge of the Taylor series expansion.

4.3 Derivation of a Wave Equation

Let us select one unit of the mass-spring system located at a distance x from the left end. When a wave is created, each mass deviates from its equilibrium position. We denote the displacement of the mass originally (in equilibrium) located at x by $\xi(x)$, that of the mass at $x + \Delta x$ by $\xi(x + \Delta x)$ and that of the mass at $x - \Delta x$ by $\xi(x - \Delta x)$. (Figure 4–3) The spring to the left of the mass at x suffers a net length change given by

$$\Delta x + \xi(x) - [\Delta x + \xi(x - \Delta x)] = \xi(x) - \xi(x - \Delta x) \tag{4.1}$$

and exerts a force on the mass given by

$$F_- = k_s[\xi(x) - \xi(x - \Delta x)] \tag{4.2}$$

directed to the left. The spring to the right of the mass at x similarly exerts a force

$$F_+ = k_s[\xi(x + \Delta x) - \xi(x)] \tag{4.3}$$

directed to the right. Thus the net force to act on the mass at x is

$$F = F_+ - F_- = k_s[\xi(x + \Delta x) + \xi(x - \Delta x) - 2\xi(x)] \tag{4.4}$$

You should be careful about the direction of the two forces, which is introduced only for mathematical formality. The actual direction of each force of course depends on the magnitude of displacements $\xi(x - \Delta x)$, $\xi(x)$, and

(a)

(b)

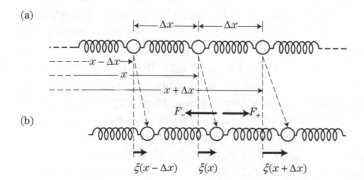

$\xi(x - \Delta x)$ $\xi(x)$ $\xi(x + \Delta x)$

FIGURE 4–3

A portion of the transmission line located at x is shown (a) without a wave (equilibrium) and (b) with a wave (perturbed).

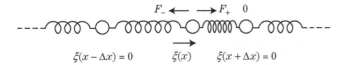

$F_- \longleftarrow \quad \longrightarrow F_+ \quad 0$

$\xi(x - \Delta x) = 0 \qquad \xi(x) \qquad \xi(x + \Delta x) = 0$

FIGURE 4–4

Sign convention of forces exerted on a mass by springs. The actual force can be opposite to the formal force.

$\xi(x + \Delta x)$. Consider the special case $\xi(x - \Delta x) = 0$ and $\xi(x + \Delta x) = 0$ as shown in Figure 4–4. A positive force directed to the left F_- is given by $k_s\xi(x)$. A negative force directed to the right F_+ is given by $-k_s\xi(x)$. This is equivalent to a positive force directed at to the left. In Figure 4–4 it can be clearly seen that both springs tend to push the mass to the left, as intuitively expected.

The new location of the mass originally located at x is given by

$$x + \xi(x) \tag{4.5}$$

But x is simply the location of the mass in equilibrium and is an independent variable. Hence our equation of motion for the mass becomes

$$M\frac{d^2\xi(x)}{dt^2} = k_s[\xi(x + \Delta x) + \xi(x - \Delta x) - 2\xi(x)] \tag{4.6}$$

However, the displacement ξ is a function of two independent variables, t and x, and we should use partial derivatives,[1]

$$M\frac{\partial^2\xi(x, t)}{\partial t^2} = k_s[\xi(x + \Delta x, t) + \xi(x - \Delta x, t) - 2\xi(x, t)] \tag{4.7}$$

We are getting closer to the wave equation described in Chapter 2. To conclude, let us expand $\xi(x + \Delta x, t)$ and $\xi(x - \Delta x, t)$ in terms of the power series of Δx using a Taylor series expansion. Assuming that Δx is small, we find

$$\xi(x + \Delta x, t) \simeq \xi(x, t) + \frac{\partial\xi}{\partial x}\Delta x + \frac{1}{2}\frac{\partial^2\xi}{\partial x^2}(\Delta x)^2 \tag{4.8}$$

$$\xi(x - \Delta x, t) \simeq \xi(x, t) - \frac{\partial\xi}{\partial x}\Delta x + \frac{1}{2}\frac{\partial^2\xi}{\partial x^2}(\Delta x)^2 \tag{4.9}$$

Then the right-hand side of Eq. (4.7) is simplified to be

$$k_s\frac{\partial^2\xi}{\partial x^2}(\Delta x)^2$$

---

[1]Equation (4.7) is called a second-order *difference* equation which is closely related to the differential equation. Under appropriate limits (such as in the long wavelength limit), a difference equation reduces to a differential equation. In the atomic or molecular scale, each molecule must be treated as a discrete particle and the difference equation is more fundamental than the differential equation, which holds only if the medium of concern can be regarded as a continuous medium. Another important application area of the difference equation can be found in the numerical integration of differential equations. In this scheme a differential equation is translated into a difference equation, which computers can handle with ease.

and the equation becomes

$$M\frac{\partial^2 \xi}{\partial t^2} = k_s(\Delta x)^2 \frac{\partial^2 \xi}{\partial x^2} \tag{4.10}$$

or

$$\frac{\partial^2 \xi}{\partial t^2} = \frac{k_s}{M}(\Delta x)^2 \frac{\partial^2 \xi}{\partial x^2} \tag{4.11}$$

This is exactly of the form of the wave equation we obtained before Eq. (2.26). The propagation velocity c_w is determined to be

$$c_w = \Delta x \sqrt{\frac{k_s}{M}} \quad (\text{m/s}) \tag{4.12}$$

which indeed contains the factor k_s/M as we had expected.

We may rewrite the velocity as

$$c_w = \sqrt{\frac{k_s \Delta x}{M/\Delta x}} \tag{4.13}$$

where $k_s \Delta x$ is called the elastic modulus of the spring and has the dimensions of force (newtons). We denote this by $K = k_s \Delta x$. This elastic modulus is actually more convenient than the spring constant k_s, since it is a normalized quantity. Let a spring have a spring constant k_s and a natural length l. To elongate (or compress) the spring by a length Δl, we have to apply a force

$$F = k_s \Delta l$$

which can be rewritten as

$$F = k_s l \frac{\Delta l}{l} = K \frac{\Delta l}{l}$$

In this form, the strain $\Delta l/l$ is a normalized quantity and the elastic modulus K is a constant determined from the material used to construct the spring and its shape but it is independent of the length of the spring.

Using the elastic modulus K, the velocity now becomes

$$c_w = \sqrt{\frac{K}{M/\Delta x}} \tag{4.14}$$

Since $M/\Delta x$ is the mass per unit length or the linear mass density (kg/m), denoted by ρ_l, we find

$$c_w = \sqrt{\frac{\text{elastic modulus}}{\text{mass density}}} = \sqrt{\frac{K}{\rho_l}} \quad (\text{m/s}) \tag{4.15}$$

In fact, all mechanical waves we will study have a similar expression for their propagation velocities. *The propagation velocities of mechanical waves are determined from the elasticity and the mass density of media.*

EXAMPLE 4.1

A spring of a total mass of 0.5 kg and a natural length of 1.5 m is elongated by 5 cm when it is stretched by a force of 20 N. Find the velocity of propagation of mechanical waves along the spring.

Solution

Since

$$F = k_l \Delta l = K \frac{\Delta l}{l}$$

we find

$$K = \frac{Fl}{\Delta l} = \frac{20\,\text{N} \times 1.5\,\text{m}}{0.05\,\text{m}} = 600\,\text{N}$$

The linear mass density is

$$\rho_l = \frac{0.5\,\text{kg}}{1.5\,\text{m}} = 0.33\,\text{kg/m}$$

The velocity of propagation is found to be

$$c_w = \sqrt{\frac{K}{\rho_l}} = 42.4\,\text{m/s}$$

Notice that in the preceding example, the mass density is that of the spring itself. You may wonder why we can use the wave equation Eq. (4.11) that was derived from a model having discrete massless springs and discrete masses. The answer is that Δx, the length occupied by the unit mass-spring combination, can be as small as we want, since the velocity is determined by the elastic modulus K and the mass density ρ_l. Both quantities remain finite no matter how small Δx is chosen. Although we started with a *discrete* medium, after taking a proper limit ($\Delta x \rightarrow 0$), we can go into a *continuous* medium, such as the spring we used in the example. The spring has a mass that is uniformly distributed along its natural length. A similar technique will be used in Chapter 9 in describing electromagnetic waves.

Energy Carried by Waves

In Chapter 2, it was briefly pointed out that any wave should carry energy and momentum with them. Let us see if this is the case for the waves on the mass-spring transmission line. We again look at the unit located at x. The displacement of the mass at this location was given by $\xi(x, t)$. Thus the velocity of the mass becomes

$$v(x, t) = \frac{\partial \xi}{\partial t} \quad \text{(velocity wave)} \tag{4.16}$$

and the kinetic energy of the mass is given by

$$\text{kinetic energy} = \frac{1}{2}M \left(\frac{\partial \xi}{\partial t}\right)^2 \quad \text{(J)} \tag{4.17}$$

The potential energy stored in the spring to the right of the mass is

$$\text{potential energy} = \frac{1}{2}k_s[\xi(x + \Delta x) - \xi(x)]^2 \tag{4.18}$$

Using the Taylor series expansion for $\xi(x + \Delta x)$, we find

$$\text{potential energy} = \frac{1}{2}k_s(\Delta x)^2 \left(\frac{\partial \xi}{\partial x}\right)^2 \tag{4.19}$$

However, since

$$\frac{\partial \xi}{\partial t} = -c_w \frac{d\xi}{dX}, \quad X = x - c_w t$$

and

$$\frac{\partial \xi}{\partial x} = \frac{d\xi}{dX}$$

we find

$$\text{kinetic energy} = \frac{1}{2}Mc_w^2 \left(\frac{\partial \xi}{\partial X}\right)^2 \tag{4.20}$$

$$\text{potential energy} = \frac{1}{2}k_s(\Delta x)^2 \left(\frac{\partial \xi}{\partial X}\right)^2 \tag{4.21}$$

Recalling

$$c_w^2 = \frac{k_s(\Delta x)^2}{M}$$

we conclude that *in traveling mechanical waves, the potential energy* (due to the elasticity) and the *kinetic energy* (due to the mass motion) *are the same everywhere and anytime.* Recall that in mechanical oscillations, the kinetic energy and potential energy are mutually exclusive.

The total energy is then

$$\text{total wave energy} = 2 \times \frac{1}{2} M c_w^2 \left(\frac{d\xi}{dX}\right)^2 = M c_w^2 \left(\frac{d\xi}{dX}\right)^2 \tag{4.22}$$

We may define the energy per unit length of the transmission line by simply dividing this equation by the length Δx,

$$\text{total wave energy density} = \rho_l c^2 \left(\frac{d\xi}{dX}\right)^2 \quad \text{(J/m)} \tag{4.23}$$

To illustrate this with an example, let us consider a sinusoidal wave,

$$\xi(x, t) = \xi_0 \sin(kx - \omega t) \tag{4.24}$$

where the phase velocity is

$$\frac{\omega}{k} = c_w = \sqrt{\frac{K}{\rho_l}} \tag{4.25}$$

Then

$$\frac{d}{dX}\xi_0 \sin[k(x - c_w t)] = \xi_0 \frac{d}{dX} \sin kX = k\xi_0 \cos[k(x - c_w t)] \tag{4.26}$$

and after substituting these terms into Eq. (4.23), we find

$$\text{wave energy density} = \rho_l c_w^2 k^2 \xi_0^2 \cos^2(kx - \omega t) \tag{4.27}$$

A snapshot of this function at $t = 0$ is plotted in Figure 4–5. These energy clumps propagate with the wave velocity c_w since Eq. (4.27) also satisfies the

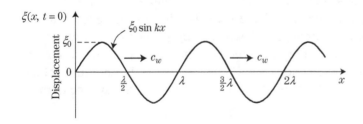

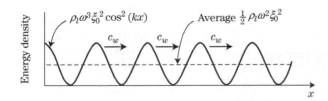

FIGURE 4–5

Displacement $\xi(x, 0)$ (top) and wave energy density (bottom). The dashed line indicates the average wave energy density $\frac{1}{2}\rho_l \omega^2 \xi_0^2$.

wave equation. The average value of the energy density is just one half of the peak value (Figure 4–5).

The average wave energy density is

$$\text{average wave energy density} = \frac{1}{2}\rho_l\omega^2\xi_0^2 \quad \text{(J/m)} \qquad (4.28)$$

Thus the energy carried by the wave in 1 second is

$$\text{rate of energy transfer} = \frac{1}{2}c_w\rho_l\omega^2\xi_0^2 \quad \text{(J/sec)} \qquad (4.29)$$

The concept of the average power resembles that described in ac circuit theory. If a sinusoidal current $I_0 \sin\omega t (A)$ is flowing through a resistor $R(\Omega)$, the instantaneous power is $RI_0^2 \sin^2\omega t$ (watts). The time-averaged power is defined by

$$\frac{1}{T}\int_0^T RI_0^2 \sin^2\omega t\, dt = \frac{1}{T}\int_0^T RI_0^2\frac{1-\cos 2\omega t}{2}dt = \frac{1}{2}RI_0^2$$

where T is the period of the oscillation, $T = 2\pi/\omega$ (s). The factor $1/2$ always appears for the average value of energy (or power) associated with oscillating physical quantities. You may recall that the rms (root-mean-square) value of the current in the preceding example is $I_0/\sqrt{2}$, which comes from

$$\frac{1}{2}RI_0^2 \equiv RI_{\text{rms}}^2$$

An oscillator connected to a transmission line creates waves on the line (Figure 4–6). To create the waves, the oscillator has to give energy to the waves. The waves now propagate down the transmission line toward the damper where the wave energy is converted into heat. How much energy can be transferred in one second is totally determined by the amplitude of the displacement waves and the oscillator frequency in a given wave medium.

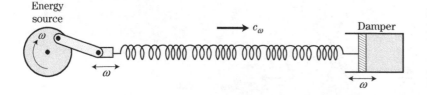

Energy source

ω

c_ω

Damper

ω

ω

FIGURE 4–6

Energy can be carried by waves from one place to another. The damper dissipates wave energy as heat.

EXAMPLE 4.2

A mechanical oscillator with a frequency of 30 Hz is connected to the spring in Example 4.1. The oscillator creates a wave with the amplitude of 1.5 cm. Find the power the oscillator has to deliver.

Solution

$$\text{Power} = \frac{1}{2} c_w \rho_l \omega^2 \xi_0^2$$

$$= \frac{1}{2} \times 42.4 \,\text{m/s} \times 0.33 \,\text{kg/m} \times \left(2\pi \times 30 \frac{1}{\text{s}}\right)^2 \times (1.5 \times 1.0^{-2} \,\text{m})^2$$

$$= 55.9 \,\text{J/s} = 55.9 \,\text{W}$$

As Eqs. (4.20) and (4.21) indicate, the kinetic and potential energies associated with mechanical waves must be the same. This reminds us of the energy relations we had for mechanical oscillations. However, there is an important difference between the two cases. In waves, the potential and kinetic energies are the same *at any time and at any location* while in an oscillation, although the amplitudes of both energies are the same, the two energies are mutually exclusive, tossing the energy back and forth. Later, in Chapter 6, we shall see that a wave medium of a finite spatial extent can become essentially an oscillation system.

Momentum Carried by Waves

The momentum carried by waves can be similarly calculated. You would conclude prematurely that since the momentum of the mass at x is

$$Mv = M\frac{\partial \xi}{\partial t} = M\frac{\partial}{\partial t}[\xi_0 \sin(kx - \omega t)] = -M\omega\xi_0 \cos(kx - \omega t) \quad (4.30)$$

which oscillates and thus whose average value is zero, the sinusoidal wave cannot carry net momentum. This, however, is a wrong argument. As a mass moving with a constant velocity has both a kinetic energy $\frac{1}{2}Mv^2$ and a momentum Mv, all waves (whether they be mechanical or electromagnetic) should carry both energy *and* momentum (and in some cases angular momentum as well).

To see this, let us consider the wave on a spring with an elastic modulus K and a linear mass density ρ_l. The quantity we have to watch carefully is the mass density ρ_l. If we stretch the spring, the mass density decreases (although the total mass M is unchanged), and if we compress it, the mass density increases. In the presence of waves any point on the spring suffers elongation and compression, and the local mass density certainly varies! Thus the momentum density (in this case, the momentum per unit length) should be written as

$$(\rho_l + \Delta\rho_l)v \tag{4.31}$$

where ρ_l is the mass density in the absence of the wave and is a constant, $\Delta\rho_l$ is the change in the mass density due to the wave, and v is the velocity of the local point on the spring,

$$v = \frac{\partial\xi}{\partial t} \tag{4.32}$$

The change in the mass density $\Delta\rho_l$ can be found as follows. Let us pick up a portion on the spring occupying a length Δx in the absence of a wave (Figure 4–7). The total mass of this section is

$$\rho_l \Delta x \ \text{(constant)} \tag{4.33}$$

Suppose the point originally at x suffers a displacement $\xi(x)$ and that at $x + \Delta x$ suffers a displacement $\xi(x + \Delta x)$. The original length Δx now becomes

$$\Delta x + \xi(x + \Delta x) - \xi(x) \tag{4.34}$$

Using the Taylor series expansion for $\xi(x + \Delta x)$ and neglecting the higher-order terms in Eq. (3.12), we write

$$\xi(x + \Delta x) \simeq \xi(x) + \frac{\partial\xi}{\partial x}\Delta x \tag{4.35}$$

and we see that the mass $\rho_l \Delta x$ is distributed over the length

$$\Delta x + \frac{\partial\xi}{\partial x}\Delta x \tag{4.36}$$

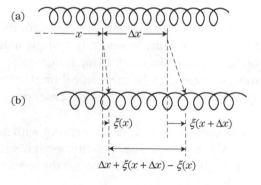

FIGURE 4–7

(a) Spring in equilibrium. (b) In the presence of a wave, the length becomes $\Delta x + \xi(x + \Delta x) - \xi(x)$, which causes the change in the mass density.

However, the total mass should be unchanged and remain equal to $\rho_l \Delta x$. Therefore we must have

$$(\rho_l + \Delta \rho_l) \left(\Delta x + \frac{\partial \xi}{\partial x} \Delta x \right) = \rho_l \Delta x \qquad (4.37)$$

or, after expanding the left-hand side of this equation, we find

$$\Delta \rho_l = -\rho_l \frac{\partial \xi / \partial x}{1 + \partial \xi / \partial x} \qquad (4.38)$$

If $|\partial \xi / \partial x| \ll 1$, we can neglect $\partial \xi / \partial x$ in comparison with 1 in the denominator and finally obtain

$$\Delta \rho_l = -\rho_l \frac{\partial \xi}{\partial x} \qquad (4.39)$$

(This appropriately defines the "density wave.") Substituting Eq. (4.39) into Eq. (4.31), we find

$$\text{momentum density} = \rho_l \frac{\partial \xi}{\partial t} - \rho_l \frac{\partial \xi}{\partial t} \frac{\partial \xi}{\partial x}$$

Assuming a sinusoidal wave

$$\xi(x, t) = \xi_0 \sin(kx - \omega t) \qquad (4.40)$$

we obtain

$$\text{momentum density} = -\rho_l \omega \xi_0 \cos(kx - \omega t) + \rho_l \omega k \xi_0^2 \cos^2(kx - \omega t) \quad (4.41)$$

As we did for the energy density, we take the average over the spatial coordinate to find

$$\text{average momentum density} = \frac{1}{2} \rho_l \omega k \xi_0^2 = \frac{1}{2} \frac{k}{\omega} \rho_l \omega^2 \xi_0^2 = \frac{1}{2} \frac{\rho_l \omega^2 \xi_0^2}{c_w} \quad (4.42)$$

Comparing this with Eq. (4.28), we find the following important conclusion:

$$\text{average momentum density} = \frac{\text{average energy density}}{\text{wave velocity}} \qquad (4.43)$$

Although we derived this conclusion for a particular mechanical wave, the conclusion is quite general, including electromagnetic waves. Remember that whenever energy is transferred, momentum must be transferred as well and there exists a simple relation between them.

EXAMPLE 4.3

Find the rate of momentum transfer (the momentum transferred in one second) in Example 4.2.

Solution

The energy transfer rate is 55.9 J/s. Thus the momentum transfer rate is

$$\frac{55.9 \text{ J/sec}}{42.4 \text{ m/sec}} = 1.32 \text{ J/m} = 1.32 \text{ N}$$

Note that the momentum transfer rate is equivalent to a force in this one-dimensional case.

EXAMPLE 4.4

A giant laser pulse with a power of 500 MW (megawatts) and pulse duration of 10 ns is completely reflected by a mirror. Find the momentum that is gained by the mirror. The velocity of light is 3.0×10^8 m/s.

Solution

The momentum transfer rate is 500 MW/$(3 \times 10^8$ m/s$) = 1.67$ N. Since the laser beam is completely reflected, the laser beam suffers a momentum change twice as much as it would when completely absorbed by a black object. Thus the momentum gained by the mirror is

$$\Delta p = 1.67 \times 10^{-8} \times 2 = 3.3 \times 10^{-8} \text{ N} \cdot \text{s}$$

Example 4.4 will be studied again later. Here this example is given only to illustrate that electromagnetic waves can indeed carry momentum with them. The force exerted by electromagnetic waves is called the *radiation pressure*. The radiation pressure can become substantial for powerful laser beams as the preceding example indicates. Sound waves in air also exert a net force on our eardrums. (Radiation force is not always pushing. When a light wave falls normal upon a glass surface, the glass surface is subject to a negative pressure and the force is directed toward the air. This pulling force was predicted by Poynting in 1905 based on Maxwell's equations.)

4.6 — ## Transverse Waves on a String

In Section 2.2, we briefly discussed the waves on a string. To create waves on a string, the string must be under a tension, such as in musical instruments. This tension provides the elasticity that is a fundamental requirement for

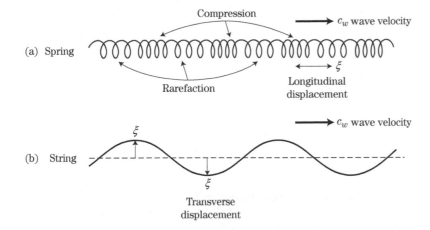

FIGURE 4–8

Longitudinal and transverse waves. In longitudinal waves, the displacement ξ is in the same direction as the wave velocity c_w. In transverse waves, the displacement ξ is normal to c_w.

creating mechanical waves and plays exactly the same role as the spring elastic modulus, $K(\text{N})$. In fact, the velocity of the waves on a string under a tension $T(\text{N})$ has a form that is very similar to Eq. (4.15) and is given by

$$c_w = \sqrt{\frac{T}{\rho_l}} \ (\text{m/s}) \tag{4.44}$$

where ρ_l (kg/m) is the linear mass density of the string.

However, there is a fundamental difference between the waves that propagate on a string and those that propagate in a spring. The difference is in the direction of motion of the perturbation of the mass. In the case of the spring (Figure 4–8[a]), segments of the spring move in the same direction (parallel or antiparallel) as the wave velocity and the wave creates a localized compression and rarefaction. Such waves are called *longitudinal waves*. Sound waves in solids, liquids, and gases are all longitudinal waves and will be studied in the following chapter. On the other hand, the waves on a string are associated with the mass motion that is perpendicular to the direction of the wave velocity as clearly shown in Figures 2–1 and 2–2. Such waves are called *transverse* waves since the mass motion is in directions that are transverse to the wave velocity. In general, waves that can be described with vector quantities perpendicular to the wave velocity are called transverse waves. In the case of the string, the string displacement ξ is normal to the string or the direction of propagation. Most *electromagnetic waves are transverse waves*, as we will study later.

Let us derive the expression for the wave propagation velocity on a string Eq. (4.44). As before, we pick up a small segment on the string having a length

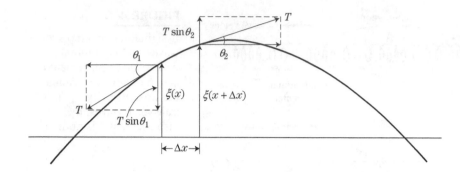

FIGURE 4–9

A segment Δx of the string suffers a displacement $\xi(x, t)$ that is normal to the string.

Δx (Figure 4–9). The segment has a mass that is given by

$$\rho_l \Delta x \quad \text{(kg)} \tag{4.45}$$

Our purpose here is to derive an equation of motion for this mass. We also assign the displacements for each end of the segment $\xi(x)$ and $\xi(x + \Delta x)$, which are now perpendicular to the x axis, the direction of wave propagation. The string is under a tension force T (N) and the net vertical force can be written as

$$F = F_+ - F_- = T \sin \theta_2 - T \sin \theta_1 \tag{4.46}$$

where θ_2 and θ_1 are the angles of the tangent lines at A and B, respectively,

$$\tan \theta_1 = \frac{\partial \xi}{\partial x} \text{ at } x$$

and

$$\tan \theta_2 = \frac{\partial \xi}{\partial x} \text{ at } x + \Delta x$$

If the displacement $\xi(x, t)$ is small, so are the angles. We may approximate $\sin \theta$ by θ and $\tan \theta$ by θ (see Chapter 3). Therefore,

$$F = T(\theta_2 - \theta_1) = T \left(\left.\frac{\partial \xi}{\partial x}\right|_{x+\Delta x} - \left.\frac{\partial \xi}{\partial x}\right|_x \right) \tag{4.47}$$

where $\partial \xi / \partial x |_{x+\Delta x}$ indicates the value of $\partial \xi / \partial x$ evaluated at $x + \Delta x$. Using the Taylor series expansion [Eq. (3.12)], we find (let $f = \partial \xi / \partial x$ in Eq. (3.12))

$$\left.\frac{\partial \xi}{\partial x}\right|_{x+\Delta x} - \left.\frac{\partial \xi}{\partial x}\right|_x \simeq \Delta x \frac{\partial^2 \xi}{\partial x^2} \tag{4.48}$$

and the force becomes

$$F = \Delta x T \frac{\partial^2 \xi}{\partial x^2} \tag{4.49}$$

This is the force that acts on the segment with the mass $\rho_l \Delta x$. Therefore, Newton's equation of motion gives

$$\rho_l \Delta x \frac{\partial^2 \xi}{\partial t^2} = \Delta x \, T \frac{\partial^2 \xi}{\partial x^2}$$

or

$$\frac{\partial^2 \xi}{\partial t^2} = \frac{T}{\rho_l} \frac{\partial^2 \xi}{\partial x^2} \tag{4.50}$$

This equation is almost identical with Eq. (4.15), except the elastic modulus K is replaced with the tension of the string T, both terms having the same dimensions N. Equation (4.50) immediately yields the velocity of propagation of the transverse waves on the string, Eq. (4.44).

EXAMPLE 4.5

A sinusoidal transverse wave train is moving along a string having a mass density of 20 g/m. The string is under a tension of 40 N. The amplitude of the wave is 5 mm, and the wave frequency is 80 cycles/s.

(a) Write an expression for the displacement wave $\xi(x, t)$.
(b) For the expression assumed in (a), find the expression for the velocity wave.
(c) Calculate the average energy density, power, and momentum transfer rate.

Solution

(a) The wave velocity is

$$c_w = \sqrt{T/\rho_l} = \sqrt{40/0.02} = 44.7 \text{ m/s}$$

Since the frequency is 80 cycles/s and the amplitude is 5 mm, we may write

$$\xi(x, t) = 5 \times 10^{-3} \sin\left(2\pi \frac{x}{\lambda} - 2\pi v t\right) = 5 \times 10^{-3} \sin(11.25x - 503t) \text{ (m)}$$

where

$$\lambda = c_w/v = 0.56 \text{ m}$$

is the wavelength.

(b) Since the velocity wave is given by

$$v(x, t) = \frac{\partial \xi}{\partial t}$$

we find

$$v(x, t) = 5 \times 10^{-3} \times (-503) \cos(11.25x - 503t)$$
$$= -2.5 \cos(11.25x - 503t) \text{ m/s}$$

(c) The expressions we obtained earlier for the energy, power, and momentum transfer rate in the longitudinal waves in a spring are also applicable for the transverse waves on a string. From Eqs. (4.28), (4.29), and (4.42), we find

$$\text{average energy density} = \frac{1}{2}\rho_l\omega^2\xi_0^2 = 6.3 \times 10^{-2} \text{ J/m}$$

$$\text{average power} = c_w \times \text{energy density} = 2.8 \text{ W}$$

$$\text{average momentum transfer rate} = \text{power}/c_w = 6.3 \times 10^{-2} \text{ N}$$

Momentum transfer accompanies energy transfer along the direction of wave propagation regardless the wave of concern is transverse or longitudinal. We will learn that electromagnetic waves are transverse in the sense that relevant electric field and magnetic field vectors are perpendicular to the wave vector (direction of wave propagation). Electromagnetic waves carry energy and momentum just like the longitudinal and transverse mechanical waves do.

We have learned that for a displacement $\xi_0 \sin(kx - \omega t)$, the energy transfer rate is

$$\text{energy transfer rate} = c_w \rho_l \left(\frac{\partial\xi}{\partial t}\right)^2 = c_w \rho_l \omega^2 \xi_0^2 \cos^2(kx - \omega t) \quad \text{(W)} \tag{4.51}$$

Since energy and momentum are both transferred at the same velocity c_w, we can find the momentum transfer rate by dividing the energy transfer rate by c_w,

$$\text{momentum transfer rate} = \rho_l \omega^2 \xi_0^2 \cos^2(kx - \omega t) \quad \text{(N)} \tag{4.52}$$

This holds for both longitudinal wave and transverse wave. Likewise, the wave energy density and momentum density are related through

$$\text{momentum density} = \frac{1}{c_w} \times \text{(energy density)}$$

$$= \frac{1}{c_w}\rho_l\omega^2\xi_0^2 \cos^2(kx - \omega t) \quad \text{(N} \cdot \text{s/m)} \tag{4.53}$$

These relationships between the wave momentum and wave energy can be derived from the wave equation more formally as follows. Mechanical waves, either longitudinal or transverse, are described by the wave equation

$$\frac{\partial^2\xi}{\partial t^2} = c_w^2 \frac{\partial^2\xi}{\partial x^2} \tag{4.54}$$

where ξ is the displacement and c_w is the wave velocity. For a transverse wave in a string, the kinetic energy density and potential energy density are identical,

$$\frac{1}{2}\rho_l \left(\frac{\partial\xi}{\partial t}\right)^2 = \frac{1}{2}T \left(\frac{\partial\xi}{\partial x}\right)^2 \tag{4.55}$$

The square root of both sides of Eq. (4.55) yields

$$\sqrt{\rho_l}\frac{\partial\xi}{\partial t} = -\sqrt{T}\frac{\partial\xi}{\partial x} \tag{4.56}$$

where the minus sign is required since for the displacement $\xi(x,t) = \xi_0 \sin(kx - \omega t)$, $\partial\xi/\partial t$ and $\partial\xi/\partial x$ have opposite signs. Then

$$-\rho_l\frac{\partial\xi}{\partial t}\frac{\partial\xi}{\partial x} = \frac{1}{c_w}T \left(\frac{\partial\xi}{\partial x}\right)^2 \tag{4.57}$$

The quantity

$$-\rho_l\frac{\partial\xi}{\partial t}\frac{\partial\xi}{\partial x} \tag{4.58}$$

is defined to be the wave momentum density which is related to the wave energy density through Eq. (4.57).

Problems

1. When a helical spring of mass 0.1 kg and natural length 2 m is stretched by a force of 30 N, an elongation of 10 cm results. Find the velocity of propagation for the longitudinal waves along the spring assuming that the spring is at its natural length.

2. A device called a *wave demonstrator* consists of rods [moment of inertia of each I (kg · m^2)] and springs [torsional constant of k_τ (N · m)] alternatively connected with a spacing Δx as shown in Figure 4–10.
   (a) Derive a wave differential equation for the angular displacement $\theta(x,t)$.
   (b) What is the propagation velocity?

3. A driver attached to one end of a long spring of mass density 0.5 kg/m and elastic modulus 300 N creates sinusoidal waves with a displacement amplitude

of 2 cm and a frequency of 40 cycles/s. Neglecting wave reflection from the other end, find (a) the power (average, or rms) and (b) the average rate of momentum transfer.

4. Consider a pulse

$$\xi(x,t) = \xi_0 e^{-(x-c_w t)^2/a^2}$$

propagating with a velocity c_w along a spring of mass density ρ_l (kg/m and elastic modulus K (N). Calculate
   (a) The kinetic energy.
   (b) The potential energy.
   (c) The momentum associated with the pulse.

5. A sinusoidal wavetrain is moving along a spring. The amplitude of the longitudinal displacement wave is 0.5 cm and the propagation velocity is 25 m/s.

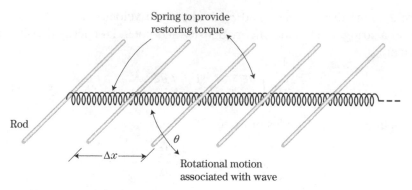

Spring to provide
restoring torque

Rod

Δx

θ

Rotational motion
associated with wave

FIGURE 4–10

Problem 2.

(a) Write down an expression to describe the displacement wave, $\xi(x, t)$

(b) For the expression assumed in (a), find the expression for the velocity wave.

6. Show that the expressions for the average energy density, the average power, and the average momentum transfer rate associated with sinusoidal transverse waves on a string are identical to those derived for longitudinal waves in a spring.

7. A steel wire of radius 0.5 mm is subject to a tension of 10 N. Steel has a volume mass density of 7800 kg/m^3. Find the velocity of propagation of the transverse waves on the wire.

8. A steel wire having a radius of 0.4 mm hangs from the ceiling. When a mass of 5.0 kg is hung from the free end, what is the velocity of propagation of transverse waves on the wire?

9. Show that if the tension is varied by a small amount ΔT, the change in the velocity of propagation of the transverse waves on a string is given approximately by

$$\Delta c_w = \frac{1}{2} \frac{\Delta T}{T} c_w$$

where T and c_w are the original tension and velocity, respectively.

10. Show that if the mass density is varied by a small amount $\Delta \rho_l$, the change in the velocity of propagation of the transverse waves on a string is given

approximately by

$$\Delta c_w = -\frac{1}{2} \frac{\Delta \rho_l}{\rho_l} c_w$$

where ρ_l is the original mass density.

11. A mechanical oscillator connected to the end of a stretched string creates a transverse displacement of the end that is given by

$$\xi = 0.01 \sin (20t) \text{ (m)}$$

The tension in the string is 10 N and the string has a linear mass density of 20 g/m. Find (a) the velocity of propagation of the transverse waves, (b) the frequency v, (c) the wavelength, and (d) the average power delivered by the oscillator.

12. Assuming that about 20% of the power of an incandescent lamp is converted into light power (the rest is wasted as heat), estimate the radiation pressure at a distance 2 m away from a 100-W light bulb. The velocity of light is 3.0×10^8 m/s and the bulb radiates light isotropically (i.e., in every spherical direction).

13. It is possible to find an exact dispersion relation for a longitudinal wave on the periodic mass-spring transmission line as shown in Figure 4–1 without the restriction that the wavelength be much longer than Δx. In other words, the equation of motion, Eq. (4.7),

can be solved exactly. Let $\xi(x, t) = \xi_0 \sin(kx - \omega t)$ in

$$m \frac{\partial^2 \xi}{\partial t^2} = k_s [\xi(x + \Delta x) + \xi(x - \Delta x) - 2\xi(x)]$$

where k is the wavenumber and k_s is the spring constant.

(a) Calculate $\xi(x+\Delta x)-\xi(x)$ and $\xi(x-\Delta x)-\xi(x)$ using

$$\sin \alpha - \sin \beta = 2 \cos \frac{\alpha + \beta}{2} \sin \frac{\alpha - \beta}{2}$$

(b) Show that the equation of motion yields the following dispersion relation:

$$\omega^2 = \frac{k_s}{m} 4 \sin^2 \left(\frac{k \, \Delta x}{2} \right)$$

(c) Check that in the long wavelength limit $k \Delta x \ll 1$, we recover

$$\omega = \sqrt{\frac{k_s}{M}} k \Delta x$$

(d) Plot ω as a function of k for $0 < k < \pi/\Delta x$

(e) Is the wave dispersive or nondispersive? Calculate the phase and group velocities.

Sound Waves in Solids, Liquids, and Gases

Introduction

Longitudinal waves in an elastic body (or medium) are generally called sound waves. The most familiar sound waves are those that propagate in air. However, sound waves can propagate even in solids or liquids. Sound waves are associated with the compressional and rarefactional motion of molecules in the direction that the wave propagates. This is similar to the longitudinal waves that propagate along a mass-spring transmission line that was discussed in the previous chapter. Earthquakes generally produce both longitudinal waves and transverse waves, the latter propagating slower than the former. When we are hit by an earthquake, we first feel a horizontal motion arising from the longitudinal waves, and some time later, a tumbling vertical movement from the transverse waves. In this chapter, we study the properties of the longitudinal sound waves in solids, liquids, and gases.

Sound Velocity Along a Solid Rod

In the previous chapter we learned that the propagation velocities of mechanical waves are in general given by

$$c_w = \sqrt{\frac{\text{elastic modulus}}{\text{mass density}}} \tag{5.1}$$

83

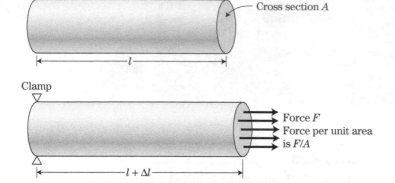

FIGURE 5–1

Relative elongation of a solid rod $\Delta l/l$ is proportional to the force per unit area F/A (Hooke's law).

The elastic modulus is a constant that relates the stress to the strain as

$$\text{stress} = \text{elastic modulus} \times \text{strain} \tag{5.2}$$

For the case of a continuous spring having a uniformly distributed mass density, we have

$$F = K\frac{\Delta l}{l} \tag{5.3}$$

where the stress is the force itself (N) and the strain is $\Delta l/l$ (dimensionless). The proportionality between the stress and the strain is known as Hooke's law. For the case of a spring, Hooke noted that after removing a few coils from the spring, the spring would become stiffer.

Hooke's law for an elastic body (rather than the spring) takes a slightly different form. Consider a solid rod with a natural length l(m) and a cross-sectional area A(m^2) (Figure 5–1). If a tension F(N) is applied along the rod, the rod will be elongated by a length Δl, and we may write

$$F = \text{constant}\frac{\Delta l}{l} \tag{5.4}$$

However, if the cross section is increased, a larger force must be applied in order to obtain the same longitudinal deformation Δl. Therefore the constant in the preceding equation is still size dependent and is not a real material constant. If we divide Eq. (5.4) by the cross section A and write

$$\frac{F}{A} = Y\frac{\Delta l}{l} \tag{5.5}$$

a more general relationship can be found. The constant Y(N/m^2) is called the Young's modulus and it is a material constant. Equation (5.5) may be regarded as the microscopic form of Hooke's law.

The stress in this case is thus given by the force per unit area N/m^2 and the elastic modulus Y has the same dimensions. The mass density of the rod should be the volume mass density, ρ_v (kg/m^3), rather than the linear mass density as in the case of the spring and string. The velocity of propagation of the longitudinal waves in the rod is thus given by

$$c_w = \sqrt{\frac{\text{elastic modulus}}{\text{mass density}}} = \sqrt{\frac{Y}{\rho_v}} \qquad (5.6)$$

It should be remembered that we are dealing with a slender rod rather than an unbounded volume of the solid. Strictly speaking, the expression for the wave velocity Eq. (5.6) is valid only for a rod, along which pure longitudinal waves propagate. For the velocity formula in Eq. (5.6) to be applicable, the rod diameter must be much smaller than the wavelength of the propagating wave. The velocity of longitudinal spherical waves in an unbounded solid contains an additional factor in the elastic modulus (shear modulus) and it is larger than that given by Eq. (5.6). Liquids and gases, on the other hand, cannot support shear stress, and thus transverse waves, and such a complication does not occur. Transverse waves on a string discussed in the previous chapter may be alternatively called shear waves.

Any longitudinal mechanical wave can appropriately be called a sound wave. The most familiar form of a sound wave is of course is the one that propagates in air. The air molecules move back and forth in the same direction as the velocity of propagation of the sound wave. The molecules in the rod do the same thing. They undergo displacements in the same direction as the wave velocity. Whenever molecules are displaced, an increase or decrease in the density corresponding to a compression or rarefaction occurs. This is similar to the longitudinal waves that propagate in the mass-spring system.

The energy and momentum densities associated with the sound waves in solids can similarly be found by just replacing the linear mass density ρ_l with the volume mass density ρ_v in Eqs. (4.28) and (4.43). We write

$$\text{energy density} = \frac{1}{2}\rho_v\omega^2\xi_0^2 \quad (\text{J/m}^3) \qquad (5.7)$$

$$\text{momentum density} = \frac{1}{2}\frac{\rho_v\omega^2\xi_0^2}{c_w} \quad (\text{N}\cdot\text{sec/m}^3) \qquad (5.8)$$

for a sinusoidal wave with a displacement amplitude ξ_0 (m). The power flow needs some further explanation. If the rod has a cross section A (m^2), the wave energy per unit length along the rod (see Figure 5–2) is

$$\frac{1}{2}\rho_v\omega^2\xi_0^2 A \quad (\text{J/m}) \qquad (5.9)$$

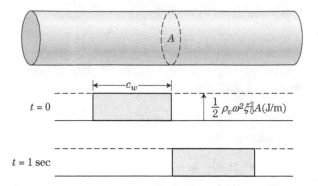

FIGURE 5–2

In 1 second, the energy $\frac{1}{2}\rho_v\omega^2\xi_0^2 A\, c_w$ is transferred through the cross section of the rod.

$t = 0$

$\frac{1}{2}\rho_v\omega^2\xi_0^2 A$(J/m)

$t = 1$ sec

In one second, the energy of an amount

$$\frac{1}{2}\rho_v\omega^2\xi_0^2 A\, c_w \quad (\text{J}) \tag{5.10}$$

goes through the cross sectional area A (m^2) at an arbitrary location. Hence, the power density (energy flow rate across a unit area in unit time) is

$$\frac{1}{2}\rho_v\omega^2\xi_0^2 c_w \quad (\text{W/m}^2) \tag{5.11}$$

This quantity is called the *intensity* of the wave.

EXAMPLE 5.1

Steel has a Young's modulus of 2×10^{11} N/m^2 and a volume mass density of 7800 kg/m^3. Assuming a sinusoidal longitudinal wave with a displacement amplitude of 1.0×10^{-6} mm and a frequency $\nu = 5$ kHz in a steel rod, find

(a) the wave velocity,
(b) the wave intensity.

Solution

(a) The velocity of the wave is

$$c_w = \sqrt{\frac{Y}{\rho_v}} = \sqrt{\frac{2 \times 10^{11}}{7.8 \times 10^3}} = 5.1 \times 10^3 \text{ m/sec}$$

(b) The intensity of the wave can be computed from Eq. (5.11),

$$\text{intensity} = \frac{1}{2}\rho_v\omega^2\xi_0^2 c_w$$

$$= \frac{1}{2} \times 7800 \times (2\pi \times 5 \times 10^3)^2 \times (10^{-9})^2 \times 5.1 \times 10^3 = 1.96 \times 10^{-2} \text{ W/m}^2.$$

In the table, Young's moduli, densities, and sound velocities in rods of typical materials at room temperature are shown.

Material	$Y(\times 10^{10} \text{ N/m}^2)$	$\rho_v \text{ (kg/m}^3)$	$c_w(\times 10^3 \text{ m/sec})$
Aluminum	6.9	2,700	5.0
Cast iron	19	7,200	5.1
Copper	11	8,900	3.5
Lead	1.6	11,340	1.2
Steel	20	7,800	5.1
Glass	5.4	2,300	5.0
Brass (70% Cu, 30% Zn)	10.5	8,600	3.5

5.3   Rigorous Derivation of Velocity of Sound in a Solid Rod

In finding the propagation velocity of sound waves in solid rods, we have just used the general expression that was obtained in Chapter 4,

$$c_w = \sqrt{\frac{\text{elastic modulus}}{\text{mass density}}} \tag{5.12}$$

Here, for redundancy, we derive Eq. (5.6) directly from Hooke's law and the equation of motion. The procedure, however, is almost exactly the same as we employed in Chapter 4.

Consider a long uniform rod with a cross section A (m^2) having an elastic or Young's modulus $Y(\text{N/m}^2)$ and a mass density ρ_v (kg/m^3). We select a thin slice of thickness Δx located at a distance x (m) from one end of the rod (Figure 5–3).

When a sound wave is excited along the rod, the thin slice will move about its original location. At the same time, the slice is deformed since the forces that act on the cross sections A_x and $A_{x+\Delta x}$ will be different. Let the displacements of the cross sections of A_x and $A_{x+\Delta x}$ be $\xi(x)$ and $\xi(x + \Delta x)$,

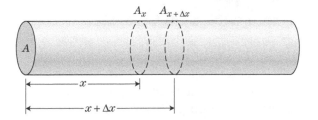

FIGURE 5–3

A solid rod in equilibrium. The small element, having a volume of $A\Delta x$, experiences no net force.

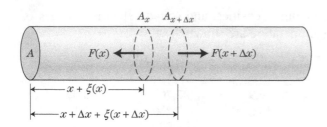

FIGURE 5–4

With the presence of the wave, the volume $A\Delta x$ is displaced. At the same time, the volume is deformed.

respectively (Figure 5–4). Then the net deformation is

$$\xi(x + \Delta x) - \xi(x) \simeq \frac{\partial \xi}{\partial x}\Delta x \tag{5.13}$$

where we have used the Taylor series expansion for $\xi(x + \Delta x)$ and retained only the lowest-order terms

$$\xi(x + \Delta x) \simeq \xi(x) + \frac{\partial \xi}{\partial x}\Delta x \tag{5.14}$$

Then Hooke's law,

$$\text{stress} = Y \times \text{strain} \tag{5.15}$$

becomes

$$\frac{F}{A} = Y\frac{(\partial \xi/\partial x)\Delta x}{\Delta x} = Y\frac{\partial \xi}{\partial x} \tag{5.16}$$

Note that Δx is the original thickness of the slice that corresponds to l in Eq (5.5), and $(\partial \xi/\partial x)\Delta x$ corresponds to Δl. Next, the equation of motion,

$$\text{mass} \times \text{acceleration} = \text{net force} \tag{5.17}$$

can be written as

$$\rho_v \Delta x A\frac{\partial^2 \xi}{\partial t^2} = F(x + \Delta x) - F(x) \simeq \frac{\partial F}{\partial x}\Delta x \tag{5.18}$$

where we have expanded $F(x + \Delta x)$ using the Taylor series expansion. Substituting Eq. (5.16) into Eq. (5.18), we find

$$\frac{\partial^2 \xi}{\partial t^2} = \frac{Y}{\rho_v}\frac{\partial^2 \xi}{\partial x^2} \tag{5.19}$$

and the velocity of propagation can immediately be found to be

$$c_w = \sqrt{\frac{Y}{\rho_v}} \tag{5.20}$$

One notes that any physical quantity associated with the sound wave should obey the same equation. If we differentiate Eq. (5.19) with respect to x, we

obtain

$$\frac{\partial^3 \xi}{\partial x \partial t^2} = \frac{Y}{\rho_v} \frac{\partial^3 \xi}{\partial x^3} \tag{5.21}$$

But the pressure or force wave satisfies

$$\frac{F}{A} = Y \frac{\partial \xi}{\partial x}$$

Hence we find that the force $F(x, t)$ obeys the same differential equation. Similarly, the velocity defined by

$$v = \frac{\partial \xi}{\partial t}$$

should also obey the same equation. However, the force F and the velocity v are not of the same functional form as the displacement ξ. For example, if we assume a sinusoidal wave given by $\xi(x, t) = \xi_0 \sin(kx - \omega t)$, the force becomes

$$F(x, t) = AY \frac{\partial \xi}{\partial x} = AY k \xi_0 \cos(kx - \omega t) \tag{5.22}$$

which is 90° out of phase with respect to the displacement ξ as shown in Figure 5–5. The velocity is also 90° out of phase with respect to the displacement ξ. Since the work is given by force × displacement, we can easily see that the rod is, on the average, not doing any work or gaining any energy. The average of the function

$$\sin(kx - \omega t) \cdot \cos(kx - \omega t) \tag{5.23}$$

is exactly zero. The rod is transferring energy from left to right but it is not dissipating or creating energy. The rod can accommodate energy that is moving or propagating in the rod and only acts as a medium that accommodates the wave.

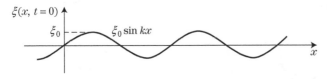

$\xi(x, t = 0)$

ξ_0 $\xi_0 \sin kx$

$F(x, t = 0)$

$AYk\, \xi_0 \cos kx$

FIGURE 5–5

The force wave is 90° out of phase with respect to the displacement wave. Both waves are shown at $t = 0$.

In fact, good wave media all have this nondissipation property. By *good* it is meant here that the medium does not dissipate energy. Otherwise, the wave energy is gradually absorbed or dissipated by the medium and by the time the wave propagates to the other end, little energy would remain. In practice, however, any wave that propagates in material media actually does undergo damping. In our studies, however, we assume that the dissipation is small and can be neglected except for one case in the chapter on electromagnetic waves where we examine wave propagation in metals and describe skin effects. In other words, the media we study are all *reactive* media. The concept of reactive media will later become clear in the chapters on electromagnetic waves.

 ## Sound Waves in Liquids

Sound waves require a compressive (and thus rarefactive) medium. Sound waves in solids can exist because solids are elastically compressive. Water is compressive also and there is a relationship between the force F and the change in volume of the water ΔV, just as in Hooke's law for solids.

Consider a liquid in a rigid cylinder occupying a volume V (m^3) in the absence of a compression force (Figure 5–6). If the cylinder has a cross-sectional area $A(m^2)$, the stress due to the force $F(N)$ is F/A (N/m^2) and the strain is $-\Delta l/l$ where $-\Delta l$ is essentially the change in the volume ΔV since $V = Al$ and the cross-sectional area does not change. The stress–strain relation for liquids is written as

$$\frac{F}{A} = -M_B \frac{\Delta l}{l} = -M_B \frac{\Delta V}{V} \tag{5.24}$$

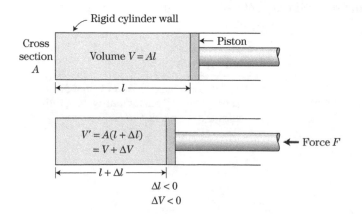

FIGURE 5–6

Liquids such as water can be compressed by an external force.

where M_B (N/m^2) plays the role of Young's modulus in solids and is called the bulk modulus of liquids. For a compressive force as shown in Figure 5–6, the change in the volume ΔV is negative or $\Delta V < 0$. The bulk modulus is a material constant and has the same dimensions as Young's modulus. Since the mass density is inversely proportional to the volume

$$\rho_v V = \text{constant}, \tag{5.25}$$

the decrease in the volume causes an increase in the mass density and we may write Eq. (5.24) in an alternative form,

$$\frac{F}{A} = M_B \frac{\Delta \rho_v}{\rho_v}, \quad \Delta \rho_v > 0 \text{ if } \Delta V < 0 \tag{5.26}$$

Once we find the elastic modulus (M_B in this case), the velocity of the longitudinal (sound) waves is immediately found to be

$$c_w = \sqrt{\frac{M_B}{\rho_v}} \tag{5.27}$$

in analogy with Eq. (5.6). Since liquids cannot support shear stress, this expression is not subject to the geometrical constraint as sound waves in solids are. The velocity of sound waves in isotropic free solids (without boundaries) is given by

$$c_s = \sqrt{\frac{K + \frac{4}{3}G}{\rho_v}} \tag{5.28}$$

where K is the bulk elastic modulus and G is the shear elastic modulus. The difference from the sound speed in rods, $c_s = \sqrt{Y/\rho_v}$ should be noted.

EXAMPLE 5.2

Water has a mass density of $\rho_v = 10^3$ kg/m^2 and a bulk modulus of $M_B = 2.1 \times 10^9$ N/m^2. Find the velocity of sound waves in water.

Solution

$$c_w = \sqrt{\frac{M_B}{\rho_v}} = \sqrt{\frac{2.1 \times 10^9 \text{ N/m}^2}{10^3 \text{ kg}}} = 1.45 \times 10^3 \text{ m/sec}$$

EXAMPLE 5.3

Derive Eq. (5.27) directly from Newton's equation of motion for a small differential volume in a liquid. Follow the procedure employed in Section 5.3 for the sound waves in solids.

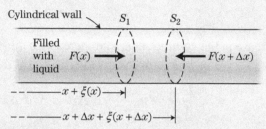

FIGURE 5–7

Displacement of two surfaces S_1 and S_2, originally separated by a distance Δx.

Solution

We consider a pipe filled with a liquid in which a sound wave is excited. A small element with a thickness Δx has a mass $\rho_v A \, \Delta x$, where ρ_v is the unperturbed liquid mass density and A is the cross section of the pipe. Let the displacements of surfaces S_1 and S_2 be $\xi(x)$ and $\xi(x + \Delta x)$, respectively (Figure 5–7). Then the net change in the volume from the original volume $A \, \Delta x$ is

$$A[\Delta x + \xi(x + \Delta x) - \xi(x)] - A\Delta x = A[\xi(x + \Delta x) - \xi(x)] \simeq A\frac{\partial \xi}{\partial x}\Delta x$$

where we have used the first two terms of a Taylor series expansion for the quantity $\xi(x + \Delta x)$,

$$\xi(x + \Delta x) \simeq \xi(x) + \frac{\partial \xi}{\partial x}\Delta x$$

This change in the volume must be caused by the internal force F acting on the surface of the element. From Eq. (5.24), we then find

$$\frac{F}{A} = -M_B \frac{1}{A\Delta x} A \, \Delta x \frac{\partial \xi}{\partial x} = -M_B \frac{\partial \xi}{\partial x} \tag{A}$$

However, the forces that act on the surface S_1 and on S_2 must be different in order to cause the displacement of the whole volume $A\Delta x$. The net force directed to the right is

$$F(x) - F(x + \Delta x) \simeq -\frac{\partial F}{\partial x}\Delta x \tag{B}$$

where we have again Taylor series-expanded $F(x + \Delta x)$. Then the equation of motion for the segment can be written as

$$\rho_v A\Delta x \frac{\partial^2 \xi}{\partial t^2} = -\frac{\partial F}{\partial x}\Delta x \tag{C}$$

After substituting the force in Eq. (A) into Eq. (C), we finally obtain

$$\frac{\partial^2 \xi}{\partial t^2} = \frac{M_B}{\rho_v}\frac{\partial^2 \xi}{\partial x^2} \qquad \text{(D)}$$

which immediately yields the sound velocity, Eq. (5.27).

Sound Waves in Gases

Sound waves in air are probably the most familiar wave phenomena. Human ears can detect sound waves with frequencies ranging from about 20 Hz to 20 kHz (the *audio* frequency range). Some animals (dogs, bats) apparently can detect sound waves with higher frequencies, called the *ultrasonic* frequencies. Frequencies below the audible limit are called *infrasonic* frequencies. Earthquakes are usually accompanied by infrasonic waves in the air in addition to the physical waves in the ground. Acoustics is a scientific branch that is devoted to the studies of sound waves and their applications.

Sound waves can be created with physical objects that are oscillating or vibrating. When we speak, our vocal cords vibrate to create compressive and rarefactive motion of the molecules in air. At room temperature 20°C, these compressive and rarefactive air molecule perturbations propagate at a speed of approximately 340 m/sec. Since sound waves in air (or gases, in general) fall into the class of waves called mechanical waves, we can find the sound velocity from the general formula

$$c_w = \sqrt{\frac{\text{elastic modulus}}{\text{mass density}}}$$

To find the mass density of air, let us recall that 1 mole of gas occupies 22.4 litres of volume in the standard condition, 0°C and atmospheric pressure. Since 1 mole of air has a mass of 29 g (about 80% of nitrogen with molecular molar mass of 28 g and 20% of oxygen with 32 g of molar mass), we find the mass density

$$\rho_v = \frac{0.029 \text{ kg}}{22.4 \times 10^{-3} \text{ m}^3} = 1.29 \text{ kg/m}^3$$

The elastic modulus of gases is defined in the same manner as that for liquids.

As before (see Figure 5–8), we consider a cylinder with a cross-sectional area $A(\text{m}^2)$ and a length l filled with a gas having a pressure P. If the piston is pushed by an external force, F (N), the pressure rises (ΔP, the change in the pressure, is positive), but the volume decreases by $A\Delta l$. If the volume of

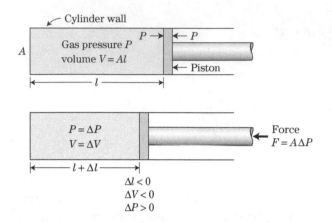

FIGURE 5–8

When a gas is compressed, the pressure increases due to the increase in both molecular density and temperature.

the gas is large compared with the wavelength of sound, the compression is adiabatic and the gas pressure P and the volume occupied by the gas V are related through *the adiabatic equation of state*:

$$PV^\gamma = \text{constant} \tag{5.29}$$

where γ is the ratio of specific heats. We find after differentiating Eq. (5.29)

$$\Delta P V^\gamma + P\gamma V^{\gamma-1}\Delta V = 0 \tag{5.30}$$

Therefore

$$\Delta P = -\gamma P \frac{\Delta V}{V} \tag{5.31}$$

Since $V = Al$, $\Delta V = A\Delta l$, we finally obtain

$$\Delta P = -\gamma P \frac{\Delta l}{l} \tag{5.32}$$

This is identical with Eq. (5.24) (stress–strain relation for solids) provided we replace F/A (which has the dimensions of pressure, N/m^2) by ΔP, and M_B by γP. Therefore we may define the bulk modulus of a gas having a pressure P and a ratio of specific heats γ by

$$M_B = \gamma P \quad (\text{N/m}^2) \tag{5.33}$$

Denoting the volume mass density of the gas with ρ_v (kg/m^3), we find the sound velocity is given as

$$c_w = \sqrt{\frac{\gamma P}{\rho_v}} \quad (\text{m/sec}) \tag{5.34}$$

The origin of the additional factor γ (compared with the sound velocity in liquids) stems from the fact that gases tend to be heated when compressed.

The pressure of a gas is given by

$$P = nk_B T \quad (\text{N/m}^2) \tag{5.35}$$

where n is the number density of gas molecules (m^{-3}), k_B is the Boltzmann constant, $k_B = 1.38 \times 10^{-23}$ J/K, and T is the absolute temperature (K). If the gas is compressed, the molecular density obviously increases. At the same time, the gas is heated if the gas is *thermally insulated* from external agents. This process is called *adiabatic compression*. Then the total pressure increase is given by

$$\Delta P = k_B(\Delta n T + n \Delta T) = P\left(\frac{\Delta n}{n} + \frac{\Delta T}{T}\right) \tag{5.36}$$

Recalling

$$\frac{\Delta n}{n} = -\frac{\Delta V}{V} \tag{5.37}$$

Eq. (5.36) becomes

$$\Delta P = -P\frac{\Delta V}{V} + P\frac{\Delta T}{T} \tag{5.38}$$

which is obviously larger than the pressure change due to the density (or volume) change alone. The coefficient γ is defined from

$$-P\frac{\Delta V}{V} + P\frac{\Delta T}{T} \equiv -\gamma P\frac{\Delta V}{V} \tag{5.39}$$

and is always larger than or equal to 1. Its numerical value depends on the actual molecules that the gas contains. For gases with *monatomic molecules* such as helium (He) and argon (Ar), $\gamma = 5/3$. For gases with *diatomic molecules* such as oxygen (O_2), nitrogen (N_2), and air (mixture of oxygen and nitrogen gases), $\gamma = 7/5$. The ratio of specific heats γ is given by $\gamma = (f + 2)/f$, where f is the number of degrees of freedom of molecular energy partition. For monatomic gases, $f = 3$ (or $\gamma = 5/3$) corresponding to three possible kinetic energies, $\frac{1}{2}mv_x^2, \frac{1}{2}mv_y^2, \frac{1}{2}mv_z^2$, in the x, y, z directions, respectively. In a diatomic gas, there are additional two degrees of freedom (Figure 5–9). These are associated with rotational energies about the two axes of bound atoms as shown. The rotation about the z axis has negligible energy because of the small moment of inertia of the atom. You may wonder why one should not include the vibrational energy along the z axis. Only quantum mechanics can answer this question. As the energy of the electron in a hydrogen atom is not arbitrary but quantized (or discrete), so is the energy of this vibration in the diatomic molecule. That is, the vibration energy cannot increase with temperature and this energy does not contribute to the number of degrees of freedom. (A bound vibration is not *free* in a sense.) A brief introduction to quantum mechanics is given in Chapter 13. It is interesting that quantum

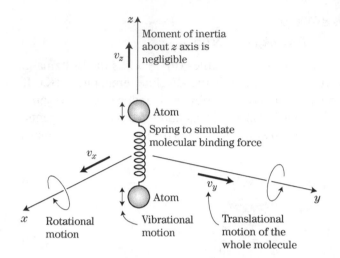

FIGURE 5–9

Translational, rotational, and vibrational energies of a diatomic molecule.

effects appear even in sound waves, which have been an important physical subject since Newton's era.

EXAMPLE 5.4

Find the velocity of sound in (a) air and (b) helium gas at atmospheric pressure and a temperature of 0°C.

Solution

(a) Since air is largely composed of nitrogen and oxygen gases with both components possessing diatomic molecules, we may choose $\gamma = 7/5$. The pressure P is 1.013×10^5 N/m^2 and the mass density ρ_v is 1.293 kg/m^3 at 0°C, 1 atmosphere pressure. Then

$$c_w = \sqrt{\frac{\gamma P}{\rho_v}} = \sqrt{\frac{1.4 \times 1.013 \times 10^5}{1.293}} = 331 \text{ m/sec}$$

(b) Since helium gas is monatomic, we choose $\gamma = 5/3$. The mass density ρ_v can be found as follows. One mole of a gas occupies a volume of 22.4 litres (L) at atmospheric pressure at 0°C and contains 6×10^{23} (Avogadro's number) molecules. Since the molar mass of He is 4 g, we find

$$\rho_v = \frac{0.004 \text{ kg}}{22.4 \times 10^{-3} \text{ m}^3} = 0.179 \text{ kg/m}^3$$

The velocity of sound in He gas in the standard condition is

$$c_w = \sqrt{\frac{\gamma P}{\rho_v}} = \sqrt{\frac{5/3 \times 1.013 \times 10^5 \text{ N/m}^2}{0.179 \text{ kg/m}^3}} = 972 \text{ m/sec}$$

The reader may have experienced the demonstration of the elevation of pitch of an individual who inhaled helium and started to speak from the same individual who was breathing normal air and speaking. This is related to the differences in the velocities of sound because of the difference of the masses between helium and air.

The expression for the velocity of sound waves in gases Eq. (5.34) can be rewritten in terms of microscopic quantities. Since the pressure P is nk_BT Eq. (5.35) and the mass density is

$$\rho_v = nm \ (\text{kg/m}^3) \tag{5.40}$$

where m is the mass of one molecule, Eq. (5.39) becomes

$$c_w = \sqrt{\frac{\gamma k_B T}{m}} \tag{5.41}$$

Observe that the velocity of sound in gases is actually independent of the gas density n and is determined by the gas temperature and the molecular mass. From an investigation of the kinetic theory of gases, one finds that the sound velocity given by Eq. (5.41) is very similar to the *random molecular velocity*,

$$\sqrt{\frac{3k_B T}{m}} \tag{5.42}$$

The only difference is in the numerical factors, γ and 3. The close resemblance is actually due to the propagation mechanism of sound waves or the origin of the bulk modulus (elasticity) of gases.

If we further multiply both numerator and denominator in Eq. (5.42) by Avogadro's number $N = 6.0 \times 10^{23}$/mole, we obtain another way to express the sound velocity.

$$c_w = \sqrt{\frac{\gamma N k_B T}{Nm}} \tag{5.43}$$

Here, $Nk_B = 8.3 \text{ J/(K mol)} \equiv R \text{ J/(K mol)}$ is known as the gas constant, and $Nm = M_{\text{mol}}$ is the mass of one mole. Then, using these, we have

$$c_w = \sqrt{\frac{\gamma R T}{M_{\text{mol}}}} \tag{5.44}$$

EXAMPLE 5.5

Find the velocity of sound in air at 20°C. One mole of air has a mass of 29 g.

Solution

Substituting $\gamma = 7/5$, $R = 8.3$ J/K·mol, $T = 273 + 20 = 293$ K, and $M_{mol} = 0.029$ kg, we find

$$c_w = \sqrt{\frac{(7/5) \times 8.3 \times 293}{0.029}} = 343 \text{ m/sec}$$

5.6   **Intensity of Sound Waves in Gases**

In Section 5.2 we derived expressions for the average energy density and power density or energy flux for sinusoidal sound waves in solids. Those expressions can also be applied for sound waves in gases. The quantity of practical importance is the power density (W/m^2), which indicates how much energy (J) passes through a unit area (1 m^2) per unit time (1 sec). The power density is alternatively called the intensity of sound waves.

If a sinusoidal wave train is described with the displacement profile

$$\xi(x, t) = \xi_0 \sin(kx - \omega t),$$

the intensity I is given by Eq. (5.11)

$$I = \frac{1}{2}\rho_v \omega^2 \xi_0^2 c_w \text{ (W/m}^2) \tag{5.45}$$

The human ear is a very sensitive organ. At the same time, the human ear is very flexible and can stand a tremendously wide range of intensities. The lower limit of the audible intensity is of the order of 10^{-12} W/m^2 and the maximum safety limit is of the order of 1 W/m^2. The ratio between these two values is 10^{12}! The intensity of ordinary conversation is of the order of 10^{-6} W/m^2, street traffic is 10^{-5} W/m^2 and jet planes is 10^{-2} W/m^2.

EXAMPLE 5.6

Sinusoidal sound waves in air have an intensity 1.0×10^{-6} W/m^2 and a frequency of 2 kHz. Assuming a room temperature (20°C) and 1 atmosphere pressure, find the amplitude ξ_0 of the displacement waves. What is the amplitude of the velocity wave? What is the amplitude of the pressure wave?

Solution

At 20°C the velocity of sound is 343 m/sec (Example 5.5). In Eq. (5.45), $I = 1.0 \times 10^{-6}$ W/m, $\rho_v = 1.29$ kg/m^3 × (273 K/293 K) = 1.20 kg/m^3, $\omega = 2\pi \times 2 \times 10^3$ rad/sec, and $c_w = 343$ m/sec. Then

$$\xi_0 = \frac{1}{\omega}\sqrt{\frac{2I}{\rho_v c_w}} = \frac{1}{2\pi \times 2 \times 10^3}\sqrt{\frac{2 \times 10^{-6}}{1.2 \times 343}} = 5.55 \times 10^{-9} \text{ (m)}$$

The amplitude of the velocity wave is $\omega\xi_0$ since

$$v = \frac{\partial \xi}{\partial t} = -\omega\xi_0 \cos(kx - \omega t)$$

Then

$$\omega\xi_0 = 2\pi \times 2 \times 10^3 \times 5.55 \times 10^{-9} = 7.0 \times 10^{-5} \text{ m/sec}$$

The amplitude of the pressure wave can be found from

$$\Delta P = -\gamma P \frac{\partial \xi}{\partial x}$$

in analogy with sound waves in liquids, as in Eq. (A) in Example 5.3. Then

$$\Delta P = -\gamma P k \xi_0 \cos(kx - \omega t)$$

and the amplitude of the pressure wave is

$$\gamma P k \xi_0$$

where the wavenumber k can be found from the dispersion relation

$$\frac{\omega}{k} = c_w \quad \text{or} \quad k = \frac{\omega}{c_w}$$

Substituting $\gamma = 7/5$, $P = 1.013 \times 10^5$ N/m^2, $\xi_0 = 5.55 \times 10^{-9}$ m, and $k = 2\pi \times 2 \times 10^3/343 = 36.6$ rad/m, we find

$$\gamma P \xi_0 = 2.85 \times 10^{-2} \text{ N/m}^2$$

As this example indicates, the air molecules hardly move (they move only 5.6×10^{-9} m!) for the intensity of 10^{-6} W/m^2, which is the typical intensity of human conversation. The pressure perturbation relative to the equilibrium pressure is only

$$\frac{2.85 \times 10^{-2}}{1.0 \times 10^5} = 2.85 \times 10^{-7}$$

We indeed realize how sensitive our ears actually are.

Because the audible intensity range is so wide (10^{12}!) and the pressure on the eardrum may have deleterious effects, it is common to specify the intensity in terms of a ratio of the actual intensity divided by a standard intensity using a logarithmic expression. The intensity I is proportional to the square of the amplitude A of the wave. The decibel (*deci* = "ten," *bel* after "Bell"), abbreviated as dB is defined as

$$dB = 10 \log_{10}\left(\frac{I}{I_0}\right) = 10 \log_{10}\left(\frac{A^2}{A_0^2}\right) = 20 \log_{10}\left(\frac{A}{A_0}\right) \qquad (5.46)$$

The standard sound intensity I_0 is chosen to be 1.0×10^{-12} W/m^2, this being approximately the threshold of human hearing. For example, the intensity of 1.0×10^{-6} W/m^2 is equivalent to 60 dB since

$$10 \log_{10}(10^{-6}/10^{-12}) = 10 \log_{10} 10^6 = 60$$

Apparently, the human feeling for sound intensity (loudness of sound) is in logarithmic form and not in a linear form. In this respect, the introduction of decibels is rather natural. The decibel representation is also used in electrical engineering to express a power relative to a standard power. The citation of 3 dB, which often appears in electrical engineering, indicates a difference of a factor of 2 between two powers. (Recall that $\log_{10} 2 \simeq 0.3$.)

 Problems

1. Ice has a density of 920 kg/m^3 and Young's modulus of 1×10^{10} N/m^2 at 0°C. Estimate the speed of sound along an ice rod.

2. Longitudinal earthquake waves typically have a velocity of 5×10^3 m/sec. Assuming the average earth density is 1500 kg/m^3, estimate the elastic modulus of the earth.

3. A long steel rod having a diameter of 5 cm is forced to transmit longitudinal waves excited with a mechanical oscillator connected to the end. The amplitude of the displacement waves is 10^{-5} m and the frequency is 400 Hz. Find
   (a) The expression to describe the displacement wave.
   (b) The average energy per unit length of the rod.
   (c) The average power transfer through a cross section of the rod.
   (d) The power delivered by the oscillator.

4. In Problem 3, find the expression for
   (a) The velocity wave.
   (b) The force wave.
   (c) The energy wave.
   Assume that the displacement wave is described by $\xi_0 \sin(kx - \omega t)$.

5. Compute the velocity of sound in
   (a) A hydrogen (H$_2$) gas.
   (b) An argon (Ar) gas. Both gases are at 0°C.
   The atomic mass of hydrogen atom is 1.0 and that of argon is 40. Be careful with your choice of the value of γ.

6. Show that the change in the sound velocity Δc_w caused by a small change in the temperature ΔT is given by

$$\Delta c_w = \frac{1}{2}\frac{\Delta T}{T}c_w$$

where $c_w = \sqrt{\gamma RT/M_{mol}}$ is the original velocity.

7. An observer detects the sound in air caused by an explosion on a lake 2 sec after he detects the sound in water caused by the same explosion. How far is the explosion point from the observer? Assume the sound velocity in water is 1500 m/sec and in air is 340 m/sec.

8. In a thunderstorm, an individual detects the thunder three seconds after seeing the lightning flash. How far away was this particular lightning stroke from the individual?

9. A sinusoidal sound wave with a frequency of $\nu = 400$ Hz in air (20°C, 1 atm pressure) has an intensity of 1×10^{-7} W/m^2. What is
   (a) The amplitude of the displacement wave?
   (b) The amplitude of the pressure wave?
   (c) The intensity of the wave expressed in dB?

10. The air molecule displacement associated with a harmonic (sinusoidal) sound wave train in air (20°C, 1 atmosphere pressure) is described by

$$\xi(x, t) = 1.0 \times 10^{-8} \sin(kx - \omega t)(\text{m})$$

where $\omega = 2\pi \times 10^3$ rad/sec.
   (a) What is the value of the wavenumber k?
   (b) How intense is the wave? Answer in terms of W/m^2 and dB (decibels).
   (c) What is the amplitude of the pressure wave? How does this compare with the equilibrium pressure?

11. Two sound waves, one in air and one in water, have the same frequency and the same intensity. What is the ratio between molecular displacement amplitudes? Assume $T = 20°$C.

12. The displacement wave of a harmonic sound wave in air is given by

$$\xi(x, t) = \xi_0 \sin(kx - \omega t), \frac{\omega}{k} = c_s$$

Show that the ratio between the pressure wave $-\gamma P \frac{\partial \xi}{\partial x}$ and the velocity wave $\frac{\partial \xi}{\partial t}$ is constant and given by $\sqrt{\gamma P \rho_v}$, where ρ_v is the volume mass density. This quantity is called the "impedance" for sound wave.

CHAPTER 6

Wave Reflection and Standing Waves

6.1 Introduction

When a ball hits a hard wall, it is reflected by the wall. This reflection phenomenon can alternatively be interpreted in terms of the reflection of energy and momentum associated with the ball. If the wall is soft, the collision is inelastic and the wall completely absorbs the energy and momentum of the ball. No reflection occurs in this case.

As we have seen, waves carry energy and momentum and whenever waves encounter an obstacle, they are reflected by the obstacle. Echoes are caused by the reflection of sound waves. Radars use the reflection of electromagnetic waves (microwaves) from metal objects such as airplanes. Wave reflection gives rise to an important wave phenomenon called standing waves, which play essential roles in most musical instruments. As the name indicates, standing waves do not propagate and therefore are not associated with energy and momentum transfer. They essentially behave as spatially distributed oscillators that only store energy. They can create waves in a surrounding medium by radiation. For example, the strings in a piano oscillate with distinct frequencies that are determined by the length, tension, and mass of each string. Each string can create sound waves in air with a particular frequency.

6.2 Reflection at a Fixed Boundary, Standing Waves

Suppose a transverse pulse is propagating along a string from left to right and it approaches the end that is rigidly clamped (Figure 6–1). When the pulse hits the end, it exerts a force on the end that is directed vertically upward.

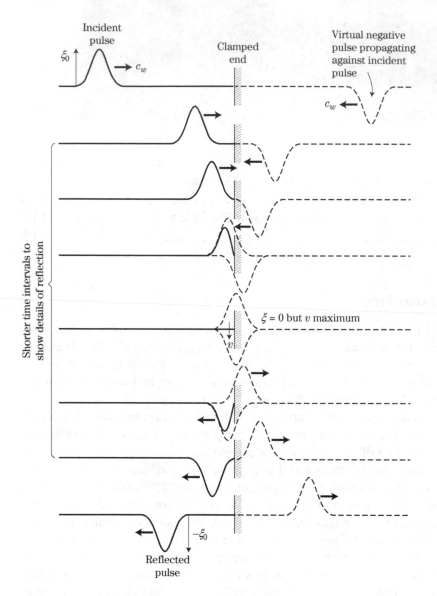

FIGURE 6–1

Wave reflection at a fixed boundary. The displacement $\xi = 0$ at the boundary and the reflected wave changes its sign. A virtual wave, which is an inverted mirror image of the wave on the string, is convenient to analyze the wave reflection.

However, the end is clamped and cannot move. Therefore the clamped end should exert a force on the string that is directed vertically downward. This new force in turn creates a pulse that is propagating from right to left with the opposite polarity. We can then conclude that at a fixed boundary, the vertical displacement ξ remains zero and the reflected wave changes its polarity. Notice that the magnitude of the displacement $|\xi|$ remains unchanged and the pulse is completely reflected and maintains the same shape. A boundary to cause the polarity change in the reflected wave (relative to the original wave) is called a *hard* boundary. The rigid clamped boundary in the preceding description is an example of a hard boundary.

Let us next consider a sinusoidal wavetrain that is propagating toward the clamped end. We choose the location of the clamped end to be at $x = 0$ and the wave to be launched at $x = -\infty$. We describe the incident wave by

$$\xi_+(x, t) = \xi_0 \sin(kx - \omega t), \quad \frac{\omega}{k} = c_w \tag{6.1}$$

where $+$ indicates a wave that is propagating to increasing values of the coordinate x or a positive-going wave. Note that x is negative in the region to the left of the clamped end that has been located at $x = 0$. Eq. (6.1) still describes the wavetrain that is propagating to the right or toward the clamped end. When the wavetrain is reflected, the direction of propagation should be in the opposite direction. As we will see, choosing the location of the boundary to be at $x = 0$ will simplify the mathematics in the following discussion. Also the polarity of the reflected wave must be reversed. Therefore the reflected wavetrain must be described by

$$\xi_-(x, t) = -\xi_0 \sin(-kx - \omega t) \tag{6.2}$$

Since

$$\sin(-\theta) = -\sin\theta$$

the reflected component ξ_- can be written as

$$\xi_-(x, t) = \xi_0 \sin(kx + \omega t) \tag{6.3}$$

The total displacement is the sum of the incident and reflected components ξ_+ and ξ_-,

$$\xi(x, t) = \xi_0[\sin(kx - \omega t) + \sin(kx + \omega t)] \tag{6.4}$$

Recalling that (see Appendix B)

$$\sin\alpha + \sin\beta = 2\sin\left(\frac{\alpha + \beta}{2}\right)\cos\left(\frac{\alpha - \beta}{2}\right)$$

we find that the sum of the incident plus the reflected components can be written as

$$\xi(x, t) = 2\xi_0 \sin kx \cos \omega t \tag{6.5}$$

(Note that $\cos(-\theta) = \cos\theta$.) Equation (6.5) is *not* of the form of a traveling wave since it does not contain a factor, $X = kx - \omega t$ or $kx + \omega t$. Although we started from two propagating waves going in the opposite directions, we ended up with something that does not propagate! The expression in Eq. (6.5) is a typical form of a standing wave.

In Figure 6–2 the snapshots of Eq. (6.5) at several instances of time are shown. We can clearly see that the sinusoidal wave patterns are not moving along the x axis but appear to be stationary. In fact, they simply oscillate up and down and a picture with a long exposure time (superposition of many snapshots) indicates the formation of clumps with *a spatial period $\lambda/2$ where λ* is the wavelength of the original wave. Such wave patterns are called *standing waves* in contrast with the traveling waves we have so far been studying.

Standing waves are formed when two sinusoidal waves with the same frequency (and thus the same wavelength) propagating in opposite directions are superposed. Simple superposition is valid for the linear waves that we are considering at this point.

Although they are not propagating, standing waves are created by traveling waves that are propagating in the opposite directions. Since standing waves are not propagating, *no energy can be carried by the standing waves*. Rather, energy is confined between the "nodes," where the displacement ξ is zero. The hard boundary will be one of the nodes as shown in Figure 6–3. The absence of energy flow in the standing wave is understandable since the positive-going wavetrain ξ_+ and the negative going wavetrain ξ_- each carry the same amount of energy but in the opposite directions pointing out that the net energy flow in a particular direction must be zero.

String musical instruments (piano, guitar, violin, etc.) all use the standing-wave phenomena. A string with a given length L that is clamped at both ends allows standing waves *with discrete* wavelengths to exist. This starts with the longest wavelength $\lambda_1 = 2L$. This determines the frequency of oscillation of the string through $\lambda\nu = c_w$ where c_w is the velocity of the transverse waves on the string that is $c_w = \sqrt{T/\rho_l}$ with T being the tension (N) and ρ_l, the linear mass density (kg/m). This lowest frequency of oscillation is called the fundamental frequency. We denote this frequency by ν_0.

The next possible wavelength is $\lambda_2 = L$ (Figure 6–3b). The frequency of oscillation for this mode is twice that of the fundamental frequency $2\nu_0$ and it is called *the second harmonic*. The third possible wavelength is $\lambda_3 = 2L/3$ yielding the third harmonic to be $3\nu_0$. This procedure can be continued to find the higher harmonics.

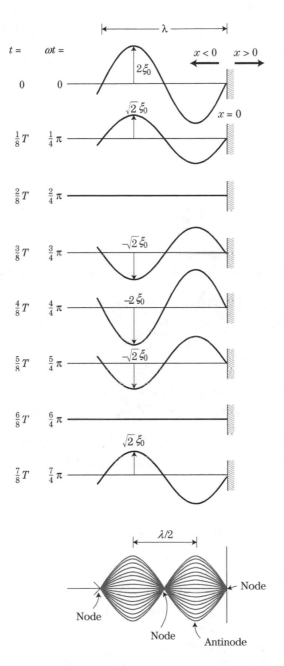

FIGURE 6–2

Snapshots of the standing wave at various times as described with Eq. (6.5). The bottom picture was taken with a long exposure time. Standing waves are essentially spatially distributed oscillators.

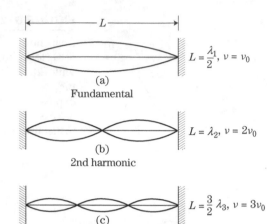

FIGURE 6–3

Allowed standing waves on a string that is clamped at both ends that are separated by a distance L.

EXAMPLE 6.1

A piano string having a mass 15 g and length 1 m (clamped at both ends) is used for creating the note that has a resonant frequency of, $\nu = 220$ Hz (fundamental mode). Find the tension to be applied to the string.

Solution

The wavelength of the fundamental mode is $\lambda = 2 \times 1 = 2$ m. Then

$$c_w = \lambda \nu = 2 \text{ m} \times 220 \text{ sec}^{-1} = 440 \text{ m/sec}$$

From $c_w = \sqrt{T/\rho_l}$ (see Eq. (4.44)), we determine that the tension should be

$$T = c_w^2 \rho_l = (440 \text{ m/sec})^2 \times 0.015 \text{ kg/m} = 2.9 \times 10^3 \text{ N}.$$

6.3 Reflection at a Free Boundary

At a free boundary, the material composing a wave medium is free to move in contrast with the case of the fixed boundary. In Figure 6–4, we again show a pulse-shaped transverse wave traveling on a string toward the end that is now free to move. When the pulse hits the free end, it is reflected, but the reflected pulse will have *the same polarity* as the incident pulse in contrast with the case of the clamped end that was discussed in the previous section. In order to explain this, let us recall that the vertical restoring force that acts on a segment of the string was obtained in Eq. (4.46).

$$F = T \frac{\partial \xi}{\partial x} \qquad (6.6)$$

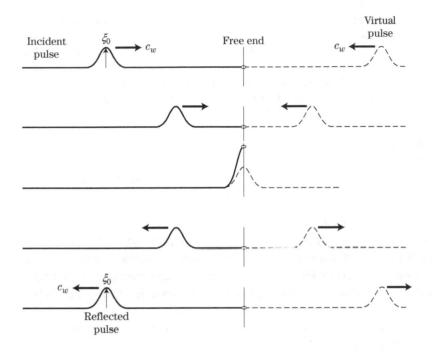

Incident pulse ξ_0 c_w Free end   Virtual pulse c_w

Incident pulse ξ_0 c_w Free end   Virtual pulse c_w

c_w ξ_0

Reflected pulse

FIGURE 6–4

Wave reflection at a free boundary. The amplitude at the free boundary is doubled and the reflected wave has the same polarity as the incident wave. The virtual wave in this case is an erect mirror image.

Let the displacements of incident wave and reflected wave be described by

$$\xi_+ = \xi_0 \sin\left(kx - \omega t\right) \tag{6.7}$$

$$\xi_- = A \sin\left(-kx - \omega t\right) = -A \sin\left(kx + \omega t\right) \tag{6.8}$$

where A is an amplitude that is yet to be determined. Since the total vertical displacement of the string is

$$\xi = \xi_+ + \xi_- = \xi_0 \sin\left(kx - \omega t\right) - A \sin\left(kx + \omega t\right) \tag{6.9}$$

The vertical force acting on the end of the string is

$$T\frac{\partial \xi}{\partial x} = T[\xi_0 k \cos\left(kx - \omega t\right) - Ak \cos\left(kx + \omega t\right)] \tag{6.10}$$

which is to be evaluated at the free end $x = 0$. Then

$$T\frac{\partial \xi}{\partial x}\bigg|_{x=0} = Tk(\xi_0 - A) \cos \omega t \tag{6.11}$$

The force is a restoring force and would tend to push the string back to the equilibrium position. However, at the free end, the string is free to have any amplitude since this force is obviously absent. We assert that $T\frac{\partial \xi}{\partial x} = 0$ for all times. Therefore, we must conclude that the amplitude A has the value

$$A = \xi_0 \tag{6.12}$$

and the reflected wave should have the same polarity as the incident wave. (You should apply this argument to the case of a fixed boundary and show that $A = -\xi_0$ results in this case.) Hence we conclude

At a free boundary, the restoring force is zero, which implies $\partial \xi / \partial x = 0$. This means that the reflected wave has the same polarity as the incident wave. The amplitude at the free boundary is twice as large as that of the incident wave.

Free boundaries are often called *soft* boundaries.

To speak of a free boundary of a string has little practical meaning since if the string were subject to a tension, it would be almost impossible to make the end move without friction. However, in musical wind instruments including a pipe organ, the reflection of sound waves at free boundaries plays an essential role.

If $A = \xi_0$ in Eq. (6.9), we find the expression for standing waves created by the free boundary,

$$\xi = 2\xi_0 \cos kx \sin(-\omega t) = -2\xi_0 \cos kx \sin \omega t. \tag{6.13}$$

The amplitude of the standing waves at $x = 0$ is indeed twice as large as that of the original incident wave. The long exposure picture is shown in Figure 6–5.

As should be obvious, the pipes of a pipe organ must have at least one open end. Otherwise standing sound waves created in a pipe would not be able to escape and we would not hear anything! An organ pipe of length L having one closed end and one open end can create a standing wave whose wavelength is

$$\lambda = 4L \tag{6.14}$$

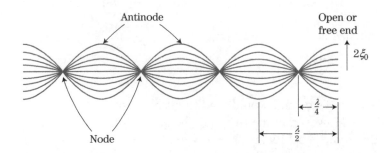

FIGURE 6–5

Long exposure picture of standing waves created by a free boundary.

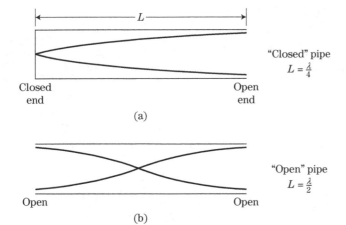

Closed
end

Open
end

(a)

"Closed" pipe
$L = \frac{\lambda}{4}$

Open

Open

(b)

"Open" pipe
$L = \frac{\lambda}{2}$

FIGURE 6–6

Standing waves in a pipe. (a) One end closed and (b) both ends open. Organ pipes are operated with this standing wave profile. The amplitude of the standing wave as the air blows is maximum and acts as an open end. Only the fundamental modes are shown.

with a distinct resonant frequency

$$\nu_1 = \frac{c_w}{4L} \tag{6.15}$$

as shown in Figure 6–6(a). Note that the closed end acts as a fixed boundary and the open end as a free boundary. At the closed end, the molecule displacement must be zero since the wall is rigidly fixed. At the open end, the air molecules are free to move. The amplitude becomes maximum at the open end which acts as a speaker that will radiate sound waves into the air.

A pipe having two open ends has its fundamental wavelength as shown in Figure 6–6b

$$\lambda_1 = 2L \tag{6.16}$$

with a corresponding fundamental frequency

$$\nu_0 = \frac{c_w}{2L} \tag{6.17}$$

You will also find higher resonant frequencies that are harmonics of these fundamental frequencies in both cases.

EXAMPLE 6.2

Find the length of a closed pipe to create sound waves with a resonant frequency of 40 Hz. What is the next resonance frequency? Assume the ambient temperature is $T = 20°C$.

112   Wave Reflection and Standing Waves

Solution

At 20°C, the sound velocity is $c_w = 343$ m/s (see Chapter 5). For the closed pipe, the fundamental wavelength is $\lambda = 4L$, with L the length of the pipe. Since $c_w = \lambda v$, we find

$$L = \frac{c_w}{4v} = \frac{343 \text{ m/s}}{4(40/\text{s})} = 2.14 \text{ m}$$

The next resonant wavelength is $4L/3$ as shown in Figure 6–7. Then the frequency is three times higher than the fundamental. The second resonant frequency is 3×40 Hz $= 120$ Hz.

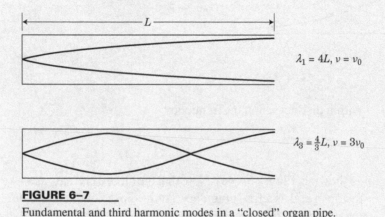

FIGURE 6–7

Fundamental and third harmonic modes in a "closed" organ pipe.

6.4

Theory of Wave Reflection, Mechanical Impedance

In previous sections, we have considered two ideal cases of wave reflection, reflection at a fixed boundary and reflection at a free boundary. In both cases, the reflection was *complete* in the sense that all the energy associated with an incident wave was reflected at the boundary. In practice, however, complete reflection rarely occurs. Consider, for example, sound waves incident upon a concrete wall. Although most of the energy is reflected from the wall since concrete is much "harder" than air, a portion of the sound wave is able to penetrate into the concrete and propagate in this solid substance. The reflection is not one hundred percent. Some of the incident energy will be absorbed in the concrete wall. A similar and probably more important phenomenon can be readily observed in optics. Light incident upon glass is partially reflected from the glass and partially transmitted through the glass. Since optical devices are important in their own right, they will be further discussed in Chapter 9.

There will always be a partial reflection of an incident wave at the interface between two different media and a partial transmission of the wave into the second medium. In the preceding example, sound waves would propagate in air and also in concrete. The sound velocity in air is quite different from the velocity in concrete. In the case of light incident upon a glass surface, the velocity of light in air and that in glass are different (the velocity of light in glass is approximately 2/3 of that in air or in a vacuum) and, therefore, air and glass are different media for light waves. If a wave medium has certain characteristics as in the two ideal cases considered in the previous sections, the wave reflection will be complete.

To understand *incomplete reflection*, let us consider two strings with different linear mass densities, ρ_{l1} (kg/m) and ρ_{l2} (kg/m), respectively. Both strings are subject to a common tension T (N) (Figure 6–8). The velocities of propagation of the transverse waves on each string are

$$c_{w1} = \sqrt{\frac{T}{\rho_{l1}}}, \quad c_{w2} = \sqrt{\frac{T}{\rho_{l2}}} \tag{6.18}$$

respectively (Chapter 4). We found there that the energy flow associated with a sinusoidal wavetrain with amplitude ξ_0 and frequency $\omega = 2\pi\nu$ along a string having a linear mass density ρ_{lj} (kg/m) is given by

$$\frac{1}{2}\rho_{lj}c_{wj}\omega^2\xi_0^2 \text{ (W)} \tag{6.19}$$

where c_{wj} is the velocity of propagation of the transverse waves where $j = 1$ or 2. We assume that the incident wave has an amplitude of ξ_i, the reflected wave has an amplitude of ξ_r, and the transmitted wave has an amplitude of ξ_t.

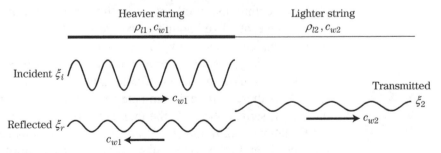

FIGURE 6–8

Wave reflection at the joining point of two strings with different mass densities occurs because of impedance mismatch. In general, reflection is not complete and some energy can cross the boundary.

The energy flow associated with the incident wave is

$$\frac{1}{2}\rho_{l1}c_{w1}\omega^2\xi_i^2(\rightarrow) \text{ incident wave} \tag{6.20}$$

that with the reflected wave is

$$\frac{1}{2}\rho_{l1}c_{w1}\omega^2\xi_r^2(\leftarrow) \text{ reflected wave} \tag{6.21}$$

and that with the transmitted wave is

$$\frac{1}{2}\rho_{l2}c_{w2}\omega^2\xi_t^2(\rightarrow) \text{ transmitted wave} \tag{6.22}$$

where the arrows indicate the direction of the energy flow. Since the energy must be conserved at the interface, we must have

$$\rho_{l1}c_{w1}\omega^2\xi_i^2 - \rho_{l1}c_{w1}\omega^2\xi_r^2 = \rho_{l2}c_{w2}\omega^2\xi_t^2 \tag{6.23}$$

or

$$\rho_{l1}c_{w1}(\xi_i^2 - \xi_r^2) = \rho_{l2}c_{w2}\xi_t^2 \tag{6.24}$$

This result is of the desired form. For example, the case $\xi_t = 0$ indicates complete reflection and yields $\xi_i^2 = \xi_r^2$. This implies that the reflected wave must have the same magnitude of the amplitude as the incident wave. (Notice $\xi_r = \pm\xi_i$, the plus sign will be chosen for the complete reflection at the free boundary and the minus sign for the complete reflection at the fixed boundary.) In addition, if there is no reflected wave $\xi_r = 0$, all the incident energy will be transmitted through the boundary.

To find the amplitudes of the reflected wave ξ_r and the transmitted wave ξ_t in terms of the incident wave ξ_i, we note that at the interface

$$\xi_i + \xi_r = \xi_t \tag{6.25}$$

which results from the requirement that the displacement at the boundary must be continuous. In other words, the string is not broken at the interface. For example, at a fixed boundary where $\xi_t = 0$, we indeed recover $\xi_r = -\xi_i$, or the complete reflection with the polarity change in the reflected wave. At a free boundary, the displacement at the boundary was twice as large as the displacement of the incident wave. Equation (6.25) indeed yields $\xi_t = 2\xi_i$ with $\xi_r = \xi_i$ or there will be a complete reflection with the reflected component having the same polarity as the incident wave.

Equations (6.24) and (6.25) are two simultaneous equations for two unknowns, ξ_r and ξ_t. Solving these in terms of the amplitude of the incident wave, we find

$$\xi_r = \frac{c_{w1}\rho_{l1} - c_{w2}\rho_{l2}}{c_{w1}\rho_{l1} + c_{w2}\rho_{l2}}\xi_i = \frac{\sqrt{\rho_{l1}T} - \sqrt{\rho_{l2}T}}{\sqrt{\rho_{l1}T} + \sqrt{\rho_{l2}T}}\xi_i \tag{6.26}$$

and

$$\xi_t = \frac{2\sqrt{\rho_{l1}T}}{\sqrt{\rho_{l1}T} + \sqrt{\rho_{l2}T}}\xi_i \tag{6.27}$$

Note that the tension T could have been canceled between numerator and denominator but we did not do this. The reason is that the quantity

$$Z_m = \sqrt{\text{mass density} \times \text{elastic modulus}} \tag{6.28}$$

has an important physical meaning and for sound waves in solid rods, the quantity $\sqrt{\rho_v Y}$ plays the same role. In the case of two strings, the elastic moduli (tension) are common. This is generally not the case.

The quantity defined by Eq. (6.28) is called the *impedance of mechanical waves*. The concept of impedance has been introduced in electrical engineering for electromagnetic waves that propagate either in space or along transmission lines and will be examined in Chapter 9, but it also can be applied to mechanical waves. Of course, the *mechanical impedance does not have the dimensions of ohms* as obtained for the electromagnetic impedance.

For sound waves in a gas, Eq. (6.28) takes the following form

$$Z_m = \sqrt{\rho_v \gamma P}$$

where ρ_v is the volume mass density and γP is the bulk modulus. As we saw in the previous chapter, the impedance can be interpreted as the ratio of the pressure (or force) wave and the velocity wave. If one wishes at this stage to make an analogy between mechanical waves and electromagnetic waves, the correspondence is illustrated in Table 6–1.

The impedance Z_m is a quantity that is characterized by the wave media. Using Z_m, we can rewrite Eq. (6.26) as

$$\xi_r = \frac{Z_1 - Z_2}{Z_1 + Z_2}\xi_i \tag{6.29}$$

where Z_1 is the impedance of medium 1 and Z_2 of medium 2. From Eq. (6.29) we can draw an important conclusion:

The condition for the absence of wave reflection at a boundary between two media is that the impedances of the two media be the same.

Table 6–1 Comparison of mechanical and electrical waves

Mechanical	Electrical
Force wave	Voltage wave
Velocity wave	Current wave
$Z_m = \dfrac{\text{force wave}}{\text{velocity wave}}$	$Z = \dfrac{\text{voltage wave}}{\text{current wave}}$

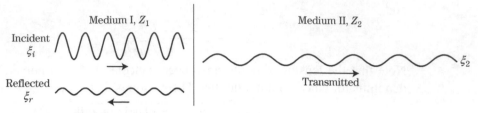

FIGURE 6–9

Wave reflection occurs whenever waves try to penetrate into a medium with a different mechanical impedance. If $Z_1 = Z_2$, there will be no reflection of the incident wave and this is called impedance matching. If $Z_2 = 0$ or ∞, the reflection will be complete.

The fraction of reflected wave energy is given by (Figure 6–9)

$$\left(\frac{\xi_r}{\xi_i}\right)^2 = \left(\frac{Z_1 - Z_2}{Z_1 + Z_2}\right)^2 \tag{6.30}$$

In mechanical wave, the displacement $\xi_0 \sin(kx - \omega t)$ is normally used to describe wave motion. As we have seen, when an incident wave ξ_i in medium 1 is reflected at a boundary where the impedance jumps to Z_2, the reflected displacement wave is

$$\xi_r = \frac{Z_1 - Z_2}{Z_1 + Z_2}\xi_i \tag{6.31}$$

As we will see in Chapter 9, reflection of voltage wave is described by

$$V_r = \frac{Z_2 - Z_1}{Z_2 + Z_1}V_i \tag{6.32}$$

while that of the current wave is given by

$$I_r = \frac{Z_1 - Z_2}{Z_1 + Z_2}I_i \tag{6.33}$$

This is because the energy conservation of electromagnetic waves can be written in terms of either voltage waves as

$$\frac{V_i^2}{Z_1} = \frac{V_r^2}{Z_1} + \frac{V_t^2}{Z_2} \tag{6.34}$$

or current waves as

$$Z_1 I_i^2 = Z_1 I_r^2 + Z_2 I_t^2 \tag{6.35}$$

The mechanical displacement wave is analogous to the electromagnetic current wave.

EXAMPLE 6.3

Sound waves in air are incident in a direction that is normal to a water surface. Calculate the percentage of energy that is reflected at the water surface. Water has a bulk modulus of 2.1×10^9 N/m^2. Assume that the temperature of air is 20°C and the pressure is 1 atm.

Solution

The impedance for sound waves in water is

$$Z_{\text{water}} = \sqrt{\rho_v M_B} = \sqrt{10^3 \ (\text{kg/m}^3) \times 2.1 \times 10^9 \ (\text{N/m}^2)} = 1.45 \times 10^6 \ \text{kg/m}^2 \cdot \text{sec}.$$

The impedance for sound waves in air is

$$Z_{\text{air}} = \sqrt{\rho \gamma P} = \sqrt{1.29 \ (\text{kg/m}^3) \times 7/5 \times 1.0 \times 10^5 \ (\text{N/m}^2)} = 4.3 \times 10^2 \ \text{kg/m}^2 \cdot \text{sec}.$$

Then the energy reflection coefficient is

$$\left(\frac{Z_{\text{air}} - Z_{\text{water}}}{Z_{\text{air}} + Z_{\text{water}}} \right)^2 = 0.9988 = 99.88\%$$

This indicates that practically all of the incident sound energy is reflected at the water surface. Only 0.1% penetrates into the water.

EXAMPLE 6.4

A pulse of amplitude 1 cm is propagating along a string toward a boundary where the string is connected to another string having a mass density four times larger. Both strings are subject to a common tension.

(a) Find the amplitudes of reflected and transmitted pulses.
(b) Find the energy reflection coefficient.
(c) Sketch qualitatively the reflected and transmitted pulses after the incident pulse reached the boundary.

Solution

(a) Since the elastic modulus (tension in this case) is common, the mechanical impedance is proportional to the square root of the mass density $Z_1/Z_2 = 1/2$. From Eq. (6.29), we find

$$\xi_r = \frac{Z_1 - Z_2}{Z_1 + Z_2} \xi_i = \frac{-0.5}{1.5} \xi_i = -\frac{1}{3} \xi_i = -0.33 \ \text{cm}$$

and from Eq. (6.27),

$$\xi_t = \frac{2Z_1}{Z_1 + Z_2} \xi_i = \frac{2}{3} \xi_i = 0.67 \ \text{cm}.$$

The reflected pulse has the opposite polarity relative to the incident pulse, as expected since the second string is heavier and thus "harder."

(b) The energy reflection coefficient, Eq. (6.30), becomes

$$\left(\frac{Z_1 - Z_2}{Z_1 + Z_2}\right)^2 = \left(-\frac{1}{3}\right)^2 = \frac{1}{9} = 11\%$$

(c) See Figure 6–10.

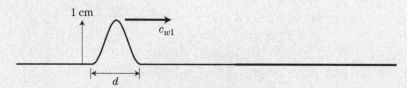

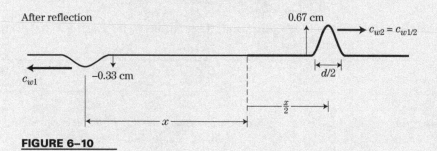

FIGURE 6–10

The opposite case in which a pulse is propagating on the heavier string toward a lighter string is given as a problem. You will find that the energy reflection coefficient remains unchanged since

$$\left(\frac{Z_1 - Z_2}{Z_1 + Z_2}\right)^2 = \left(\frac{Z_2 - Z_1}{Z_2 + Z_1}\right)^2$$

To summarize, we have seen that waves are reflected at a boundary where the discontinuity in the mechanical impedance exists. A similar reflection will occur for electromagnetic waves as will be discussed later. The criteria of requiring the same propagation velocity in the two media does not necessarily ensure the absence of reflection at the boundary between the two media. The impedances must also be the same. The subject treated in this section is an extremely important concept for all wave phenomena.

Problems

6.5

1. A certain violin string is 50 cm long between its fixed ends and has a mass of 0.5 g. It sounds the A note (440 Hz) without fingering the string.
   (a) What is the tension that should be applied?
   (b) Where is the fingering position if it is to play the C note (528 Hz)?

2. A string 2 m long has a mass of 500 g. If the string is fixed at the ends with a tension of 200 N, what is the fundamental frequency of oscillation?

3. What is the fundamental resonance frequency of a 50-cm-long organ pipe with one end closed? What are the higher resonance frequencies? Take $c_s = 340$ m/sec.

4. What are the resonance frequencies of a well 20 m deep? Assume the velocity of sound is $c_s = 340$ m/sec.

5. When wave reflection is not complete, incomplete standing waves are formed. Let the incident harmonic wave be $\xi_0 \sin(kx - \omega t)$ and the reflected wave be $-\Gamma \xi_0 \sin(kx + \omega t)$, where Γ is the reflection coefficient defined by
$$\Gamma = \frac{\text{reflected wave at } x = 0}{\text{incident wave at } x = 0}$$
   If $|\Gamma| = 1$, we have complete reflection and thus the largest amplitude standing wave:
   (a) Where does the amplitude maximum occur? Answer in terms of the distance from the discontinuity, $x = 0$.
   (b) Where does the amplitude minimum occur?
   (c) Show that the ratio between the amplitude maximum and minimum is given by
$$\frac{1 + |\Gamma|}{1 - |\Gamma|}.$$
   This quantity is called the *standing-wave ratio* and is a measure of how much reflection takes place.

6. A pulse 1 cm high on a string having a mass density of 50 g/m is propagating toward the boundary where the string is connected to another with a mass density 20 g/m. Both strings are under the same tension.

   (a) Is the boundary a hard or soft boundary for the incident pulse?
   (b) Find the amplitude (height) of the reflected pulse.
   (c) Find the amplitude of the transmitted pulse.

7. The transverse wave on a string discussed so far is a one-dimensional wave having only one spatial variable x. Generalization to multidimensional waves can be done by suitable changes in the elastic moduli and mass densities. As an example, let us consider transverse waves on a rectangular membrane whose four edges are rigidly clamped such as the membrane of a drum or a trampoline. The membrane has *a surface* mass density ρ_s (kg/m^2) and is subject to a uniform, isotropic surface tension T_s (N/m).
   (a) The wave differential equation for the vertical displacement $\xi(x, y, t)$ is given by
$$\frac{\partial^2 \xi}{\partial t^2} = \frac{T_s}{\rho_s}\left(\frac{\partial^2 \xi}{\partial x^2} + \frac{\partial^2 \xi}{\partial y^2}\right)$$
   What is the velocity of the wave on the membrane? The derivation will be similar to that used to derive the wave equation for a string.
   (b) The standing waves on a string whose ends are clamped are described by
$$\xi(x, t) = \xi_0 \sin\left(\frac{m\pi x}{L}\right)\sin \omega t$$
   where m is a nonzero integer and L is the length of the string. Show that for the standing waves on the membrane, the appropriate expression for the standing waves is
$$\xi(x, y, t) = \xi_0 \sin\left(\frac{m\pi x}{a}\right)\sin\left(\frac{n\pi y}{b}\right)\sin \omega t$$
   where m and n are nonzero integers. Explain why $\cos((m\pi x/a))$, $\cos((n\pi y/b))$ are not allowed for the clamped membrane.
   (c) Show that the resonant frequencies of the membrane are given by
$$\omega = \sqrt{\frac{T_s}{\rho_s}}\,\pi\left[\left(\frac{m}{a}\right)^2 + \left(\frac{n}{b}\right)^2\right]^{1/2} \quad \text{(rad/sec)}$$

Are the higher harmonics integer multiples of the lowest-order resonant frequency,

$$\omega(m = 1, n = 1) = \sqrt{\frac{T_s}{\rho_s}\pi}\left(\frac{1}{a^2} + \frac{1}{b^2}\right)^{1/2} ?$$

(*Note:* The harmonics are not always integer multiples of the fundamental frequency in contrast with the case of strings. The ratio $\omega(m, n)/\omega(m = 1, n = 1)$ may not even be a rational number. This explains why sounds created by a drum are rather uncomfortable for the listener. An analysis of a circular membrane reveals a similar conclusion. In this case, the higher harmonics are always an irrational number × the fundamental frequency. A complete analysis of the waves on a circular membrane would require Bessel functions.

8. The three-dimensional sound wave is described by

$$\frac{\partial^2\xi}{\partial t^2} = c_s^2\left(\frac{\partial^2\xi}{\partial x^2} + \frac{\partial^2\xi}{\partial y^2} + \frac{\partial^2\xi}{\partial z^2}\right), c_s = \text{sound velocity.}$$

Following the procedure outlined for Problem 7, find the resonance frequencies of sound waves in a rectangular box having dimensions of $a \times b \times c$.

9. Make an order of magnitude estimate for the period of the earth resonance. The earth has a radius of 6400 km. Assume the sound velocity of 5×10^3 m/sec.

10. A steel rod is joined to a copper rod at a smooth, flat surface. Steel has a mass density of 7800 kg/m³ and a Young's modulus of 2.0×10^{11} N/m² and copper has a mass density of 8900 kg/m³ and a Young's modulus = 1.1×10^{11} N/m². A sinusoidal sound wavetrain in the steel rod is incident on the boundary.
  (a) Calculate the fraction of reflected and transmitted wave energies relative to the incident wave energy.
  (b) Calculate the amplitudes of reflected and transmitted waves relative to the amplitude of the incident wave.
  (c) What are the mechanical impedance of the steel and copper rod for sound waves?

11. Glass and aluminum have approximately the same sound speed, 5.0×10^3 m/sec. From this, can we conclude that sound waves incident on a glass–aluminum boundary suffer very little reflection? Explain.

12. Show that the resonant frequency of a sound wave in a wine bottle is approximately given by

$$\omega = \sqrt{\frac{a}{Vl}}\,c_s$$

where V is the main volume, l is the length of the narrow neck, a is the cross-sectional area of the neck, and c_s is the sound speed. Hint: What oscillates in this case is the air in the neck. Write an equation of motion for the mass of air in the neck. The formula based on a quarter wavelength standing wave cannot be applied because the cross-sectional area changes, that is, we are dealing with a nonuniform wave medium. The equation of motion for the air in the neck is

$$\rho_v a l \frac{\partial^2\xi}{\partial t^2} = -\gamma P \frac{a^2}{V}\xi.$$

13. A nonuniform string has a linear mass density distribution ρ_{l1} (kg/m) in the region $0 < x < L_1$ and ρ_{l2} in the region $L_1 < x < L_1 + L_2$. The string is under a tension T. Find an equation to determine the frequencies of the standing waves. Assume the following displacement,

$$\xi_1(x, t) = \xi_{10}\sin(k_1 x)\sin\omega t, \quad 0 < x < L_1$$
$$\xi_2(x, t) = \xi_{20}\sin[k_2(L_1 + L_2 - x)]\sin\omega t,$$
$$L_1 < x < L_1 + L_2$$

where $k_1 = \omega/c_1 = \omega/\sqrt{T/\rho_{l1}}, k_2 = \omega/\sqrt{T/\rho_{l2}}$. Use the boundary condition that at the point $x = L_1$, both the displacement and its spatial derivatives are to be continuous.

14. In analyzing wave reflection and transmission at an impedance discontinuity, the energy conservation was imposed. Is the momentum conservation satisfied too? (In general, it is not, because the wave medium can absorb momentum. When a sound wave is reflected by a wall, the sound wave tends to pull the wall, not push. Similar phenomenon occurs when light wave falls on a glass surface.)

CHAPTER 7

Spherical Waves, Waves in a Nonuniform Media, and Multidimensional Waves

Introduction

We have frequently used the following representative waveform

$$\xi(x, t) = \xi_0 \sin(kx - \omega t) \tag{7.1}$$

to describe several kinds of waves (waves on a string, sound waves in gases and solids, etc.) where ξ_0 is the amplitude of the sinusoidal displacement wave. The amplitude ξ_0 is a constant that is specified by the source of the waves which may be a speaker for sound waves in air. (Recall that the power associated with mechanical waves is proportional to ξ_0^2.) For a wave described by Eq. (7.1), we observe the same amplitude ξ_0 everywhere along the wave that is propagating in a lossless media. Such waves are called *one-dimensional* or *plane waves*. The terminology of a plane wave results from the fact that all points transverse to the direction of propagation reside in a plane.

However, you can quickly observe that you would hear louder sounds as you get closer to a radio receiver. The radio receiver, on the other hand, can receive radio waves better at locations that are closer to the broadcasting station. A stone thrown into a pond creates water waves whose amplitude becomes smaller and smaller as the waves propagate radially outward from the initial perturbation. These examples indicate that wave amplitudes become smaller as the waves in extended media propagate away from localized wave sources. In this chapter we study these geometrical effects on the amplitude of the wave. Also, we briefly study how waves behave in a nonuniform medium in which the wave velocity slowly varies as a function of the coordinates. For example, we may wish to understand why water waves increase their amplitude as they approach the shore after they have propagated for long distances with an amplitude that appears to be almost constant.

Conservation of Energy Flow as Applied to Spherical Waves

As frequently stated, waves carry energy. The amount of energy passing through a unit area (1 m^2) in a unit time (1 sec) was defined as the power density or intensity I with the units of J/sec·m^2, or W/m^2. For example, a plane sound wave described by Eq. (7.1) in either solids or gases has an average power density or intensity I,

$$I = \frac{1}{2}\rho_v c_w \omega^2 \xi_0^2 \quad (\text{W/m}^2) \tag{7.2}$$

where ρ_v (kg/m^3) is the volume mass density of the medium, c_w is the sound velocity, and $\omega = 2\pi v$ (rad/sec) is the angular frequency of the wave. For plane waves, this power density is a constant since ξ_0 does not depend on the spatial coordinate x measured from the source of the waves. In Figure 7–1, we show the power

$$P = IA \quad (\text{W}) \tag{7.3}$$

passing through the area A (m^2) that is normal to the direction of the propagation of the wave. We assume that the waves are confined within the area A. The power P is provided by a wave source and the total power can be assumed to be a constant. Therefore, we may conclude

The intensity of the wave times the cross-sectional area is a constant and is equal to the power provided by the wave source. The cross-sectional area A is normal to the direction of the propagation of the wave and covers the region in which waves are present.

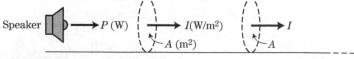

FIGURE 7–1

Sound waves that are confined in a uniform pipe have the same intensity I everywhere. The power P is given by the product of the intensity of the wave times the cross-sectional area of the pipe.

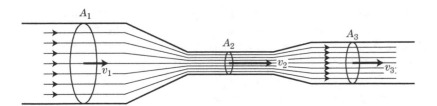

FIGURE 7–2

Water flows through a pipe with a nonuniform cross section. The flow rate Av is constant.

This is reminiscent of water flow through a pipe (Figure 7–2). The flow rate (L/sec) through any cross section must be the same because of mass conservation. For the energy flow with waves, the principle of energy conservation applies.

Consider now a point source that is spherically radiating waves and hence energy radially outward. The total power at each radius must be equal to the total power at a different radius. It must also be equal to the value that was emitted by the source

$$I_1 A_1 = I_2 A_2 = \text{constant}$$

We consider two areas, A_1 and A_2, located at the radii r_1 and r_2, respectively. The surface area of a sphere is proportional to the radius squared and we write

$$\frac{A_1}{A_2} = \frac{4\pi r_1^2}{4\pi r_2^2} \qquad (7.4)$$

The wave intensity at each radial position will therefore be inversely proportional to the radius squared:

$$\frac{I_1}{I_2} = \frac{r_2^2}{r_1^2} \qquad (7.5)$$

Obviously, the wave intensity is larger as we go closer to the wave source as intuitively expected.

Waves characterized by Eq. (7.5) are called *spherical waves*. Sound waves created by a loudspeaker, a radio, and the TV wave signals emitted from antennas are typical examples. It is noted here that spherical waves are not necessarily radiated isotropically or uniformly in every direction. For example, electromagnetic waves emitted by a rod antenna have an intensity peak in the direction perpendicular to the antenna.

Since the intensity is proportional to the wave amplitude squared, Eq. (7.2), we find that the amplitude of the radiated spherical waves is inversely proportional to the distance r from the wave source,

$$\xi_0(r) \propto \frac{1}{r} \qquad (7.6)$$

Any vector quantity (velocity, force, field, etc.) associated with a spherical wave propagating in a uniform media must have this $1/r$ dependence as required by the energy conservation principle and geometry. Similar arguments will also apply in cylindrical coordinates where one finds that the amplitude of a cylindrical wave will change approximately as $1/\sqrt{\rho}$ where ρ is the distance between a long wave source and the point of observation. The decrease of the wave amplitude that is excited with a stone thrown into a pond is another example of cylindrical waves.

EXAMPLE 7.1

A loudspeaker is emitting sound waves in air with a power 5 W over a solid angle of 0.1π steradian (Figure 7–3). Find

(a) The wave intensity at a distance 25 m from the speaker.
(b) The amplitude of air molecule displacement waves at the same position. Assume $T = 20°C$, $\rho_v = 1.2$ kg/m^3 (air density), and the source frequency is 500 Hz.

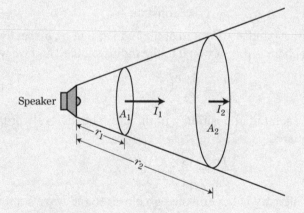

FIGURE 7–3

Spherical sound waves. The speaker can be regarded as a point sound source at sufficiently large distances from the speaker.

Solution

(a) The area at distance r (m) away from the speaker is $0.1\pi r^2$ (m^2). Then the intensity is

$$I = \frac{P}{A} = \frac{5\,\text{W}}{0.1\pi \times (25)^2} = 0.0255\,\text{W/m}^2 = 10\log_{10}\left(\frac{0.0255}{10^{-12}}\right) = 104\,\text{dB}$$

(b) In the expression for the intensity

$$I = \frac{1}{2}\rho_v c_w \omega^2 \xi_0^2 \quad (\text{W/m}^2)$$

where ρ_v (mass density) = 1.3 kg/m^3, c_w = 343 m/sec (20°C), $\omega = 2\pi \times 500$ rad/sec, and $I = 0.0255$ W/m^2, we find

$$\xi_0 = \frac{1}{\omega}\sqrt{\frac{2I}{\rho_v c_w}} = 3.4 \times 10^{-6}\,\text{m}$$

As we will see in the problems, the wave equation for spherical waves is given by

$$\frac{\partial^2}{\partial t^2}(r\xi) = c_w^2 \frac{\partial^2}{\partial r^2}(r\xi) \tag{7.7}$$

or

$$\frac{\partial^2 \xi}{\partial t^2} = c_w^2 \left(\frac{\partial^2 \xi}{\partial r^2} + \frac{2}{r}\frac{\partial \xi}{\partial r} \right) \tag{7.8}$$

which are fundamentally different from our previous one-dimensional wave equation. The general harmonic solution of these equations is

$$\xi(r, t) = \frac{A}{r} \sin(kr - \omega t), \quad \frac{\omega}{k} = c_w \tag{7.9}$$

as can be proved by direct substitution of Eq. (7.9) into Eq. (7.8). Here c_w is the wave velocity and A is a constant. In cylindrical geometry, the wave equation is

$$\frac{\partial^2 \xi}{\partial t^2} = c_w^2 \left(\frac{\partial^2}{\partial \rho^2} + \frac{1}{\rho}\frac{\partial}{\partial \rho} \right) \xi$$

This does not have a simple solution. (The solution involves the Bessel function.) However, at large distances ρ from the source such that $k\rho \gg 1$, the solution approaches the expected form,

$$\xi \propto \frac{1}{\sqrt{\rho}}$$

Nonuniform Wave Medium

We have studied the transverse waves on a string under tension T in Chapter 4. The tension is of course the same everywhere on the string and the velocity of the transverse waves given by

$$c_w = \sqrt{\frac{T}{\rho_l}}$$

is also the same everywhere on the string. The string, subject to a tension, is a *uniform medium* for the transverse waves. Consider now a string hanging from the ceiling by its own weight (see Figure 7–4). In practice, we have to use a relatively heavy string so that the string hangs straight. Let the linear mass density of the string be ρ_l (kg/m). At a point that is a distance x (m)

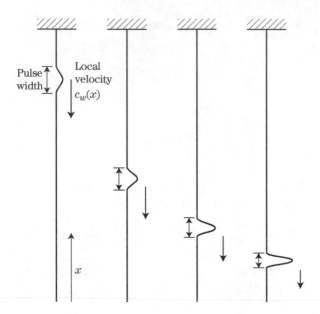

FIGURE 7–4

Transverse waves on a string hanging from the ceiling. The tension T is proportional to x, the distance measured from the bottom of the string. The pulse sent out downward will slow down as it propagates. Also the pulse width decreases as the pulse propagates.

away from the lower end, the tension T is given by

$$T = g\rho_l x \tag{7.10}$$

which now depends on the coordinate x! The velocity for the transverse waves then also depends on the coordinate x and is approximately given by

$$c_w(x) = \sqrt{\frac{g\rho_l x}{\rho_l}} = \sqrt{gx} \tag{7.11}$$

The string hanging vertically is a *nonuniform medium* for the propagation of transverse waves.

In Figure 7–4 the propagation of a pulse created at the top end is qualitatively shown. Observe the following: As the pulse approaches the lower end,

1. The pulse propagates more slowly.
2. The pulse width becomes narrower.
3. The pulse height (amplitude) becomes larger.

The pulse is squeezed and the amplitude increases as it approaches the lower end of the string. This is understandable since the portion of the pulse behind the peak tends to catch up with the peak and thus the squeezing takes place.

EXAMPLE 7.2

Water waves approaching a beach. The velocity of surface waves on water is given by

$$c_w = \sqrt{gh}$$

where h is the depth of water and g is the gravitational acceleration. Discuss how a wave pulse behaves as it approaches the beach. Assume that the bottom surface has no irregularities and can be modeled with a tilted plane. (See Figure 7–5.)

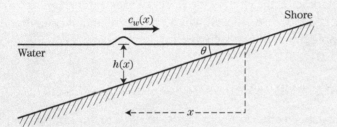

FIGURE 7–5

The velocity of surface water waves depends on the localized depth h. As the wave approaches the shore, it slows down, and the amplitude increases.

Solution

If the bottom is planar, the depth is proportional to the distance x from the beach and is given by

$$h(x) = x \tan \theta$$

where θ is the angle between the water surface and the bottom. Then the velocity of the surface wave is

$$c_w = \sqrt{gx \tan \theta}$$

which is mathematically identical to Eq. (7.11). Therefore exactly the same things happen as in the case of the vertical string: The pulse increases its amplitude and becomes narrower as it approaches the beach.

Surfers use this pulse-steepening phenomenon. The final fate of the pulse when it hits the beach is catastrophic in that the simple pulse shape of the wave is destroyed, an event with which we are probably familiar. The amplitude becomes so large that the waves break down as shown in Figure 7–6.

Surfers enjoy big waves, but for those living near the seashore, tidal waves (tsunami) created by earthquakes are terrifying. Although the initial amplitude of the wave may be small when created by an earthquake at the bottom of an ocean, the amplitude of the resultant wave can increase as it approaches the shore. Tidal waves as high as 30 m have been recorded. Recall the 2004 tsunami that occurred in the Pacific Ocean and all the human and material

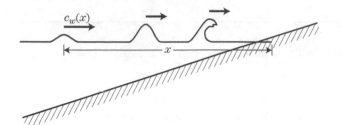

FIGURE 7–6

Breaking of water waves near the shore takes place when the amplitude becomes too large.

devastation that it inflicted throughout the world. Waves of this large amplitude with respect to the depth of the ocean will also require an analysis involving nonlinearity, a topic that will be discussed in Chapter 15.

Multidimensional Waves

Waves that we have been analyzing have only one spatial variable (x for plane waves, ρ for cylindrical waves, and r for spherical waves) and are called one-dimensional waves. However, it often becomes necessary to introduce more than one spatial variable in analyzing wave phenomena. Examples of "two-dimensional" waves are the standing wave on the membrane of a drum or the water waves bouncing around in a bathtub. In contrast to the one-dimensional counterpart, namely a string with both ends clamped, the resonant frequencies of standing waves on a membrane under tension or in the bathtub are not, in general, integer multiples of the fundamental frequency.

Since the analysis of a circular membrane requires some knowledge of Bessel functions that are beyond the scope of this book, we model the drum with a rectangular membrane having an area $a \times b$ with its four edges rigidly clamped. We also assume that the membrane is under a uniform surface tension T_s (N/m).

The wave equation for the transverse displacement $\xi(x, y, t)$, which depends upon the two spatial coordinates, is now given by

$$\frac{\partial^2 \xi}{\partial t^2} = c_w^2 \left(\frac{\partial^2 \xi}{\partial x^2} + \frac{\partial^2 \xi}{\partial y^2} \right) = c_w^2 \nabla^2 \xi \tag{7.12}$$

where $c_w^2 = T_s/\rho_s$, with ρ_s being the surface mass density (kg/m^2) of the membrane. The Laplacian operator ∇^2 is also introduced in this equation. The derivation of this equation is left as a problem. We wish to find standing-wave solutions to this wave equation.

The standing wave solution of a clamped string can be written as

$$\xi(x, t) = \xi_0 \sin\left(\frac{n\pi}{L}x\right) \cos(\omega_n t) \tag{7.13}$$

where $n = 1, 2, 3, \ldots$ (integer), L is the length of the string, and $\omega_n = n\omega_1$ with $\omega_1 = \pi c_w / L$ being the fundamental frequency of oscillation (see Section 6.2.) The appearance of $\sin[(n\pi/L)x]$ is understandable since the string is clamped at both ends: $x = 0$ and L. This requires that the vertical deflection ξ must be equal to 0 at $x = 0$ and L for all time. In the case of the rectangular membrane whose edges are all clamped, we must have

$$\xi = 0 \text{ at } x = 0, a \text{ and } \xi = 0 \text{ at } y = 0, b$$

at any time. A trampoline would satisfy these boundary conditions. A two-dimensional version of Eq. (7.13) is

$$\xi(x, y, t) = \xi_0 \sin\left(\frac{m\pi}{a}x\right) \sin\left(\frac{n\pi}{b}y\right) \cos(\omega t) \tag{7.14}$$

and determine the oscillation frequency ω that satisfies the wave equation Eq. (7.14). Noting that the various derivatives can be written as

$$\frac{\partial^2 \xi}{\partial x^2} = -\left(\frac{m\pi}{a}\right)^2 \xi^2$$

$$\frac{\partial^2 \xi}{\partial y^2} = -\left(\frac{n\pi}{b}\right)^2 \xi^2$$

$$\frac{\partial^2 \xi}{\partial t^2} = -\omega^2 \xi^2$$

and substituting these into Eq. (7.14), we find that the resonant frequency can be found from

$$\omega^2 = c_w^2 \left[\left(\frac{m\pi}{a}\right)^2 + \left(\frac{n\pi}{b}\right)^2\right] \tag{7.15}$$

where m, n are nonzero integers. This indicates that the higher harmonic frequencies of the standing waves in a rectangular membrane are not in general integer multiples of the fundamental frequency of oscillation ($m = 1, n = 1$), that is,

$$\omega(m = 1, n = 1) = c_w \pi \sqrt{\frac{1}{a^2} + \frac{1}{b^2}}$$

If $a > b$, the next resonant frequency is given by

$$\omega(m = 2, n = 1) = c_w \pi \sqrt{\frac{4}{a^2} + \frac{1}{b^2}}$$

This is clearly not an integer multiple of the fundamental frequency of oscillation. The next resonant frequency is given by

$$\omega(m = 1, n = 2) = c_w \pi \sqrt{\frac{1}{a^2} + \frac{4}{b^2}}, \quad \left(\sqrt{\frac{8}{3}}b > a > b\right)$$

and so on. Although there are certain resonant frequencies that are integer multiples of the fundamental oscillation frequency that are based on certain choices of the integers ($m = n = 2, m = n = 3, \ldots$), most resonant frequencies are not related to the fundamental oscillation frequency through rational numbers. This explains why sounds coming from a metal plate hit with a hammer are rather uncomfortable to our ears.

In the case of a circular membrane, such as the one found in the drum, the ratio between the higher-order resonant frequencies and the fundamental is an irrational number. The lowest frequency of oscillation for a clamped circular membrane with a radius a is

$$\omega_1 \simeq 2.405 \frac{c_w}{a}$$

The next resonant frequency is $\omega_2 = 3.832\,(c_w/a)$, the third is $\omega_3 = 5.136(c_w/a)$, and so on. The specific numbers that appear in these frequencies are obtained from an analysis of the Bessel function. Here $c_w = \sqrt{T_S/\rho_s}$ is the velocity of the transverse waves on the membrane.

7.5   **Problems**

1. A speaker is radiating spherical sound waves with a power 5 W. The radiation is limited within a cone with a 20° angle as shown in Figure 7–7. Within the cone, the radiation can be assumed to be uniform.

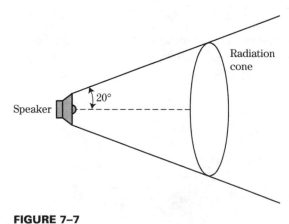

FIGURE 7–7

Problem 1.

(a) What is the power density 10 m away from the speaker?

(b) At what distance does the intensity become 10^{-6} W/m²?

2. Assuming that water waves obey the law of cylindrical waves, find the amplitude of water waves 50 m away from a source. The waves have an amplitude of 15 cm when 10 m away from the source.

3. A radio station is emitting spherical waves at 50 kW.
(a) Find the intensity 1 mile away from the station assuming isotropic radiation.
(b) Electromagnetic waves are characterized by electric and magnetic fields as we will see later. What is the ratio between the electric field at 1 mile and that at 10 miles?
(c) A radio has the lowest receivable intensity of 3μW/m². How far can one bring the radio to listen to the station?

4. (a) Following the procedure used for the transverse waves on a string (Chapter 4, Section 4.6), show that the differential equation the transverse waves on a vertical free string should satisfy is given by

$$\frac{\partial^2 \xi}{\partial t^2} = gx \frac{\partial^2 \xi}{\partial x^2} + g \frac{\partial \xi}{\partial x}$$

where $\xi(x, t)$ is the displacement and x is measured from the lower end of the string.

(b) Under what conditions can we use Eq. (7.10) for the velocity of the transverse waves on the string?

5. A whip used by animal tamers has a nonuniform cross-sectional area. Explain how this can be understood in terms of the amplitude of a pulse created on a whip.

6. Show that $\xi(r, t) = (A/r) \sin(kr - \omega t)$ satisfies the spherical wave equation in Eqs. 7.7 and 7.8.

7. Show that $\xi(\rho, t) = (A/\sqrt{\rho}) \sin(k\rho - \omega t)$ approximately satisfies the cylindrical wave equation,

$$\frac{\partial^2 \xi}{\partial t^2} = c_w^2 \left(\frac{\partial^2}{\partial \rho^2} + \frac{1}{\rho} \frac{\partial}{\partial \rho} \right) \xi$$

What condition(s) must be imposed for this solution to be a sufficiently accurate one?

8. Discuss the wave-steepening mechanism of water waves on a beach using the energy conservation principle.

9. We saw that water waves slow down as they approach a beach. How then can a surfer get accelerated by water waves?

10. Derive Eq. (7.12), the two-dimensional wave equation for a membrane.

11. A rectangular membrane has edges 20 cm and 30 cm, a surface tension of 5 N/m, and a surface mass density of 40 g/m^2.

(a) What is the velocity of transverse waves on the membrane?

(b) Calculate the lowest-order standing-wave frequency.

(c) What are the higher-order resonance frequencies?

12. (a) Show that the wave equation for three-dimensional sound waves is given by

$$\frac{\partial^2 \xi}{\partial t^2} = c_w^2 \nabla^2 \xi$$

where

$$\nabla^2 = \frac{\partial^2}{\partial x^2} + \frac{\partial^2}{\partial y^2} + \frac{\partial^2}{\partial z^2} \quad \text{(Laplacian)}$$

(b) Discuss the resonance frequencies of sound waves in a rectangular box having a volume $a \times b \times c$ (m^3).

CHAPTER 8

Doppler Effect of Sound Waves and Shock Waves

8.1 Introduction

The tone or frequency of a siren from a police car or fire engine seems to change from higher to lower value when the vehicle passes an observer even though the actual siren frequency is a constant. The driver of the vehicle, however, always hears the same frequency as if it were stationary. The apparent change in the frequency caused by the motion of the source of the wave (the siren in this case) relative to an observer is called the *Doppler effect*. C. J. Doppler (1803–1853) was an Austrian physicist who discovered the effect in light waves. Earlier in the seventeenth century, Danish astronomer Ole Roemer estimated the velocity of light from the apparent change in the revolution time of one of the moons of Jupiter. His estimate was $c = 2 \times 10^8$ m/s, which is amazingly close to the modern value of 3.0×10^8 m/s considering the quality of the data available at the time. It turns out that the method used by Roemer was based on the same principle as the Doppler effect. For the sound waves that are propagating in air, we have three separate entities to consider: the source that emits the waves with a certain frequency, the observer that detects the waves, and the medium in which the wave propagates. All of these entities can be moving with respect to any of the others.

Some objects, such as a supersonic plane, can travel faster than sound. The so-called "sonic booms" that are created by these planes are alternatively known as shock waves. Shock waves contain large amounts of energy that are concentrated over a narrow spatial range and could even cause mechanical damage. In this chapter we study the phenomena caused by the motion of sound sources relative to air and an observer.

Stationary Sound Source and Moving Observer

Let us first consider a stationary sound source emitting a sinusoidal wavetrain with a frequency ν_0. We assume that the air is stationary; that is, we assume that there is no wind. An observer is moving away from the sound source with a velocity u_0 relative to the wave medium–air (Figure 8–1). We denote the sound velocity by c_s.

Since the observer is moving in the same direction as the sound wave, the apparent velocity of the sound relative to the observer is

$$c_s - u_0 \tag{8.1}$$

We assume that the observer is moving with a subsonic velocity, $u_0 < c_s$, so that Eq. (8.1) is a positive quantity. If the wave source is stationary, it is emitting sound waves in every direction with the same wavelength λ that can be determined from (Figure 8–2)

$$\lambda = \frac{c_s}{\nu_0} \tag{8.2}$$

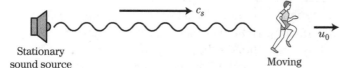

FIGURE 8–1

Stationary sound source and moving observer. The effective sound velocity for the observer is $c_s - u_0$.

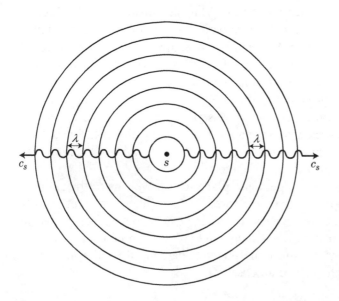

FIGURE 8–2

Wave patterns created by a stationary sound source S are symmetric in every direction or the radiation is "isotropic."

The observer, whether in motion or not, should always observe the *same wavelength*. However, if the observer is moving, the apparent sound velocity becomes $c_s - u_0$. Therefore the frequency v' that the observer hears will be different and can be found from

$$\lambda = \frac{c_s}{v_0} = \frac{c_s - u_0}{v'}$$

or solving for v', we obtain

$$v' = \frac{c_s - u_0}{c_s} v_0 \tag{8.3}$$

As expected, the apparent frequency heard by the observer is lower than the frequency that was launched from the source if the observer is moving away from the source. The case of an observer approaching the source can be given simply by changing the sign of the velocity u_0 ($u_0 < 0$), and the apparent frequency in this case is higher than that launched from the source. You should watch the sign convention that is chosen for the velocity u_0. The convention is $u_0 > 0$ if the observer is moving away from the source and $u_0 < 0$ if the observer is moving toward the source. If the observer is traveling with the sound speed, $u_0 = c_s$, the frequency detected by the observer becomes zero since the observer appears to be "riding on the waves" and does not detect anything varying with time.

Let us derive Eq. (8.3) from an alternative point of view (Figure 8–3). Assume that the observer is a distance l(m) separated from the sound source at a certain instant of time, say $t = 0$ sec. At one second later, the observer has moved a distance $u_0 \times 1$ sec $= u_0$ (m) and the distance between the source and the observer is $l + u_0$ (m) at this instant. The time t_0 required for the sound wave that was emitted at $t = 0$ to reach the observer can be found from

$$c_s t_0 = l + u_0 t_0$$

or

$$t_0 = \frac{l}{c_s - u_0} \text{ (sec)} \tag{8.4}$$

Similarly, the time t_1 at which the wave emitted at $t = 1$ sec reaches the observer is found from

$$c_s(t_1 - 1) = l + u_0 + u_0(t_1 - 1)$$

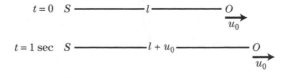

FIGURE 8–3

At the instant $t = 0$, the observer is a distance l from the source. One second later, the observer is at a distance $l + u_0$ away.

or

$$t_1 = \frac{l + c_s \times 1 \text{ sec}}{c_s - u_0} \qquad (8.5)$$

During the period of 1 sec, the source has emitted v_0 waves but the observer hears v_0 waves in the time duration of

$$t_1 - t_0 = \frac{c_s \times 1 \text{ sec}}{c_s - u_0} \text{ (sec)},$$

which is longer than 1 sec. In 1 sec the observer then hears

$$\frac{c_s - u_0}{c_s} v_0$$

waves. This is identical to Eq. (8.3).

EXAMPLE 8.1

A car moving at a velocity 50 mi/hr $\sim$ 22 m/sec is passing a stationary sound source that emits a frequency 500 Hz (Figure 8–4). The closest distance between source and the car is 20 m. Find the apparent frequency heard by the driver as a function of the distance x. Assume $c_s = 340$ m/sec.

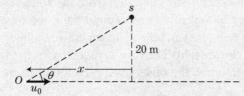

FIGURE 8–4

Observer moving along a line not intersecting the source.

Solution

In this case, the velocity of the car is not aimed directly toward the sound source and we have to find the component of the velocity vector that is directed toward the source (Figure 8–5). It is given by

$$u_0 \cos \theta = u_0 \frac{x}{\sqrt{x^2 + 20^2}} = 22 \frac{x}{\sqrt{x^2 + 20^2}} \text{ m/sec}$$

where we have converted the velocity of 50 mi/hr into 22 m/sec. Then the Doppler-shifted frequency is

$$v'(x) = v_0 \frac{c_s + u_0 \cos \theta}{c_s} = 500 \left(1 + \left(\frac{22}{340} \right) \frac{x}{\sqrt{x^2 + 20^2}} \right) \text{ Hz}$$

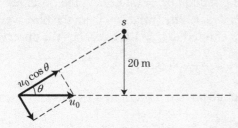

FIGURE 8–5

The velocity component toward (or away from) the source is responsible for the Doppler shift.

This is plotted below for $-100 \leq x \leq 100$ m. At $x = 0$, the car is moving purely perpendicular to the wave and at the location where the car passes this point, the driver will hear the true frequency emitted by the source 500 Hz.

$$500 \left(1 + \frac{22}{340} \frac{x}{\sqrt{x^2 + 400}} \right)$$

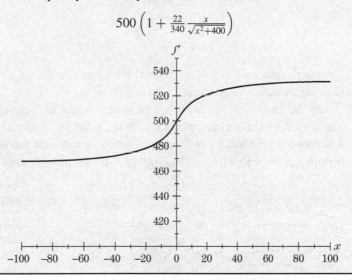

Moving Sound Source and Stationary Observer

Next we consider the opposite case in which the sound source is moving toward a stationary observer with a velocity u_s. You may wonder why this case should be different from the case discussed in Section 8.2, since it seems that the only thing that matters is the relative velocity between the source and the observer. One may prematurely conclude that the observer approaching a stationary sound source should hear the same frequency as a stationary observer hearing the same frequency from a moving sound source approaching the observer. This argument is wrong, however, for sound waves *although it is correct for electromagnetic waves.*

Let the sound source be a distance l (m) away from the observer at a certain instant in time, say $t = 0$ sec. Since the observer is not moving, it takes

$t_0 = l/c_s$ sec for the sound wave emitted at $t = 0$ to reach the observer. One second later the source is $l - u_s \times 1 = l - u_s$ (m) away from the observer and the time at which the sound wave emitted at $t = 1$ sec reaches the observer is

$$t_1 = 1 + \frac{l - u_s}{c_s} \text{ (sec)}$$

During the period of 1 sec, the sound source has emitted ν_0 waves, which are heard by the observer in

$$t_1 - t_0 = 1 - \frac{u_s}{c_s} \text{ (sec)}$$

Therefore the apparent (Doppler-shifted) frequency to be heard by the observer is

$$\nu' = \nu_0 \frac{1}{t_1 - t_0} = \frac{c_s}{c_s - u_s} \nu_0 \qquad (8.6)$$

Compare this with Eq. (8.3). The velocity of the sound source u_s appears in the denominator in contrast to the result in Eq. (8.3).

In the case of a stationary sound source and a moving observer, the sound velocity appears to change because of the relative motion, but the wavelength λ remains the same. Here the sound velocity remains the same but the wavelength appears to change. In Figure 8–6, the circles indicate the locations

FIGURE 8–6

Moving source and stationary observer. The wavelength changes are due to the motion of the source relative to the observer.

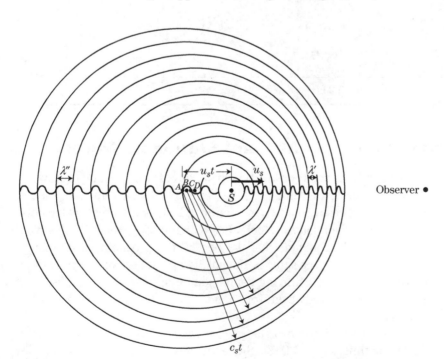

Observer •

of the waves emitted by the source when it was at $A, B, C, \ldots$, and so on that are equally spatially separated. We can clearly see that the waves are squeezed into the region between the sound source and the observer. In the region behind the source, the wavelength is elongated.

EXAMPLE 8.2

A sound source with a frequency 1000 Hz is approaching an observer who also has a sound source with the same frequency. When the observer hears 5 beats a second, what is the velocity of the moving sound source? Assume the sound velocity is 340 m/sec.

Solution

We should recall that the beat phenomenon is caused by two waves with slightly different frequencies and this difference in the frequencies appears as detectable beats [Eq. (2.49)]. Let the velocity of the moving source be u_s. From Eq. (8.6), the Doppler-shifted frequency v' is

$$v' = \frac{c_s}{c_s - u_s} v_0$$

where $v_0 = 1000$ Hz and $c_s = 340$ m/sec. Then the beat frequency or the frequency difference is

$$\Delta v = v' - v_0 = \frac{u_s}{c_s - u_s} v_0$$

Solving for u_s, we find

$$u_s = \frac{\Delta v}{v_0 + \Delta v} c_s = 1.7 \text{ m/sec.}$$

8.4   General Expression for Doppler-Shifted Frequency

We can generalize the two cases: (a) a stationary sound source and moving observer, Eq. (8.3) and (b) a moving sound source and stationary observer, Eq. (8.6). If both the source and observer are moving (Figure 8–7), the apparent frequency to be observed is given by

$$v' = \frac{c_s - u_0}{c_s - u_s} v_0 \tag{8.7}$$

Again you should be careful about the sign convention adopted here. We have assigned positive velocities (u_s and u_0) for those that are in the same direction

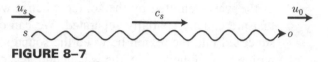

FIGURE 8–7

Both the source and the observer are moving relative to the ambient medium. The velocities u_s and u_0 are positive if they both are in the same direction as the sound velocity.

as the sound velocity c_s. If either of these velocities are in the opposite direction of the sound velocity, a negative value for the velocity must be used.

EXAMPLE 8.3

A police car is approaching a solid wall with a velocity 30 m/sec. If it is sounding a siren with a frequency of 1500 Hz, what is the frequency of the reflected sound waves that will be heard by the police officer? Assume the velocity of sound to be $c_s = 340$ m/sec.

Solution

The wall can be regarded as an observer who hears the Doppler-shifted frequency of the approaching siren given by Eq. (8.6),

$$v' = \frac{c_s}{c_s - u_s} v_0$$

where $u_s = 30$ m/sec and $v_0 = 1500$ Hz. When the sound waves are reflected, *the wall now acts as a new sound source with the preceding frequency v'.* The police officer is now the observer who is *approaching* the sound source with the velocity u_s. Then Eq. (8.3) applies with the sign of u_s flipped and the frequency observed by the police officer is

$$v'' = \frac{c_s + u_s}{c_s} v'$$

Substituting v', we find

$$v'' = \frac{c_s + u_s}{c_s} \frac{c_s}{c_s - u_s} v_0 = \frac{c_s + u_s}{c_s - u_s} v_0 = \frac{340 + 30}{340 - 30} \times 1500 \, \text{Hz} = 1790 \, \text{Hz}$$

Note that the frequency of the reflected wave is Doppler-shifted twice in this example.

EXAMPLE 8.4

Show that if a wind with a velocity u_w exists in the direction of the sound velocity c_s, the Doppler-shifted frequency Eq. (8.7) should be rewritten as

$$v' = \frac{c_s + u_w - u_0}{c_s + u_w - u_s} v_0 \tag{8.8}$$

Solution

We recall the fact that the sound velocity c_s is relative with respect to the sound wave medium, which in this case is air. Therefore if the medium is moving, or a wind exits, the sound velocity relative to the ground becomes $c_s + u_w$, where u_w is the wind velocity chosen to be positive if the wind is directed in the same direction as the sound wave. Equation (8.7) should then be modified to be

$$v'' = \frac{c_s + u_w - u_0}{c_s + u_w - u_s} v_0$$

where u_0 and u_s are the velocities relative with respect to the ground.

The Doppler effect in electromagnetic waves is given by

$$v' = \frac{c}{c - u} v_0 \tag{8.9}$$

where $c = 3.0 \times 10^8$ m/sec is the speed of light in free space and u is the *relative* velocity between a wave source with frequency v_0 and an observer. This velocity can be assumed to be much smaller than the velocity of light, $|u| \ll c$. *For electromagnetic waves, only the relative velocity matters and it is immaterial whether the source or the observer is moving.* This surprising fact stems from Albert Einstein's relativity theory. Einstein discovered that electromagnetic waves in free space have entirely different properties from waves in a material media such as sound waves in air. According to the theory of relativity, the velocity of light in a vacuum is constant ($c = 3.0 \times 10^8$ m/sec) irrespective of the velocity of the light source or of the observer. If the relative velocity u approaches the speed of light c, Eq. (8.9) should be generalized to be

$$v' = \frac{\sqrt{1 - (u/c)^2}}{1 - u \cos \theta} v_0$$

where the factor in the numerator is due to the time dilation effect and θ is the angle between the light velocity $\mathbf{c}$ and the relative velocity $\mathbf{u}$. Even at $\theta = 90°$, there is a relativistic Doppler shift that is given by

$$v' = \sqrt{1 - (u/c)^2}\, v_0$$

due to the time dilation effect.

Shock Waves

If the velocity of the source of the sound approaches the velocity of sound in air, Eq. (8.6) predicts that the Doppler-shifted frequency ν' becomes larger and finally diverges when the velocity of the source is further increased up to the sound velocity. What happens if the velocity of the source exceeds the velocity of sound? When an object is traveling through air with a velocity u that is larger than the velocity of sound c_s, the object is said to have a supersonic velocity. The ratio u/c_s is called the Mach number, after Ernst Mach (1838–1916), the Austrian physicist and philosopher. A supersonic object has a Mach number larger than 1.0.

Let us see if we can draw wave patterns similar to Figure 8–6 when the source velocity u_s, is larger than the velocity of sound c_s (Figure 8–8). Suppose that the source is at the point A at a time $t = 0$. After a time t (sec), the waves emitted at A are on a spherical surface with a radius $c_s t$ (m). Since $u_s > c_s$, the distance traveled by the source $AS = u_s t$ is larger than the distance traveled by the sound waves. The waves emitted at successive points, B, C, and so on, are on the line $A'S$, where the circles are most crowded. We can see that sound waves are confined within a cone. External to this cone, no sound waves are present and are thus undetectable. The sound velocity c_s is normal to the surface of the cone.

The energy of the sound waves is mostly concentrated on the conical surface where the wave pattern circles are most crowded. When the conical surface hits an observer, the sudden arrival of a large amplitude pulse is detected. This is known as a shock wave or a sonic boom.

EXAMPLE 8.5

A supersonic plane is traveling horizontally 1 km above the ground. An observer on the ground experiences a shock wave approximately 2 sec after the plane passes directly overhead. Estimate the Mach number of the plane. Assume $c_s = 340$ m/sec.

Solution

In Figure 8–8, the angle θ between the axis AS and the cone surface $A'S$ is given by

$$\sin\theta = \frac{c_s}{u_s}, \quad \text{or} \quad \theta = \sin^{-1}\left(\frac{c_s}{u_s}\right)$$

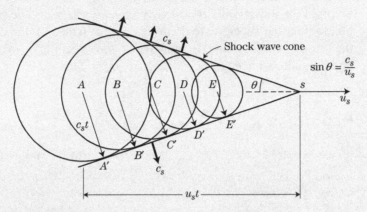

FIGURE 8–8

Shock waves created by a sound source moving faster than the velocity of sound.

However, $\sin \theta$ is also equal to $\dfrac{h}{\sqrt{(2u_s)^2 + h^2}}$ (Figure 8–9).

FIGURE 8–9

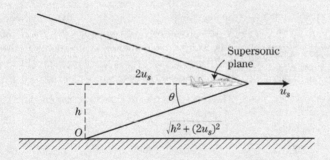

Then

$$\frac{c_s}{u_s} = \frac{h}{\sqrt{(2u_s)^2 + h^2}}$$

Solving for u_s, we find

$$u_s = \frac{c_s h}{\sqrt{h^2 - 4c_s^2}} \quad \text{or} \quad \frac{u_s}{c_s} = \frac{h}{\sqrt{h^2 - 4c_s^2}}$$

Substituting $h = 10^3$ m, $c_s = 340$ m/sec, we find the Mach number to be

$$\text{Mach number} = \frac{u_s}{c_s} = 1.36$$

Supersonic objects are necessarily associated with shock waves. The excitation of the shock wave enormously increases a drag force on the supersonic object since the shock wave drains energy from the moving object. Even with the best of intentions and designs, shock waves cannot be avoided for moving objects such as airplanes.

8.6 Problems

Assume that the sound velocity in air is 340 m/sec in the problems.

1. A police car sounding a siren of a frequency of 800 Hz is approaching a car going in the same direction. The velocity of the police car is 90 mi/hr and that of the car 55 mi/hr. Assume that the air is still.
   (a) What frequency does the car driver hear before the police car passes the car?
   (b) What frequency does the car driver hear after the police car passes the car?

2. Repeat Problem 1 assuming that both cars are moving against a steady wind of 30 mi/hr.

3. A whistle with a frequency 600 Hz is placed at the edge of an LP record (radius 15 cm, $33\frac{1}{3}$ rpm). What are the lowest and the highest frequencies heard by an observer who is sufficiently far from the record? *Note:* The whistle frequency varies sinusoidally between these frequencies being frequency modulated.

4. A boy sounding a whistle with a frequency of 800 Hz is approaching a wall that reflects sounds well. When he hears three beats a second, what is his velocity?

5. A bullet is fired with a velocity of 500 m/sec. Find the angle between the shock wave surface and the line of motion of the bullet.

6. Whenever an object travels faster than the wave characterized by a medium, a shock wave is created. Surface waves on water are slow and you can easily create a shock wave by moving a finger dipped into the water fast enough in a horizontal direction. Try this experiment.

7. If a relative velocity u between the source of electromagnetic waves and an observer is sufficiently smaller than the speed of light (which is an electromagnetic wave) $c = 3.0 \times 10^8$ m/sec, the Doppler-shifted frequency is given by

$$\nu' = \frac{c}{c - u} \nu_0,$$

where $u > 0$ when the source and the observer are approaching each other and $u < 0$ when receding. When microwaves with a frequency 4.0×10^9 Hz are reflected from a jet fighter, the frequency is changed by 10 kHz. What is the speed of the plane? (This principle is used in highway patrol radar.)

8. The exact formula for the Doppler-shifted frequency of electromagnetic waves is

$$\nu' = \frac{\sqrt{1 - \beta^2}}{1 - \beta} \nu_0$$

where $\beta = u/c$ and u is the relative velocity between the light source and the observer ($u > 0$ for approaching, $u < 0$ for receding). Starlight, known to have a wavelength of 5500 Å ($1Å = 10^{-10}$ m) and observed in the laboratory appears to have a wavelength of 6500 Å:
   (a) What is the speed of the star relative to the earth?
   (b) Is the star receding or approaching? (This is known as "redshift" of spectrum lines.)

CHAPTER 9

Electromagnetic Waves

9.1 Introduction

Two aspects of electromagnetic waves are particularly important to appreciate. The first is that the frequency of an electromagnetic wave spans an enormous range. At the lower end, 50 and 60 Hz waves are used to transmit electric power and toward the higher end, 10^{14} Hz waves are associated with visible light with X-rays and gamma rays at even higher frequencies. The second is that, unlike mechanical waves such as sound and heat which can only propagate in a material, electromagnetic waves can propagate in a vacuum. That is, vacuum can be a medium for electromagnetic waves. This seemingly obvious fact was long disbelieved and the concept of "ether" prevailed for a long time. In mechanical waves such as sound waves in gases and solids, and transverse waves on a string, we have no difficulty in "visualizing" wave motion. In sound waves, for example, molecules move about their equilibrium positions. We have seen that the motion of the molecules determines the kinetic energy and the displacement of molecules from equilibrium positions determines the potential energy associated with wave motion. Said conversely, in any media capable of storing kinetic and potential energies, mechanical waves can be produced and will propagate.

Close analogy can be found in a vacuum. Take a capacitor first. A capacitor can store electrical energy in its volume. Although most capacitors are filled with dielectric materials, this is not essential. Dielectrics can be replaced by air or vacuum; that is, a vacuum can store electrical energy. Next, take an inductor which is capable of storing magnetic energy. Again, the magnetic energy is stored in the volume occupied by the inductor, which can be air or vacuum. Thus we draw an important conclusion. The vacuum is capable of storing electric and magnetic energies. This corresponds to the potential and kinetic energies that were found in the case of mechanical waves. In any media capable of storing electrical and magnetic energies, electromagnetic waves can be produced and propagated.

In an ideal vacuum, electromagnetic waves are completely dispersionless. The phase velocity equals the group velocity for any frequency,

$$\frac{\omega}{k} = \frac{d\omega}{dk} = c = 3.0 \times 10^8 \text{ m/sec}$$

Can you imagine what would happen to FM radio if this were not the case? If the electromagnetic waves were dispersive, two waves with different frequencies would be received at different times and the signals would be all mixed up!

Electromagnetic waves can be dispersive, however, in media other than vacuum. In fact, the velocity of visible light in glass is slower than that in vacuum. More important, the velocity of light in glass depends, although slightly, on the wave frequency; that is, the electromagnetic waves in glass are not dispersionless. Another example is the propagation of shortwave radio around the earth's surface, being reflected by both the earth's surface and the ionospheric plasma surrounding the earth. As we have seen in mechanical waves, waves can be reflected whenever they enter another medium in which the propagation velocity (more precisely, the impedance) is different. As we will see later, the ionospheric plasma acts as a "soft" boundary. Visible light has no difficulty in penetrating into the ionospheric plasma since the plasma becomes essentially nondispersive and transparent at such high frequencies.

In this chapter we derive the wave equation for electromagnetic waves using basic knowledge. In fact, all we need is Kirchhoff's voltage and current laws. Then we generalize the primitive method using macroscopic Maxwell's equations (namely, Faraday's induction law and Maxwell's displacement current). Finally, you will be introduced to the differential (or microscopic) form of Maxwell's equations, which govern all electromagnetic phenomena.

Wave Equation for an *LC* Transmission Line

A ladder network composed of series inductances and parallel capacitances is called an *LC* transmission line or delay line (Figure 9–1). This transmission line can also be used in modeling sections of integrated circuits and is

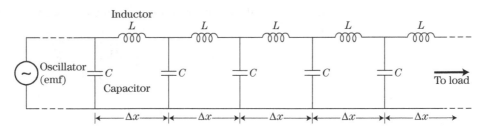

FIGURE 9–1

LC transmission line.

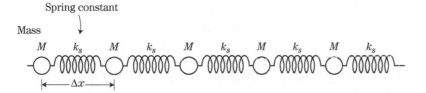

FIGURE 9–2

Distributed mass-spring system.

investigated in electromagnetics. It is also an excellent analogue of the distributed mass-spring system (Figure 9–2) we studied in Chapter 4 and in fact can describe electromagnetic waves under many practical situations.

We select one section of the *LC* transmission line located at x, and assign currents and voltages as shown in Figure 9–3. Kirchhoff's voltage law that states that the sum of the voltage drops around a closed loop is equal to 0 yields

$$V(x, t) = L\frac{\partial I(x, t)}{\partial t} + V(x + \Delta x, t) \tag{9.1}$$

Kirchhoff's current law that states that the sum of the currents that enter a node is equal to 0 yields

$$I(x, t) = C\frac{\partial V(x, t)}{\partial t} + I(x + \Delta x, t) \tag{9.2}$$

The first term in the right-hand side of Eq. (9.2) should have been written as

$$C\frac{\partial V(x + \Delta x, t)}{\partial t}$$

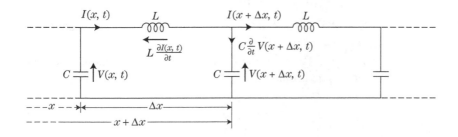

FIGURE 9–3

Voltages and currents at one section of an *LC* transmission line.

but if Δx is sufficiently small which we assume, we may Taylor series expand $V(x + \Delta x, t)$ as

$$V(x + \Delta x, t) \simeq V(x, t) + \Delta x \frac{\partial V(x, t)}{\partial x} \tag{9.3}$$

and $(\partial/\partial t)V(x + \Delta x, t)$ can be approximated with $(\partial/\partial t)V(x, t)$. The current at $x + \Delta x$ can also be Taylor series expanded as

$$I(x + \Delta x, t) \simeq I(x, t) + \Delta x \frac{\partial I(x, t)}{\partial x} \tag{9.4}$$

After substituting Eqs. (9.3) and (9.4) into Eqs. (9.1) and (9.2), we obtain

$$-\Delta x \frac{\partial V(x, t)}{\partial x} = L \frac{\partial I(x, t)}{\partial t} \tag{9.5}$$

$$-\Delta x \frac{\partial I(x, t)}{\partial x} = C \frac{\partial V(x, t)}{\partial t} \tag{9.6}$$

which may be considered as two coupled first-order differential equations for $V(x, t)$ and $I(x, t)$. Next, we differentiate Eq. (9.5) with respect to the spatial coordinate x and Eq. (9.6) with respect to time t to obtain

$$-\Delta x \frac{\partial^2 V(x, t)}{\partial x^2} = L \frac{\partial^2 I(x, t)}{\partial x \partial t} \tag{9.7}$$

$$-\Delta x \frac{\partial^2 I(x, t)}{\partial t \partial x} = C \frac{\partial^2 V(x, t)}{\partial t^2} \tag{9.8}$$

Since $\partial^2 I(x, t)/\partial x \partial t = \partial^2 I(x, t)/\partial t \partial x$, the current $I(x, t)$ can be eliminated and we finally obtain a partial differential equation for the voltage $V(x, t)$,

$$\frac{\partial^2 V(x, t)}{\partial t^2} = \frac{(\Delta x)^2}{LC} \frac{\partial^2 V(x, t)}{\partial x^2} \tag{9.9}$$

A similar equation for the current $I(x, t)$ can be obtained

$$\frac{\partial^2 I(x, t)}{\partial t^2} = \frac{(\Delta x)^2}{LC} \frac{\partial^2 I(x, t)}{\partial x^2} \tag{9.10}$$

by differentiating Eq. (9.5) with respect to t and Eq. (9.6) with respect to x. (You should work this out yourself.)

Equations (9.9) and (9.10) are of the form of a wave differential equation described in Chapter 2. The propagation velocity is immediately found to be

$$c_w = \frac{\Delta x}{\sqrt{LC}} \tag{9.11}$$

or

$$c_w = \frac{1}{\sqrt{L/\Delta x \cdot C/\Delta x}} \tag{9.12}$$

But $L/\Delta x$ and $C/\Delta x$ are, respectively, the inductance and capacitance per unit length of the transmission line. It should be noted that both of these parameters depend upon the materials encased within them. This implies that the numerical value for the velocity of propagation can be significantly different than the velocity of light in a vacuum. Thus we have found that:

The propagation velocity of voltage and current waves on this transmission line is determined by the inductance and capacitance per unit length having the units of henry/m, and farad/m.

This conclusion is in fact quite general and once we know the inductance and capacitance per unit length of any transmission line, the wave propagation velocity can readily be found.

Let us go back a little bit in the derivation of the wave equation. We assumed Δx to be small, but did not specify how small it should be. When we wrote Eq. (9.2), we used $C[\partial V(x, t)/\partial t]$ instead of $C[\partial V(x + \Delta x, t)/\partial t]$. This can be done if

$$\left| \Delta x \frac{\partial V(x, t)}{\partial x} \right| \ll |V(x, t)|$$

as is clear from the Taylor series expansion for $V(x + \Delta x, t)$. For a sinusoidal waveform of the voltage, $V(x, t) = V_0 \sin(kx - \omega t)$, with $\omega/k = c_w$, we find

$$\frac{\partial V(x, t)}{\partial x} = k V_0 \cos(kx - \omega t) = \frac{2\pi}{\lambda} V_0 \cos(kx - \omega t)$$

Therefore $|\Delta x(\partial V(x, t)/\partial x)| \ll |V(x, t)|$ requires that

$$\Delta x \frac{2\pi}{\lambda} \ll 1$$

or roughly speaking, Δx must be smaller than the wavelength λ. Thus when we have a discrete *LC* transmission line as shown in Figure 9–1, the propagation of the voltage and current signals can be described with a dispersionless, linear wave equation only if the preceding condition ($\Delta x \ll \lambda/2\pi$) is satisfied. This is the major limitation of this model although it is possible to find an exact dispersion relation for the discrete *LC* transmission line (Example 9.2).

This limitation seems very severe, but we do not have to worry about it at all in practical transmission lines, which in most cases are *continuous*. We saw that the propagation velocity is determined by the inductance and capacitance *per unit length*; that is, Δx can be taken as small as we wish in a

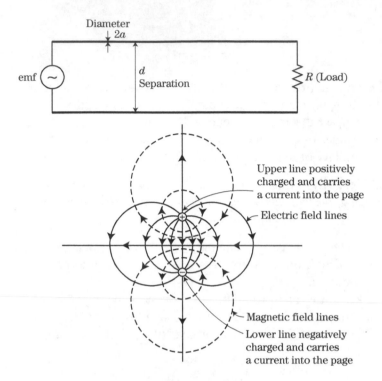

FIGURE 9–4

Parallel wire transmission line for low-frequency electromagnetic waves. The lower figure indicates the electric and magnetic fields lines.

continuous medium. The inductance and capacitance per unit length remain as finite quantities no matter how small Δx is chosen. The most familiar transmission line is the one composed of just two parallel wires (Figure 9–4). The inductance and capacitance per unit length can easily be calculated for such a system (see Example 9.4).

Another important transmission line is the coaxial cable which we will study in detail later. Usually, the coaxial cable is filled with a dielectric material in order to provide mechanical strength as well as electrical insulation. The dielectric in turn increases the capacitance per unit length. Consequently, the propagation velocity of electromagnetic signals in coaxial cables is smaller than that of those filled with air. As we will see, coaxial cables filled with air would have the propagation velocity 3.0×10^8 m/sec, which is the velocity of light in air. If it is filled with an insulator having a permittivity $2\varepsilon_0$, the velocity is 2.1×10^8 m/sec.

Let us now reexamine the dimensions of the vacuum permittivity

$$\varepsilon_0 = 8.85 \times 10^{-12} \simeq \frac{1}{36\pi} \times 10^{-9} \, \frac{\text{C}^2}{\text{m}^2\text{N}}$$

and permeability

$$\mu_0 = 4\pi \times 10^{-7} \, \frac{\text{N}}{\text{A}^2}$$

The dimensions of ε_0, $\text{C}^2/\text{m}^2 \cdot \text{N}$, can be rewritten as farad/meter since the farad has the dimensions of $\text{C}^2/\text{N} \cdot \text{m}$. Thus the physical meaning of ε_0 is the capacitance per unit length in vacuum. Similarly, μ_0 can be understood as the inductance per unit length in vacuum,

$$\varepsilon_0 = \frac{C}{\Delta x} \text{ (farad/meter)}$$

$$\mu_0 = \frac{L}{\Delta x} \text{ (henry/meter)}$$

Therefore the velocity of electromagnetic waves in vacuum can immediately be found

$$c = \frac{1}{\sqrt{\varepsilon_0 \mu_0}} = 3.0 \times 10^8 \text{ m/sec}$$

This is the velocity of electromagnetic waves in unbounded vacuum or in an ideal coaxial cable with a vacuum separating the inner and outer conductors. The wave velocity of electromagnetic waves in bounded media, such as air-filled waveguides is not given by the preceding expression and the waves may become dispersive. In dielectric materials, such as glass and water, the permittivity is larger and the velocity of electromagnetic waves in dielectric materials is correspondingly smaller since

$$c = \frac{1}{\sqrt{\varepsilon_r \varepsilon_0 \mu_r \mu_0}}$$

where ε_r and μ_r are the relative permittivity and the relative permeability of the material.

Thus we have seen that the simple *LC* transmission line can model several important media for electromagnetic waves. In the introduction, we emphasized that any media capable of storing both electric and magnetic energies can permit electromagnetic waves to propagate through them. The bulk expression for both energies are $\frac{1}{2}CV^2$ and $\frac{1}{2}LI^2$ with C and L being the capacitance and inductance, respectively. We can alternatively state that any medium having both capacitance and inductance is capable of accommodating electromagnetic waves.

Consider a highly conductive metal such as copper. We know that an electrostatic field cannot penetrate into copper. In other words, metals cannot store electric energy. Thus electromagnetic waves cannot exist in metals with high electrical conductivity and waves incident upon metals are strongly reflected. The conductivity in conductors plays a much more dominant role than the permittivity for electromagnetic waves. In Section 9.7, we will see that electromagnetic waves in conductors obey a differential equation that is entirely different from the usual wave equation such as Eq. (9.10). A static field should be distinguished from a direct current field or dc field, which is dynamic and associated with a current flow in conductors while the static field will not cause motion of charged particles A dc field can penetrate into a conductor as we will see in the section dealing with the skin effect.

EXAMPLE 9.1

Find the velocity of propagation of the electromagnetic waves in a coaxial cable filled with Teflon, which has $\varepsilon = 2.0\varepsilon_0$ and $\mu = \mu_0$ (Figure 9–5).

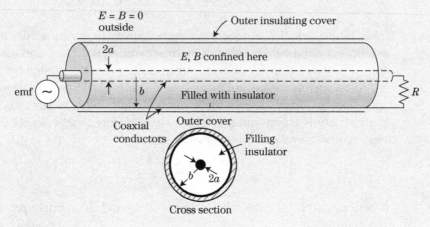

FIGURE 9–5

Coaxial cable for high-frequency electromagnetic waves. There are no fields outside the cable.

Solution

From

$$c = \frac{1}{\sqrt{\varepsilon \mu_0}}$$

we find

$$c = \frac{1}{\sqrt{2 \times 8.85 \times 10^{-12} \times 4\pi \times 10^{-7}}} = 2.1 \times 10^8 \text{ m/s}$$

EXAMPLE 9.2

(a) For the *LC* transmission line shown in Figure 9–1, derive a difference equation similar to Eq. (4.7) that we previously encountered in Chapter 4 for a mass-spring transmission line.

(b) Show that the exact dispersion relation is given by

$$\omega^2 = 4\omega_0^2 \sin^2\left(k\frac{\Delta x}{2}\right)$$

for a harmonic wave $V(x, t) = V_0 \sin(kx - \omega t)$. Here $\omega_0^2 = \frac{1}{LC}$.

Solution

(a) Let us consider two adjacent units as shown in Figure 9–6. Applying Kirchhoff's voltage law repeatedly, we obtain

$$V(x - \Delta x, t) = L\frac{\partial}{\partial t}I(x - \Delta x, t) + V(x, t) \tag{i}$$

$$V(x, t) = L\frac{\partial}{\partial t}I(x, t) + V(x + \Delta x, t) \tag{ii}$$

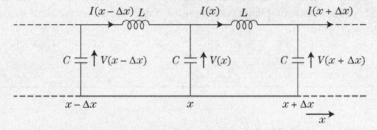

FIGURE 9–6

Subtracting and rearranging gives

$$L\frac{\partial}{\partial t}[I(x, t) - I(x - \Delta x, t)] = 2V(x, t) - V(x + \Delta x, t) - V(x - \Delta x, t) \tag{iii}$$

However, Kirchhoff's current law yields

$$I(x - \Delta x, t) = C\frac{\partial}{\partial t}V(x, t) + I(x, t)$$

or

$$I(x, t) - I(x - \Delta x, t) = -C\frac{\partial}{\partial t}V(x, t) \tag{iv}$$

Substituting (iv) into (iii),

$$LC\frac{\partial^2}{\partial t^2}V(x, t) = V(x + \Delta x, t) + V(x - \Delta x, t) - 2V(x, t) \tag{v}$$

This is the required difference equation for the voltage. Note that this is mathematically identical to Eq. (4.7) for the mechanical transmission lines.

(b) Let $V(x, t) = V_0 \sin(kx - \omega t)$. Noting that

$$\frac{\partial^2}{\partial t^2} V(x, t) = -\omega^2 V_0 \sin(kx - \omega t)$$

$$V(x + \Delta x, t) - V(x, t) = 2V_0 \sin\left(k\frac{\Delta x}{2}\right) \cos\left[\left(kx + k\frac{\Delta x}{2}\right) - \omega t\right]$$

$$V(x - \Delta x, t) - V(x, t) = -2V_0 \sin\left(k\frac{\Delta x}{2}\right) \cos\left[\left(kx - k\frac{\Delta x}{2}\right) - \omega t\right]$$

and thus

$$V(x + \Delta x, t) - V(x - \Delta x, t) - 2V(x, t) = -4V_0 \sin^2\left(k\frac{\Delta x}{2}\right) \sin(kx - \omega t)$$

we find

$$\omega^2 = \frac{1}{LC} 4\sin^2\left(k\frac{\Delta x}{2}\right) = 4\omega_0^2 \sin^2\left(k\frac{\Delta x}{2}\right)$$

Note that the long wavelength limit corresponds to satisfying the relation that $k\Delta x \ll 1$. Approximating $\sin(k\Delta x/2)$ with $k\Delta x/2$ yields our previous dispersion relation,

$$\omega^2 = \frac{\Delta x}{LC} k^2$$

 9.3 ## Coaxial Cable

A cable composed of two coaxial cylindrical conductors is called a coaxial cable and is frequently used for transmitting electromagnetic signals from one device to another. Coaxial cables can confine electromagnetic waves between the two conductors and under ideal conditions, the electromagnetic waves do not leak out. This is in contrast with the open, two-parallel-wire transmission line. Parallel-wire transmission lines are typically used for high-power, low-frequency (60-Hz) electromagnetic waves. At such a low frequency, the radiation loss can be considered to be negligible. As the frequency becomes higher, however, the parallel-wire transmission lines become very ineffective because of the radiation loss; that is, the wave energy can easily leak out from the transmission line. Therefore at high frequencies, the transmission lines must be of closed type, such as a coaxial cable and microwave waveguide. (It is a common experience that sound waves that are confined in a pipe can propagate a further distance than in free space.)

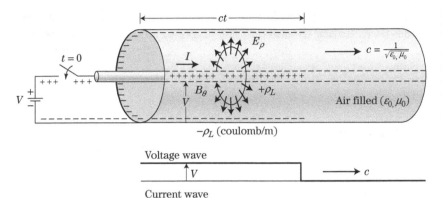

FIGURE 9–7

When a dc battery is suddenly connected to a coaxial cable, both the voltage and current waves start propagating along the cable.

Suppose that we have a long coaxial cable connected to a dc battery at $t = 0$ at one end (Figure 9–7). The cable cannot be filled with charge instantaneously, since the electromagnetic waves should travel with a large, but finite speed. We assume that the space between the inner and outer conductor is filled with air (or vacuum), which has a permittivity ε_0 and a permeability μ_0. Our purpose here is to find the propagation velocity c of the charge pulse, initially pretending that we do not know the wave equation that we found in the previous section. (Actually, we are cheating ourselves since when we assume that a square charge pulse can exist in the cable, we have to assume that the electromagnetic waves in the coaxial cable are dispersionless; that is, the propagation velocity is independent of the wave frequency. This can only be assured by the wave equation that we are going to obtain!)

Let ρ_L be the linear charge density in the region to the left of the pulse front. (The use of ρ_L should not be confused with the linear mass density ρ_l we used for mechanical waves.) Since we have cylindrical symmetry, the radial electric field in the space between the inner and outer conductor can easily be found from Gauss's law (Figure 9–8). To save space, in the following

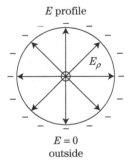

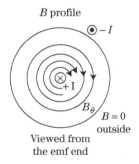

FIGURE 9–8

Electric and magnetic field profiles in the coaxial cable.

we remove the explicit space and time independent variables in the equations that we had used previously although all of the inhomogeneous fields that we obtain are space and time dependent. The electric field component of the wave is determined from Eq. (9.13).

$$E_\rho = \frac{\rho_L}{2\pi\varepsilon_0}\frac{1}{\rho} \tag{9.13}$$

Next we notice that the current I should be related to the linear charge density ρ_L through

$$I = \rho_L c$$

where c is the velocity at which the charge front is moving. (The wave propagation velocity c should not be confused with the velocity of the charge carriers, namely, electrons in the conductors. What propagates at the velocity c is the *perturbation* in the charge density and this has nothing to do with the electron velocity, which is in fact extremely small. A similar situation has already been encountered in sound waves in which the wave velocity c_s and the wave velocity $\partial\xi/\partial t$ were entirely different physical quantities.) Then the azimuthal magnetic field can be found from Ampere's law (Figure 9–8) as

$$B_\theta = \frac{\mu_0 I}{2\pi\rho} = \frac{\mu_0\rho_L}{2\pi\rho}c \tag{9.14}$$

From Eqs. (9.13) and (9.14), we find

$$\frac{E_\rho}{B_\theta} = \frac{1}{\varepsilon_0\mu_0 c} \tag{9.15}$$

Thus if we find one more relationship between the electric field E_ρ and magnetic field B_θ, we can find the propagation velocity c.

This can be done if we apply Faraday's law to a thin rectangle at the pulse front at a certain instant (Figure 9–9). (We neglect unimportant edge effects at the pulse front.) As the pulse propagates to the right, the magnetic flux enclosed by the rectangle will also increase. Thus a voltage induced along the

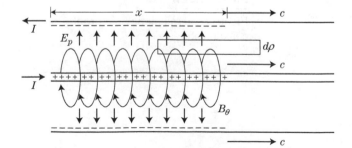

FIGURE 9–9

Faraday's law is applied to the rectangle. The magnetic flux through the rectangle is increasing with time.

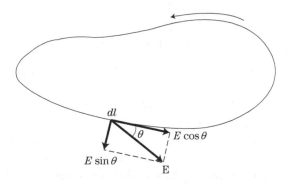

FIGURE 9–10

Closed-line integral.

rectangle is

$$\oint \mathbf{E} \cdot d\mathbf{l} = -\frac{d}{dt}(Bxd\rho) \tag{9.16}$$

The integral with a circle at the center is called a closed line integral. The integral $\oint \mathbf{E} \cdot d\mathbf{l}$ with reference to Figure 9–10 can be written as

$$\oint E \cos\theta \, dl \tag{9.17}$$

where θ is the angle between $\mathbf{E}$ and $d\mathbf{l}$, and the integration is over a closed but arbitrary shape (Figure 9–10). In the present case, the only contribution to the integral comes from the edge AB, since the electric field is perpendicular to the edges BC and DA ($\cos 90° = 0$), and we do not have an electric field along the edge CD (Figure 9–11). Then

$$\oint \mathbf{E} \cdot d\mathbf{l} = -E_\rho d\rho$$

(Note that along AB, the angle $\theta = 180°$.) The right-hand side of Eq. (9.16) becomes

$$-\frac{d}{dt}(B_\theta d\rho x) = -B_\theta d\rho \frac{dx}{dt} = -B_\theta d\rho c$$

Then

$$E_\rho = B_\theta c \tag{9.18}$$

From Eqs. (9.15) and (9.18), we find

$$c^2 = \frac{1}{\varepsilon_0 \mu_0} = 9.0 \times 10^{16} \ (\text{m/sec})^2$$

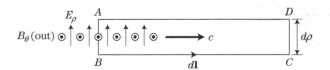

FIGURE 9–11

The only contribution to the line integral comes from the edge AB where $\mathbf{E}$ is parallel with $d\mathbf{l}$.

or

$$c = 3.0 \times 10^8 \text{ m/sec}$$

which is the desired velocity of propagation of an electromagnetic pulse in a coaxial cable filled with air.

It is obvious that if the cable were filled with a dielectric material other than air, we would have to replace ε_0 by $\varepsilon = \varepsilon_r \varepsilon_0$, where ε_r is the relative dielectric constant, and the velocity becomes correspondingly smaller. *Caution:* The relative dielectric constant ε_r is usually a function of the frequency of electromagnetic waves. For example, water has $\varepsilon_r \simeq 80$ for static fields ($\nu = 0$), but at the frequency of visible light ($\nu \simeq 5 \times 10^{14}$ Hz), $\varepsilon_r = 1.8$.

Equation (9.15) can alternatively be derived from the Ampere–Maxwell law, which states that a time-varying electric flux should induce a time-varying magnetic field. In other words, a time-varying electric field is equivalent to a current that is in reality the displacement current postulated originally by James Clerk Maxwell. This is exactly how a current can flow through a capacitor. Let us consider again a thin rectangle that is perpendicular to the one we considered before. The electric flux enclosed by the rectangle $ABCD$ induces a magnetic field along $ABCD$ (Figure 9–12).

$$\oint \mathbf{B} \cdot d\mathbf{l} = \varepsilon_0 \mu_0 \frac{d}{dt}(E_\rho x \, dl) = \varepsilon_0 \mu_0 E_\rho dl \frac{dx}{dt} = \varepsilon_0 \mu_0 E_\rho dl c \qquad (9.19)$$

Now the only contribution to the integral comes from the edge AB and we have

$$\oint \mathbf{B} \cdot d\mathbf{l} = B_\theta dl \qquad (9.20)$$

We then find

$$B_\theta = \varepsilon_0 \mu_0 E_\rho c$$

which is identical to Eq. (9.15).

Thus we have been able to derive the velocity of an electromagnetic pulse in a coaxial cable without using the wave equation. Once we find the velocity

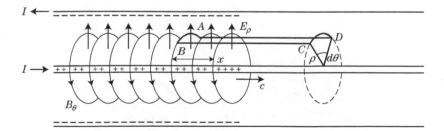

FIGURE 9–12

Electric flux through $ABCD$ is increasing with time, which induces a magnetic field through the displacement current.

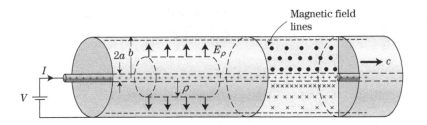

FIGURE 9–13

Gauss's law applied to a cylinder with radius ρ to find the electric field. Magnetic field lines are shown for comparison.

we can calculate the current I for a given value of the voltage source (volt) or find the impedance of the coaxial cable. In order to do this, we have to assume a finite radius of the inner conductor a (m) and that of the outer conductor b (m). The electric field at radius ρ is again found from Gauss's law (Figure 9–13),

$$2\pi\rho E_\rho = \frac{\rho_L}{\varepsilon_0}, \quad \text{for } a < \rho < b$$

The voltage difference between the inner and outer conductor is then given by

$$V = \int_a^b E_\rho d\rho = \frac{\rho_L}{2\pi\varepsilon_0}\int_a^b \frac{1}{\rho}d\rho = \frac{\rho_L}{2\pi\varepsilon_0}\ln\left(\frac{b}{a}\right)$$

But $\rho_L = I/c$. Then

$$Z_c = \frac{V}{I} = \frac{1}{2\pi\varepsilon_0 c}\ln\left(\frac{b}{a}\right) \ (\Omega)$$

Using $c = 1/\sqrt{\varepsilon_0\mu_0}$, we find

$$Z_c = \sqrt{\frac{\mu_0}{\varepsilon_0}}\frac{1}{2\pi}\ln\left(\frac{b}{a}\right) \ (\Omega) \tag{9.21}$$

Since $\ln(b/a)$ is dimensionless, we see that $\sqrt{\mu_0/\varepsilon_0}$ has the dimension of an impedance Ω. This ratio is called the *characteristic impedance* of the coaxial cable. You should check this directly from the dimensions for μ_0 and ε_0.

The characteristic impedance we just found can alternatively be written as

$$Z_c = \sqrt{\frac{\text{inductance per unit length}}{\text{capacitance per unit length}}} \tag{9.22}$$

To see this, let us find the inductance and capacitance per unit length of the coaxial cable. For every 1 m of the cable, the cable has ρ_L coulombs of charge. The voltage difference between the conductors is still V. Then

$$C = \frac{\rho_L}{V} = 2\pi\varepsilon_0\frac{1}{\ln(b/a)} \ (\text{F/m}) \tag{9.23}$$

The inductance of the cable can be found if we calculate the total magnetic flux between the two conductors, $a < \rho < b$. Since

$$B_\theta = \frac{\mu_0 I}{2\pi\rho}$$

the magnetic flux per unit length of the cable is

$$\Phi_B = \int_a^b B_\theta d\rho = \frac{\mu_0 I}{2\pi} \ln\left(\frac{b}{a}\right) \quad \text{(weber/meter)}$$

The inductance per unit length is then

$$L = \frac{\Phi_B}{I} = \frac{\mu_0}{2\pi} \ln\left(\frac{b}{a}\right) \quad \text{(H/m)} \tag{9.24}$$

We have not included the magnetic flux in the inner conductor in the inductance calculation; in other words, we have neglected the internal inductance of the inner conductor. This is allowed if the current flows only on the surface of the conductor (skin effect). Substituting these values for C and L into Eq. (9.22), we recover

$$Z_c = \sqrt{\frac{\mu_0}{\varepsilon_0}} \frac{1}{2\pi} \ln\left(\frac{b}{a}\right) \quad (\Omega)$$

You may wonder why we cannot simply use ε_0 and μ_0 as the capacitance and inductance per unit length, particularly since we are considering electromagnetic waves in *air* filling the space between the two conductors. The term $Z_0 = \sqrt{\mu_0/\varepsilon_0}$ indeed has the meaning of the characteristic impedance of vacuum (or air), but we cannot use this for the coaxial cable. The reason is that the electromagnetic wave in the coaxial cable is *not* a plane wave. We saw that both the electric and magnetic fields depend on the radial position as well as, of course, x, the direction of propagation. For plane electromagnetic waves, the impedance indeed becomes $\sqrt{\mu_0/\varepsilon_0}$. Whenever the waves are confined geometrically, as in the case of coaxial cable, the characteristic impedance must be accordingly modified.

The reason we call Z_c the characteristic impedance rather than the resistance is that the coaxial cable does not dissipate energy. The cable itself is a *reactive medium* composed of a capacitance and an inductance. It only provides a medium for electromagnetic wave propagation. If we terminate one end of the coaxial cable with a resistor having the same value as the characteristic impedance, the energy is most effectively transferred from the cable to the load resistor. There will be no reflection of the electromagnetic waves at the terminating end. At this stage, you should be reminded of the concept of impedance matching for the most efficient energy transfer from a battery having an internal resistance R_i. If the load resistance matches the characteristic impedance of the coaxial cable, the energy dissipation in the load resistance becomes a maximum. The physics behind this matching should become obvious. If the load resistance matches the characteristic impedance of the coaxial cable, no wave (and thus energy) reflection occurs, and the energy is most efficiently transferred.

Let us now take a look at the energy stored in the coaxial cable. Since we know both the electric and magnetic fields as functions of the radius ρ, we can immediately find the local energy densities:

$$\text{electric energy density} = \frac{1}{2}\varepsilon_0 E_\rho^2 = \frac{\rho_L^2}{8\pi^2\varepsilon_0\rho^2} \quad (\text{J/m}^3) \qquad (9.25)$$

$$\text{magnetic energy density} = \frac{1}{2\mu_0}B_\theta^2 = \frac{\mu_0\rho_L c^2}{8\pi^2\rho^2} \quad (\text{J/m}^3) \qquad (9.26)$$

But $c^2 = 1/\varepsilon_0\mu_0$. Thus we find that the electric energy density is equal to the magnetic energy density, as in the case of mechanical waves in which the kinetic energy density equals the potential energy density. We conclude that for nondispersive electromagnetic waves, the same amount of energy is stored in both the electric and magnetic fields.

The total energy density is

$$W = 2 \times \frac{\rho_L^2}{8\pi^2\varepsilon_0\rho^2} = \frac{\rho_L^2}{4\pi^2\varepsilon_0\rho^2} \quad (\text{J/m}^3) \qquad (9.27)$$

Then the energy per unit length of the coaxial cable is

$$\int_a^b W \cdot 2\pi\rho d\rho = \frac{\rho_L^2}{2\pi\varepsilon_0}\ln\left(\frac{b}{a}\right) \quad (\text{J/m})$$

and the power is

$$P = c\frac{\rho_L^2}{2\pi\varepsilon_0}\ln\left(\frac{b}{a}\right) = \frac{V^2}{Z_c} = I^2 Z_c \quad (\text{W}) \qquad (9.28)$$

which is equal to the power supplied by the voltage source VI.

EXAMPLE 9.3

Determine the ratio between the outer and inner conductor radii b/a of a 50-Ω coaxial cable filled with Teflon ($\varepsilon = 2\varepsilon_0$). The permeability of Teflon is approximately the same as the vacuum value.

Solution

The characteristic impedance is given by

$$Z_c = \sqrt{\frac{\mu_0}{\varepsilon}} \frac{1}{2\pi} \ln\left(\frac{b}{a}\right) \text{ (}\Omega\text{)}$$

Then

$$\ln\left(\frac{b}{a}\right) = 2\pi \sqrt{\frac{\varepsilon_r \varepsilon_0}{\mu_0}} Z_c = 2\pi \sqrt{\frac{2\varepsilon_0}{\mu_0}} Z_c$$

Substituting $\sqrt{\mu_0/\varepsilon_0} = 377\Omega$ and $Z_c = 50\Omega$, we obtain

$$\ln\left(\frac{b}{a}\right) = 2\pi \sqrt{2} \times \frac{50}{377} = 1.178$$

or

$$\frac{b}{a} = 3.25$$

As long as this ratio is used, a Teflon-filled cable will have a 50 Ω characteristic impedance irrespective of its actual size, large or small.

EXAMPLE 9.4

Calculate the characteristic impedance of a parallel-wire transmission line. Assume that the conductors have a common radius a and are separated by a distance d that is much larger than the radius a.

Solution

To find the capacitance per unit length of the transmission line, we assume that the linear charge densities on the two wires are $+\rho_L$(C/m) and $-\rho_L$(C/m), respectively. In the plane containing the wires, the electric field is given by (Figure 9–14)

$$E_\rho = \frac{\rho_L}{2\pi\varepsilon_0}\left(\frac{1}{\rho} + \frac{1}{d-\rho}\right)$$

where ρ is the distance from the positively charged wire. Then the potential difference between the wires is

$$V = \int_a^{d-a} E_\rho d\rho = \frac{\rho_L}{2\pi\varepsilon_0}\int_a^{d-a}\left(\frac{1}{\rho} + \frac{1}{d-\rho}\right)d\rho = \frac{\rho_L}{\pi\varepsilon_0}\ln\left(\frac{d-a}{a}\right) \simeq \frac{\rho_L}{\pi\varepsilon_0}\ln\left(\frac{d}{a}\right)$$

and the capacitance per unit length of the transmission line becomes

$$\frac{C}{l} = \frac{\rho_L}{V} = \frac{\pi\varepsilon_0}{\ln(d/a)} \text{ (F/m)}$$

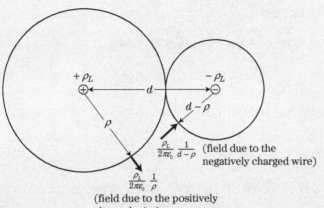

FIGURE 9–14

Electric fields.

$\frac{\rho_L}{2\pi\epsilon_0}\frac{1}{d-\rho}$ (field due to the negatively charged wire)

$\frac{\rho_L}{2\pi\epsilon_0}\frac{1}{\rho}$

(field due to the positively charged wire)

The inductance per unit length can be found in a similar manner. For this purpose, we let the wires carry currents $+I(\text{A})$ and $-I(\text{A})$, respectively. The magnetic field at a distance ρ from the positive current (Figure 9–15) is

$$B = \frac{\mu_0 I}{2\pi}\left(\frac{1}{\rho} + \frac{1}{d-\rho}\right)$$

FIGURE 9–15

Magnetic fields.

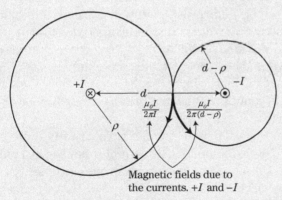

Magnetic fields due to the currents. $+I$ and $-I$

Then the magnetic flux linked to unit length of the transmission line is

$$\frac{\Phi_B}{l} = \frac{\mu_0 I}{2\pi}\int_a^{d-a}\left(\frac{1}{\rho} + \frac{1}{d-\rho}\right)d\rho \simeq \frac{\mu_0 I}{\pi}\ln\left(\frac{d}{a}\right)\ (\text{W/m})$$

The inductance per unit length is thus

$$\frac{L}{l} = \frac{\Phi_B}{lI} = \frac{\mu_0}{\pi}\ln\left(\frac{d}{a}\right)\ (\text{H/m})$$

Finally, the characteristic impedance is given by

$$Z_c = \sqrt{\frac{L/l}{C/l}} = \frac{1}{\pi}\sqrt{\frac{\mu_0}{\varepsilon_0}}\ln\left(\frac{d}{a}\right)\ (\Omega)$$

This formula is subject to the condition $d \gg a$.

Poynting Vector

In the previous section we found the energy density in the coaxial cable is

$$W = \frac{\rho_L^2}{4\pi^2 \varepsilon_0 \rho^2} \quad (\text{J/m}^3) \tag{9.29}$$

Then the power density (energy flow per unit time and per unit area) is

$$cW = \frac{\rho_L^2 c}{4\pi^2 \varepsilon_0 \rho^2} = \frac{\rho_L}{2\pi \varepsilon_0 \rho} \times \frac{\mu_0 \rho_L c}{2\pi \rho} \frac{1}{\mu_0} = E_\rho \times \frac{B_\theta}{\mu_0} = E_\rho \times H_\theta. \quad (\text{W/m}^2) \tag{9.30}$$

The quantity $\mathbf{E} \times \mathbf{H} \,(= \mathbf{E} \times \mathbf{B}/\mu_0)$ is called the Poynting vector and is denoted by $\mathbf{S}$. Its physical meaning is simply the power density (W/m^2). Furthermore, since $\mathbf{S}$ is a vector, it also tells us in which direction the electromagnetic energy is flowing. In the coaxial cable, the electric field is radially outward and the magnetic field is in the azimuthal direction. The vector direction of $\mathbf{E} \times \mathbf{B}$ is indeed toward the load and this is consistent with the direction of the energy flow (Figure 9–16).

From the arguments given above, it is now apparent that energy is *not* transferred *in* the conductors. Rather, the energy is transferred through the space between the conductors. The role of the conductors, other than to provide mechanical stability, is to provide a confinement region for the electromagnetic energy in the space between the conductors, thus preventing leakage. This is not surprising if you remember that vacuum (or air) is an excellent medium for electromagnetic waves. In contrast, conductors are an extremely poor medium for electromagnetic waves.

Let us reexamine the energy flow in some simple cases in terms of the Poynting vector, $\mathbf{S} = \mathbf{E} \times \mathbf{B}/\mu_0$. Suppose a dc current is flowing in a rod with a resistance R (Figure 9–17). The potential drop from one end of the rod to the other is $V = IR$ and the electric field in the rod is $E = V/l$. The magnetic field is azimuthal and resides at the surface of the rod (use Ampere's law),

$$B = \frac{\mu_0 I}{2\pi a}$$

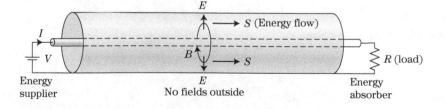

FIGURE 9–16

Electric field, magnetic field, and Poynting vector in a coaxial cable. The Poynting vector is directed from the battery to the load everywhere in the cable.

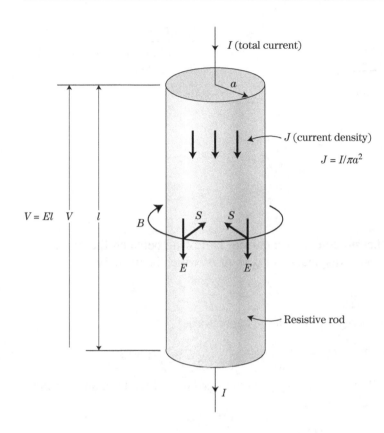

FIGURE 9–17

Poynting vector **S** around a resistor. **S** is radially inward everywhere.

Then the Poynting vector **S** is directed radially inward, and its magnitude is

$$S = \frac{EB}{\mu_0} = \frac{V}{l}\frac{I}{2\pi a} \; (\text{W/m}^2)$$

The total power flow through the surface is

$$P = S \times 2\pi al = VI \; (\text{W})$$

which is consistent with the power being dissipated in the resistive rod. In this case again, the energy is not carried in the conductor. Rather it is carried by the dc electromagnetic fields through the space surrounding the conductor and the resistor.

Next consider a parallel plate capacitor C that is being charged (Figure 9–18). The current is determined from the temporal change of the charge

$$I = \frac{dq}{dt}$$

and the charge stored in the capacitor is related to the electric potential

$$q = CV$$

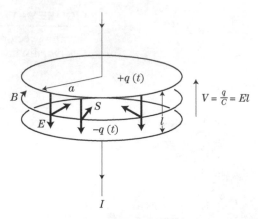

FIGURE 9–18

Poynting vector in a capacitor being charged.

We also know that no conduction current is flowing between the two plates of the capacitor, but a displacement current is, which is given by

$$I = \frac{dq}{dt} = C\frac{dV}{dt}$$

Then the rate of energy storage in the capacitor is

$$P = IV = C\frac{dV}{dt}V = \frac{d}{dt}\left(\frac{1}{2}CV^2\right) \text{ (W)}$$

Ampere's law still holds for displacement currents. Then the magnetic field at the edge of the capacitor is

$$B = \frac{\mu_0 I}{2\pi a} = \frac{\mu_0}{2\pi a}C\frac{dV}{dt}$$

The electric field is $E = V/l$. Then the Poynting vector at the edge is

$$S = \frac{EB}{\mu_0} = \frac{V}{l}\frac{1}{2\pi a}C\frac{dV}{dt} \text{ (W/m}^2)$$

and we again have

$$2\pi a l S = \frac{d}{dt}\left(\frac{1}{2}CV^2\right)$$

The preceding argument is correct only if the rate of charging is sufficiently slow. If the capacitor is rapidly charged, $E = V/l$ does not hold anymore. However, the Poynting theorem still applies.

From these examples, it is now clear that electromagnetic energy transfer is achieved and can be understood in terms of the Poynting vector. An electric field or magnetic field alone cannot cause the flow of electromagnetic energy. We must have both fields. Furthermore, since the Poynting vector is the vector product between **E** and **B**, the Poynting vector vanishes if **E** is purely parallel to **B** even though both are nonzero. In most electromagnetic phenomena we study, **E** is usually perpendicular to **B** and you do not have to worry about this complication.

Plane Electromagnetic Waves in Free Space

We know that radio or TV signals can be transmitted through air (or vacuum). We also know that solar energy is transmitted from the sun to the earth through vacuum in the form of radiation of electromagnetic waves with different frequencies or wavelengths. As noted previously, electromagnetic waves have an enormous range of frequencies. Surprisingly enough, they all propagate with the same velocity, $c = 3.0 \times 10^8$ m/sec, the velocity of light in vacuum (or air, with negligible error).

In Section 9.2, we almost obtained the wave equation for electromagnetic waves in free space. If we take $L/\Delta x = \mu_0$, and $C/\Delta x = \varepsilon_0$, we may indeed conclude that electromagnetic waves in free space should propagate with the velocity

$$c = \frac{\Delta x}{\sqrt{LC}} = \frac{1}{\sqrt{\varepsilon_0 \mu_0}} = 3.0 \times 10^8 \text{ m/sec}$$

However, it is far from obvious that we can analyze the propagation of electromagnetic waves in free space, which is a continuous medium for the waves in terms of discrete inductances and capacitances. Also, the analogy does not tell us what the electric and magnetic field should be which we will need when we calculate the Poynting vector associated with electromagnetic waves. Here we directly derive the wave equation for electromagnetic waves in free space using the fundamental laws in electromagnetism, namely, Faraday's induction law and the Ampere–Maxwell displacement current law.

For simplicity, we assume a one-dimensional or plane wave that is propagating in the $+x$ direction which is also the direction of energy flow. From the arguments concerning the Poynting vector and assuming that the electric field is polarized in the y direction, we must have the magnetic field polarized in the z direction as shown in Figure 9–19. Let us assume a thin rectangle in the x-y plane having a length of 1 m and width Δx (Figure 9–20). We apply Faraday's law to the rectangle. Since the magnetic flux enclosed by the rectangle is $\Phi_B = B(x)\Delta x \times 1$, we have

$$\oint \mathbf{E} \cdot d\mathbf{l} = E(x + \Delta x) \times 1 - E(x) \times 1 \simeq \Delta x \frac{\partial E}{\partial x}$$

This must be equal to $-\partial \Phi_B / \partial t = -\Delta x (\partial B / \partial t)$. Then

$$-\frac{\partial E}{\partial x} = \frac{\partial B}{\partial t} \tag{9.31}$$

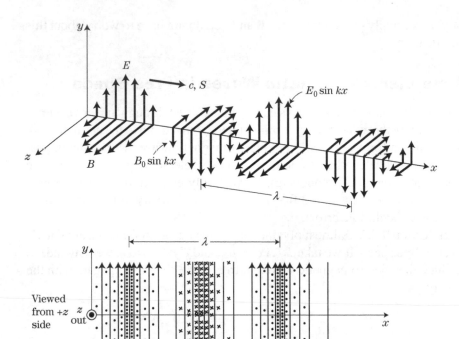

FIGURE 9–19

Plane electromagnetic wave propagating in the $+x$ direction ("snapshot"). In the upper part of the figure, the vector wavelength corresponds to the local field intensity. In the lower part, the number of field lines per unit length corresponds to the local field intensity.

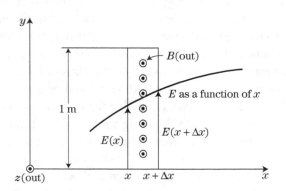

FIGURE 9–20

The magnetic flux through the rectangle is varying with time.

Next consider a thin rectangle in the x-z plane through which an electric flux

$$\Phi_E = E\,\Delta x \times 1$$

penetrates (Figure 9–21). The Ampere–Maxwell law requires that

$$\oint \mathbf{B} \cdot d\mathbf{l} = B(x,t) \times 1 - B(x + \Delta x, t) \times 1 = \varepsilon_0 \mu_0 \frac{\partial \Phi_E}{\partial t} = \varepsilon_0 \mu_0 \Delta x \frac{\partial E}{\partial t}$$

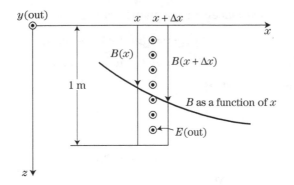

FIGURE 9–21

The electric flux through the rectangle is time varying, which induces a magnetic field along the edges (Ampere-Maxwell law for the displacement current).

or

$$-\frac{\partial B}{\partial x} = \frac{\partial E}{\partial t} \tag{9.32}$$

Differentiating Eq. (9.31) and Eq. (9.32) with respect to x and t respectively and eliminating B, we find

$$\frac{\partial^2 E}{\partial t^2} = \frac{1}{\varepsilon_0 \mu_0} \frac{\partial^2 E}{\partial x^2} \tag{9.33}$$

Similarly, we can obtain

$$\frac{\partial^2 B}{\partial t^2} = \frac{1}{\varepsilon_0 \mu_0} \frac{\partial^2 B}{\partial x^2} \tag{9.34}$$

Equations (9.33) and (9.34) are our desired wave equations for the electric and magnetic fields in free space.

The velocity of propagation is immediately found to be

$$c = \frac{1}{\sqrt{\varepsilon_0 \mu_0}} = 3.0 \times 10^8 \text{ m/sec}$$

If the medium has an electric permittivity $\varepsilon = \varepsilon_r \varepsilon_0$ and magnetic permeability $\mu = \mu_r \mu_0$, the velocity should accordingly be modified to

$$c = \frac{1}{\sqrt{\varepsilon \mu}}$$

Let us determine whether for the electromagnetic waves, the condition that the electric energy be equal to the magnetic energy is satisfied. For this, it is sufficient to see if the energy densities are the same. The electric energy density is

$$\frac{1}{2} \varepsilon_0 E^2 \ (\text{J/m}^3)$$

and the magnetic energy density is

$$\frac{1}{2\mu_0} B^2 \ (\text{J/m}^3)$$

Assuming a sinusoidal electromagnetic wave of the form

$$E(x, t) = E_0 \sin(kx - \omega t)$$
$$B(x, t) = B_0 \sin(kx - \omega t)$$

we find from Eq. (9.31) that

$$-k E_0 \cos(kx - \omega t) = -B_0 \cos(kx - \omega t)$$

or

$$k E_0 = \omega B_0$$

Since $\omega / k = c$, we have

$$E(x, t) = c B(x, t)$$

which was found earlier in Section 9.3. Then

$$\frac{1}{2}\varepsilon_0 E^2 = \frac{1}{2}\varepsilon_0 c^2 B^2 = \frac{1}{2\mu_0} B^2$$

Indeed both energy densities are equal. In the preceding derivation, we assumed a sinusoidal wave. However, this is not essential and the equipartition of energy holds for any propagating electromagnetic wave of any arbitrary form.

Let us next calculate the Poynting vector for the case of sinusoidal waves. Since E is polarized in the y direction and B is polarized in the z direction

$$\mathbf{S} = \mathbf{E} \times \frac{\mathbf{B}}{\mu_0}$$

is directed along the $+x$ axis which is the direction of the propagation of the wave and thus in the direction of the energy flow. The magnitude of $\mathbf{S}$ is

$$\frac{EB}{\mu_0}$$

Substituting $E = E_0 \sin(kx - \omega t)$, and $B = B_0 \sin(kx - \omega t)$, we find

$$S = \frac{E_0 B_0}{\mu_0} \sin^2(kx - \omega t) = c\varepsilon_0 E_0^2 \sin^2(kx - \omega t) \ (\text{W/m}^2)$$

The root-mean-square value of S is

$$S_{\text{rms}} = \frac{1}{2}c\varepsilon_0 E_0^2 = c\varepsilon_0 E_{\text{rms}}^2 = \sqrt{\frac{\varepsilon_0}{\mu_0}} E_{\text{rms}}^2 \tag{9.35}$$

which is composed of $\frac{1}{2}c\varepsilon_0 E_{\text{rms}}^2$ and $\frac{1}{2}c(B_{\text{rms}}^2/\mu_0)$ as required from the equipartition of energy between electric and magnetic fields.

The characteristic impedance for electromagnetic waves in vacuum can still be defined by

$$Z_c = \sqrt{\frac{\text{inductance/m}}{\text{capacitance/m}}} = \sqrt{\frac{\mu_0}{\varepsilon_0}} = 377\Omega \tag{9.36}$$

This quantity is also important for antenna engineering. As we will see in Chapter 10, the impedance of antennas used for electromagnetic radiation (called the *radiation resistance*) is proportional to this quantity.

Since the energy density associated with the electromagnetic wave is

$$\varepsilon_0 E_{rms}^2 = \frac{1}{2}\varepsilon_0 E_{rms}^2 + \frac{1}{2\mu_0} B_{rms}^2 \ (J/m^3)$$

we conjecture that the electromagnetic waves have a pressure ($J/m^3 = N/m^2$) given by the same expression. The pressure associated with electromagnetic waves is called the *radiation pressure*. (In gas dynamics the pressure exerted by a gas on a wall *that can absorb all the particles impinging on it* is given by $\frac{1}{2}nk_BT$ where $n(m^{-3})$ is the molecule density, k_B is the Boltzmann constant, and T (K) is the absolute temperature. For a wall that can elastically reflect particles, the pressure exerted is $2 \times \frac{1}{2}nk_BT = nk_BT$. The radiation pressure we found above corresponds to the one-way momentum transfer or the pressure exerted on an absorbing wall. For a perfect reflector, the pressure exerted is twice as large, in complete analogy to the gas pressure.) Usually, this radiation pressure is negligibly small. However, the radiation pressure is actually the measure of the rate of momentum transfer, and it should be realized that electromagnetic waves carry momentum as well as energy. A circularly polarized electromagnetic wave can even carry an angular momentum.

EXAMPLE 9.5

A monochromatic light source radiates isotropically with an rms power of 30 W.

(a) What is the rms electric field at a distance of 5 m from the source?
(b) What is the rms magnetic field at the same distance?
(c) Find the radiation force exerted on a mirror placed normal to radiation at the same distance. Assume that the mirror has an area of 10×10 cm^2.

Solution

(a) The rms Poynting vector is

$$S_{rms} = \frac{power}{4\pi r^2} = \frac{30}{4\pi (5)^2} = 0.0955 \ W/m^2$$

This should be equal to E_{rms}^2/Z_c, where $Z_c = \sqrt{\mu_0/\varepsilon_0} = 377 \ \Omega$. Then

$$E_{rms} = \sqrt{S_{rms}Z_c} = \sqrt{0.0955 \times 377} = 6.0 \ V/m$$

(b) The rms magnetic field can be found from

$$B_{rms} = \frac{E_{rms}}{c} = \frac{6.0}{3 \times 10^8} = 2.0 \times 10^{-8} \ Tesla$$

(c) The radiation pressure exerted on a reflector is

$$2 \times \varepsilon_0 E_{\text{rms}}^2 \ \text{N/m}^2$$

Then the force is

$$F = 2\varepsilon_0 E_{\text{rms}}^2 A = 2 \left(\frac{1}{36\pi} \times 10^{-9} \right) \times (6.0)^2 \times 0.01 = 6.4 \times 10^{-12} \ \text{N (negligibly small)}$$

EXAMPLE 9.6

The Poynting vector $S = \mathbf{E} \times \mathbf{B}/\mu_0$ is interpreted as the intensity of electromagnetic waves W/m^2 which is equivalent to the electromagnetic power density. How do you interpret the following quantities?

(a) $\mathbf{S}/c$, (b) $\mathbf{r} \times \mathbf{S}/c$, (c) $\mathbf{r} \times \mathbf{S}/c^2$ where $\mathbf{r}$ is the distance from the source of the radiation to the point of observation.

Solution

(a) Since $S = c\varepsilon_0 E_{\text{rms}}^2$, S/c is the energy density. However, this is alternatively interpreted as the radiation pressure (note J/m^3 = N/m^2) and furthermore as the rate of momentum transfer per unit area, that is, the momentum flux density.

(b) The angular momentum is $\mathbf{r} \times$ (momentum vector). Therefore $\mathbf{r} \times \mathbf{S}/c$ may be interpreted as the angular momentum flux density. For pure plane waves, this quantity is identically zero since $\mathbf{r}$ and $\mathbf{S}$ are parallel.

(c) The quantity $\mathbf{S}/c^2$ may be interpreted as the momentum density. Therefore $\mathbf{r} \times \mathbf{S}/c^2$ becomes the angular momentum density.

 9.6 **Reflection of Electromagnetic Waves**

The concept of the characteristic impedance introduced in previous sections can greatly simplify our understanding of the reflection phenomena of electromagnetic waves. It allows us to quantize the concept of "hard" and "soft" boundaries that we introduced for mechanical waves.

Suppose a transmission line having a characteristic impedance $Z_c(\Omega)$ and a load resistance $R(\Omega)$ is connected to a battery at $t = 0$ (Figure 9–22). As we have seen before, both voltage and current step functions start propagating toward the load with a constant velocity c, which of course is determined by the medium. If the line is l m long, it takes the pulse l/c sec to reach the load. The question here is: what would happen when and after the pulse front reaches the load? In Figure 9–22, we consider two extreme cases of an open

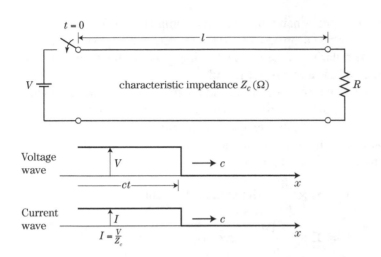

FIGURE 9–22

A transmission line having a characteristic impedance Z_c and a load resistance R is suddenly connected to a battery V. Both voltage and current step functions start propagating. The battery supplies energy that is stored in the transmission line and also dissipated in the load.

circuit $R = \infty$ and a short circuit $R = 0$. If $R = Z_c$, then there should be no reflection at the load, and in this case the power transfer is most efficiently accomplished (impedance matching).

Open Circuit ($R = \infty$)

In the case of an open circuit, no current can flow from point A to B since the resistance is infinitely large. That is, the current at $x = l$ must be zero at all times. For this to be possible, we must have a negative current of equal amplitude propagating toward the battery after the step function front reaches the open circuit (Figure 9–23).

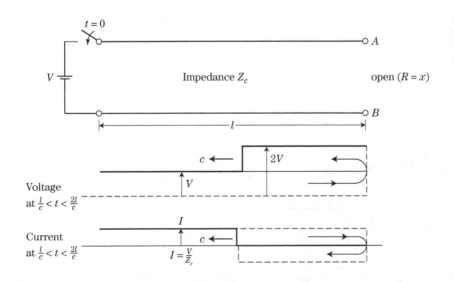

FIGURE 9–23

Reflection at an open circuit, $R = \infty$.

The voltage, on the other hand, behaves in a completely different manner. At the open end, no voltage drop occurs and the reflected voltage step function must have the same polarity as the incident voltage. Thus, after reaching the open end, a voltage of $2V$ appears at the end.

This can be argued in terms of electric and magnetic fields associated with electromagnetic waves. Since the reflected pulse must have the Poynting vector directed *toward* the battery, the electric field (corresponding to the voltage) must have the same polarity as the incident step function if the magnetic field (corresponding to the current) is reversed,

$$\mathbf{S}_i = \mathbf{E} \times \frac{\mathbf{B}}{\mu_0} \quad \text{directed to the right}$$

$$\mathbf{S}_r = \mathbf{E} \times \frac{(-\mathbf{B})}{\mu_0} \quad \text{directed to the left}$$

Thus we may conclude that at the open end, the reflected voltage wave has the same polarity as the incident wave, while the reflected current wave has the opposite polarity with respect to the incident current wave. (This statement holds as long as the termination is a resistance R that is larger than the characteristic impedance Z_c.) Of course, all the incident energy is reflected back at the boundary and we have no waves beyond $x = 1$.

Short Circuit ($R = 0$)

In the case of a short circuit (Figure 9–24), the polarity of the voltage is reversed and the polarity of the current remains unchanged. Recall that the

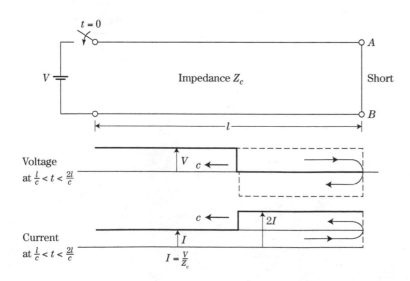

Voltage
at $\frac{l}{c} < t < \frac{2l}{c}$

Current
at $\frac{l}{c} < t < \frac{2l}{c}$

FIGURE 9–24

Reflection at the short circuit, $R = 0$.

voltage difference across a short circuit must be zero at any time. The Poynting vector of the reflected wave must also be directed toward the battery end.

Something seems wrong with this argument. If we short-circuit the terminals of a battery, a tremendously large current should flow, but Figure 9–24 indicates that the current only doubles. How do we explain this?

To answer this we have to find out what happens when the reflected current pulse reaches the battery after a time $2l/c$. If the internal resistance of the battery is small, which is usually the case, the reflected current step again sees an impedance that is a short circuit at the battery. Thus, after $t = 2l/c$, we have a current amplitude $3V/Z_c$ propagating toward the closed end. This multiple reflection process continues; the current amplitude builds up and eventually becomes dangerously large enough to destroy either the battery or the transmission line itself. This is the physics behind short-circuiting, which in any case should be avoided in the laboratory. Of course, the time required for the current buildup depends on the length of the transmission line, but you can easily see that the time scale is extremely short because of the fast propagation velocity which is close to or equal to the speed of light.

The ratio of the voltage wave that is reflected from the load impedance divided by the incident wave is called the *voltage reflection coefficient*. Similarly, the ratio of the current wave that is reflected from the load impedance divided by the incident wave is called the *current reflection coefficient*. If the incident wave is sinusoidal, there will then be a summation of the incident and reflected wave components resulting in a *standing wave* whose amplitude appears to just oscillate in time.

A convenient technique to display the propagation characteristics of the transient signal is to use a *bounce diagram* as shown in Figure 9–25. The horizontal axis of the bounce diagram will be the normalized position of the front of the wave and the vertical axis is a normalized time. The normalizations are in terms of the position divided by the length of the transmission line and of the wave velocity times the actual time divided by the length of the transmission line. With these normalizations, the slopes of the lines will then be ± 1 and the amplitude of each component can be stated on the sloped lines. Drawing a vertical line at a certain location on the transmission line will intersect these sloped lines at various locations. Drawing horizontal lines at these intersections will indicate the actual voltages that will occur at this location. Since the voltage that follows the front remains constant after it passes, one need only add up the contributions of each of the terms.

In both cases ($R = \infty$ and 0) discussed above, the reflection is complete or one hundred percent. No energy can go beyond the open circuit or closed

(a)   (b)

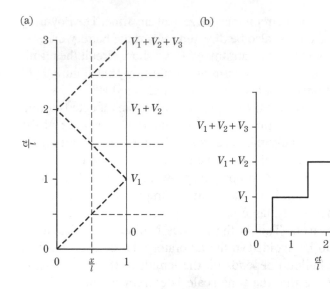

FIGURE 9–25

(a) The bounce diagram plots the normalized distance of the front of the wave as a function of normalized time. (b) The voltage as a function of normalized time at the normalized location of $x/l = 1/2$.

circuit except for small amounts of radiation that can be observed as leakage. For a finite load resistance other than the characteristic impedance, incomplete reflection occurs; that is, some energy is dissipated in the load and the rest is reflected. For example, if $R = 2Z_c$, the voltage reflection is $V/3$ and the current reflection is $-V/3Z_c$. In general, the voltage reflection coefficient is given by

$$\Gamma_V = \frac{R - Z_c}{R + Z_c} \tag{9.37}$$

and the current reflection coefficient is given by

$$\Gamma_I = \frac{Z_c - R}{R + Z_c} = -\Gamma_V \tag{9.38}$$

EXAMPLE 9.7

Derive Eq. (9.37), the voltage reflection coefficient, by using the energy method we employed for the reflection coefficient of mechanical waves.

Solution

Let the incident voltage be V_i and the reflected voltage be V_r. The power associated with the incident voltage is V_i^2/Z_c and that associated with the reflected wave is V_r^2/Z_c. The resistor R dissipates energy at the rate $(V_i + V_r)^2/R$, where $V_i + V_r$ is the voltage to appear at the terminated end. Then the power conservation principle requires that

$$\frac{V_i^2}{Z_c} = \frac{V_r^2}{Z_c} + \frac{(V_i + V_r)^2}{R}$$

Solving for V_r, we obtain

$$V_r = \frac{R - Z_c}{R + Z_c} V_i$$

which defines the voltage reflection coefficient

$$\Gamma_V = \frac{R - Z_c}{R + Z_c}$$

EXAMPLE 9.8

In Figure 9–26, discuss how the voltage wave develops after the switch is closed. The characteristic impedance Z_c of the transmission line is 50 ohms.

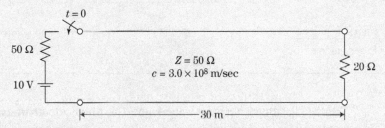

FIGURE 9–26

Solution

When the switch is closed, a voltage step function wave having an amplitude $10 \times 50/(50 + 50) = 5$ V starts propagating down the transmission line. (Note that the 50-Ω transmission line essentially acts as a 50-Ω resistor for the *initial* voltage wave, and the voltage dividing principle applies.) It takes the initial pulse $30 \text{ m}/c = 10^{-7}$ sec to reach the load end. The voltage reflection coefficient is

$$\Gamma_V = \frac{20 - 50}{20 + 50} = -0.43$$

Then, after 10^{-7} sec, a negative voltage wave of 5 V $\times(-0.43) = -2.1$ V propagates toward the battery end. When this negative wave reaches the battery end where impedance matching exists, it is completely absorbed by the 50-Ω resistor and no more reflection occurs. Then a steady state is achieved at a transmission line voltage of $5 - 2.1 = 2.9$ V. This voltage is consistent with what we expect from the dc theory, which yields the resistor voltage of

$$10 \times \frac{20}{50 + 20} \text{ V} = 2.9 \text{ V}$$

However, as we have seen, it takes a finite (although short) time to establish this dc, or steady state. The evolution of the voltage is shown in Figure 9–27. (The evolution of the current wave is left as an exercise.)

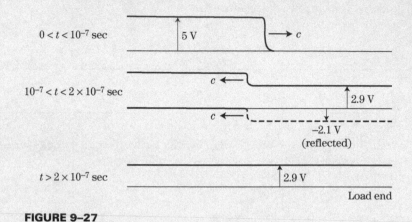

FIGURE 9–27

Propagation of the voltage wave.

Since the characteristic impedance for plane electromagnetic waves is given by

$$Z_c = \sqrt{\frac{\mu}{\varepsilon}}$$

the waves are reflected whenever they enter a medium with a different permittivity and/or permeability (Figure 9–28). In most media, we may assume that the permeability is unchanged, being equal to μ_0. However, the permittivity ε can easily have different values. The velocity of visible light in water and glass is smaller than that in vacuum. (In water, the velocity is $c/1.33$, and in a typical glass, it has a value of $c/1.5$.) This is due to the larger values of permittivity in water and glass than ε_0. Thus the characteristic impedance of water and glass

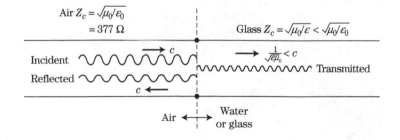

FIGURE 9–28

Transmission line analog of light reflection from water (or glass), which has a lower value of the characteristic impedance than that in air.

with respect to visible light is smaller than that in vacuum, $Z_c = 377\Omega$. It is then obvious that the electric field associated with the reflected light has the opposite polarity with respect to the incident light, since the reflection coefficient for the voltage (and thus the electric field) is negative. In other words, water (or glass) acts as a hard medium for the electric field associated with light. This polarity change at a hard boundary is quite analogous to the case of mechanical waves, particularly the transverse waves on a string under tension. The polarity changes of the electric field at hard boundaries play important roles as discussed in Chapter 11, causing interference between the incident and reflected waves.

Electromagnetic Waves in Matter

As we have seen, electromagnetic waves in vacuum are dispersionless and the propagation velocity c is independent of the wave frequency. This nondispersive nature breaks down for electromagnetic waves in matter. We have already seen that the velocity of electromagnetic waves in matter is different from $c = 3.0 \times 10^8$ m/sec. Usually, it becomes smaller than c, as for example, the velocity of light in glass. We also learned that the change in the velocity causes the change in the characteristic impedance, which in turn causes the reflection of electromagnetic waves whenever they try to penetrate into a different medium.

Let us consider shortwave radio, which can be received at a place to which the waves cannot travel directly (along a straight line), in other words, at a point that is not in the direct "line of sight." Shortwave radio actually uses the reflection of electromagnetic waves from the ionospheric plasma surrounding the earth and from the surface of the earth itself, which is a good conductor (Figure 9–29). Plasmas are ionized gases in which equal amounts

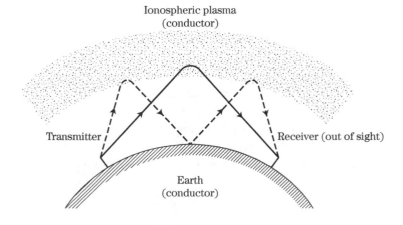

Ionospheric plasma
(conductor)

Transmitter

Receiver (out of sight)

Earth
(conductor)

FIGURE 9–29

Reflection of radio waves from the ionospheric plasma. The earth is a good conductor, too, for radio waves, and multiple reflections can occur. The plasma and the earth form a waveguide that guides the electromagnetic waves for long distances at suitable frequencies.

of negative (electrons) and positive (ions) charges coexist, maintaining gross charge neutrality. The ionospheric plasma is produced by the solar radiation (ultraviolet light and X-ray). Charged particles in the solar wind can also interact with the naturally occurring magnetic field surrounding the earth to produce the aurora.

In plasmas, charged particles are highly mobile and can easily be accelerated by the electric field associated with the electromagnetic waves. The motion of charged particles (mainly electrons because their mass is much smaller than that of ions) create a conduction current, which, as we know, induces a magnetic field in addition to the magnetic field associated with the waves. The induced magnetic field in turn induces an electric field according to Faraday's law. As we will see, the electric field induced by the electron motion always tends to oppose the electric field of the incident wave. In other words, the incident wave would encounter difficulties whenever it tries to penetrate into the plasma. Under certain conditions the wave just gives up and is reflected from the plasma.

The same argument can be applied to electromagnetic wave reflection from a metal surface. Metals are characterized by the presence of large numbers of conduction electrons. For example, copper has about 10^{23} conduction electrons in 1 cm^3. These mobile electrons can easily respond to an incident electromagnetic field, preventing it from penetrating into the conductor.

As far as the wave equation for electromagnetic waves is concerned, all that is necessary is to add another current to the Ampere–Maxwell law, which had only the displacement current density

$$\varepsilon_0 \frac{\partial E}{\partial t} \quad (\text{A/m}^2)$$

in vacuum. In the presence of a conduction current density, J (A/m^2), the total current density thus becomes

$$\varepsilon_0 \frac{\partial E}{\partial t} + J$$

and the Ampere–Maxwell equation now becomes

$$-\frac{\partial B}{\partial x} = \mu_0 \left(\varepsilon_0 \frac{\partial E}{\partial t} + J \right) \tag{9.39}$$

Then if we can somehow express the conduction current J in terms of either E or B, our problem will be solved.

To do this, let us consider how the electrons are accelerated by the electric field E. Since the electric field E exerts a force $-eE$ ($e = 1.6 \times 10^{-19}$ C) on an electron, the equation of motion for the electron can be written as

$$m_e \frac{\partial u}{\partial t} = -eE \quad (u = \text{electron velocity}) \tag{9.40}$$

We used the partial derivative $\partial/\partial t$ since we expect that the electrons would follow the wave motion of the electric field, which is a function of both the spatial coordinate x and time t. If we multiply Eq. (9.40) by the electron density n_0 (m^{-3}) and the charge of the electron $-e$, we obtain

$$m_e \frac{\partial J}{\partial t} = n_0 e^2 E \tag{9.41}$$

since the conduction current density J is given by

$$J = -n_0 e u \ (\text{A/m}^2)$$

On the other hand, the time derivative of Eq. (9.39) is

$$-\frac{\partial^2 B}{\partial t \, \partial x} = \mu_0 \left(\varepsilon_0 \frac{\partial^2 E}{\partial t^2} + \frac{\partial J}{\partial t} \right) \tag{9.42}$$

Substituting $\partial J/\partial t$ from Eq. (9.41) into Eq. (9.42), we find

$$-\frac{\partial^2 B}{\partial t \, \partial x} = \mu_0 \varepsilon_0 \left(\frac{\partial^2 E}{\partial t^2} + \frac{n_0 e^2}{m_e \varepsilon_0} E \right) \tag{9.43}$$

Faraday's induction equation is unchanged,

$$-\frac{\partial E}{\partial x} = \frac{\partial B}{\partial t} \tag{9.44}$$

or taking the spatial derivative,

$$-\frac{\partial^2 E}{\partial x^2} = \frac{\partial^2 B}{\partial x \, \partial t} \tag{9.45}$$

After eliminating $\partial^2 B/\partial x \, \partial t$ between Eqs. (9.43) and (9.45), we finally obtain

$$\frac{\partial^2 E}{\partial x^2} = \mu_0 \varepsilon_0 \left(\frac{\partial^2 E}{\partial t^2} + \frac{n_0 e^2}{m_e \varepsilon_0} E \right) \tag{9.46}$$

This is our desired differential equation for electromagnetic waves in a plasma. Of course, it reduces to the wave equation we derived before if there are no plasma electrons, $n_0 = 0$. Also note that in the equation of motion, we have assumed the free acceleration of electrons and have neglected collisions between electrons and other particles in the plasma. We still call Eq. (9.46) a wave equation although it has one additional term due to the conduction current.

To see that the term due to the plasma electrons indeed tends to reduce the total electric field, let us assume a sinusoidal wave,

$$E(x, t) = E_0 \sin(kx - \omega t)$$

Obviously,

$$\frac{\partial^2 E}{\partial t^2} = -\omega^2 E_0 \sin(kx - \omega t)$$

But the last term in Eq. (9.46) always has the opposite sign with respect to the term $\partial^2 E / \partial t^2$, since the quantity $n_0 e^2 / m_e \varepsilon_0$ is positive definite. Thus the reaction effect of the plasma electrons is clearly seen. Plasma electrons tend to reduce the effective amplitude of the electric field and wave propagation from vacuum into a plasma must encounter some difficulty.

The quantity

$$\frac{n_0 e^2}{m_e \varepsilon_0}$$

has the dimension of $\sec^{-2}$, as can easily be checked. The square root of this is called the angular electron plasma frequency and is written as

$$\omega_p = \left(\frac{n_0 e^2}{m_e \varepsilon_0} \right)^{1/2} \quad \text{(rad/sec)} \tag{9.47}$$

The corresponding electron plasma frequency in hertz is

$$\nu_p = \frac{1}{2\pi} \left(\frac{n_0 e^2}{m_e \varepsilon_0} \right)^{1/2} \quad \text{(Hz)} \tag{9.48}$$

or

$$\nu_p = \frac{\omega_p}{2\pi} = 8.97 \sqrt{n_0} \quad \text{(Hz)} \tag{9.49}$$

where the electron density n_0 is in m^{-3}. The ionospheric plasma has the electron density in the range of $10^{11} \sim 10^{13}$ m^{-3}. The corresponding plasma frequency range is $2.8 \sim 28$ MHz. In metals the electron density is much higher and is of the order of 10^{29} m^{-3}. The plasma frequency of metals is then around 3×10^{15} Hz which is higher than the frequency of visible light, 5×10^{14} Hz.

Let us now find the dispersion relation, the relationship between ω and k of electromagnetic waves in a plasma. For a sinusoidal wave of the form

$$E(x, t) = E_0 \sin(kx - \omega t)$$

Eq. (9.46) yields

$$\omega^2 = k^2 c^2 + \omega_p^2 \tag{9.50}$$

A rough sketch of the frequency ω as a function of the wavenumber k is shown in Figure 9–30. Note that we now have a minimum frequency of electromagnetic waves that can exist in a plasma. The cutoff frequency is given by the electron plasma frequency itself. We have no solutions for ω below ω_p; that is, waves with a frequency below ω_p cannot exist in a plasma. Now we see why a plasma can reflect certain electromagnetic waves. Waves with a

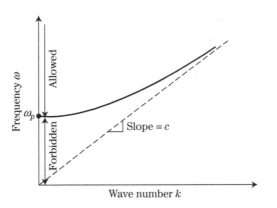

FIGURE 9–30

Dispersion relation for electromagnetic waves in a plasma.

frequency well above the plasma frequency have no difficulties in penetrating into a plasma, but those with a frequency below the plasma frequency must be completely reflected at the boundary. Visible light has a frequency around 5×10^{14} Hz, which is much higher than the electron plasma frequency of the ionospheric plasma. Thus light should have no difficulties in penetrating and propagating through the ionospheric plasma. On the other hand, shortwave radio communication uses the frequency band between 3 and 30 MHz, which is subject to reflection by the ionosphere (Figure 9–31).

The dispersion relation we obtained has a peculiar property. The phase velocity ω / k becomes

$$\frac{\omega}{k} = \frac{\sqrt{k^2 c^2 + \omega_p^2}}{k} = c\sqrt{1 + \frac{\omega_p^2}{c^2 k^2}} \tag{9.51}$$

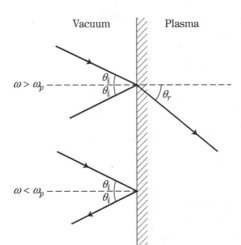

FIGURE 9–31

Wave reflection from a plasma. Note that the refraction angle θ_r is *larger* than the incident angle θ_i for a plasma in contrast with the refraction of visible light at the surface of water or glass. This means that we cannot make a converging lens using a plasma.

which is faster than the speed of light c. Is this not in grave contradiction to what Einstein found? According to him, nothing can travel faster than light. What travels with the electromagnetic waves? Energy does. But we know that energy is carried at the group velocity $d\omega/dk$ rather than the phase velocity. The group velocity for $\omega = \sqrt{k^2c^2 + \omega_p^2}$ is

$$\frac{d\omega}{dk} = \frac{kc^2}{\sqrt{k^2c^2 + \omega_p^2}} = \frac{kc^2}{\omega} \tag{9.52}$$

which can never exceed the velocity of light c. Thus the dispersion relation does not lead to a contradiction with Einstein's relativity theory.

EXAMPLE 9.9

The dispersion relation in Eq. (9.50) can be used to measure the average plasma electron density. Let a high-frequency (microwave) electromagnetic wave whose frequency is higher than the plasma frequency go through a plasma slab having an electron density n and a width d (Figure 9–32). In the absence of the plasma, the phase change due to the propagation over the distance d is $k_0d = (\omega/c)d$, where $k_0 = \omega/c$.

(a) What is the phase change in the presence of the plasma?
(b) Show that when $\omega^2 \gg \omega_p^2$, the difference between $(\omega/c)d$ and the phase change found in (a) become proportional to the electron density.

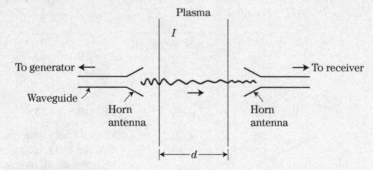

FIGURE 9–32
Plasma diagnostics.

Solution

(a) From Eq. (9.50), we obtain

$$k = \frac{1}{c}\sqrt{\omega^2 - \omega_p^2}$$

Then

$$kd = \frac{d}{c}\sqrt{\omega^2 - \omega_p^2} = \frac{d\omega}{c}\sqrt{1 - (\omega_p/\omega)^2}$$

(b) If $\omega^2 \gg \omega_p^2$, we may expand $\sqrt{1 - (\omega_p/\omega)^2}$ as

$$\sqrt{1 - \left(\frac{\omega_p}{\omega}\right)^2} \simeq 1 - \frac{1}{2}\left(\frac{\omega_p}{\omega}\right)^2 \quad \text{(binomial expansion)}$$

Then

$$kd = \frac{d\omega}{c}\left[1 - \frac{1}{2}\left(\frac{\omega_p}{\omega}\right)^2\right]$$

The difference between $k_0 d$ and kd becomes

$$k_0 d - kd = \frac{1}{2}\frac{d}{c}\frac{\omega_p^2}{\omega}$$

in which d, c, and ω are fixed parameters. Then the phase difference is proportional to ω_p^2, or the electron density. (This phase difference can be measured electronically, thus enabling us to determine the plasma electron density. This diagnostic technique is widely used in plasma research.)

We saw that electromagnetic fields cannot penetrate into metals. Then how can we use copper wires as electrical conductors at all? For the conduction electrons in metals to carry a current, we must have an electric field in metals. The dispersion relation we derived tells us no electromagnetic fields with frequencies below the plasma frequency of metals can exist in metals. For copper, $\nu_p \simeq 3 \times 10^{15}$ Hz, but we frequently use copper wires for transmitting 60-Hz (or 50-Hz) electromagnetic waves (commercial electrical power) from power plants to wherever it is used. When we state that electromagnetic waves cannot penetrate into metals, we should have in mind perfectly conducting metals or ideal conductors with zero resistivity. Such conductors exist and are used to produce a high magnetic field. Copper at room temperature has a small but finite resistivity. Its value is 1.7×10^{-8} $\Omega \cdot$ m at $20°$C and is one of the smallest among conventional metals. If the resistivity is finite, a dc electric field can fully penetrate into a metal. The electric field E_{dc} is related to the current density through

$$E_{dc} = \eta J_{dc} \tag{9.53}$$

where η is the resistivity of the material (Ω m). (You should check that this equation is dimensionally correct.) This is the microscopic form of Ohm's law. If we multiply Eq. (9.53) by the length l of a rod, we find (Figure 9–33)

$$E_{dc}l = V_{dc} = \eta l J_{dc} = \frac{\eta l}{A} A J_{dc} = R I_{dc}$$

which is the macroscopic form of the well-known Ohm's law.

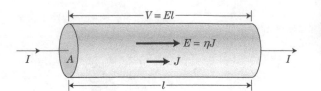

FIGURE 9–33

Current density J and the electric field intensity E in a resistive rod.

Now we will see what happens to electromagnetic waves in a metal if it has a finite resistivity. Equation (9.41) was obviously for the case of infinite conductivity or zero resistivity. But we now know that in the case of a dc field or when $\partial J/\partial t = 0$ ($J \neq 0!$) Eq. (9.53) must hold. Then, Eq. (9.41) should be modified as

$$m_e \frac{\partial J}{\partial t} = n_0 e^2 (E - \eta J) \qquad (9.54)$$

This equation is also derivable directly from the equation of motion in which we assume a finite collision frequency for the electrons,

$$m_e \frac{\partial u}{\partial t} = -eE - m_e v_c u$$

where v_c (s^{-1}) is a measure of how frequently electrons collide with either ions or neutrals to lose momentum. Multiplying this by $-n_0 e$, we obtain

$$m_e \frac{\partial J}{\partial t} = n_0 e^2 E - m_e v_c J = n_0 e^2 \left(E - \frac{m_e v_c}{n_0 e^2} J \right)$$

which, when compared with Eq. (9.54), indicates that the resistivity η is given by

$$\eta = \frac{m_e v_c}{n_0 e^2} \; (\Omega \cdot m)$$

Therefore the resistivity is a direct consequence of electron collisions. For copper, $\eta = 1.7 \times 10^{-8} \Omega \cdot m$, and $n_0 \simeq 10^{29}$ m^{-3}. Substituting $m_e = 9.1 \times 10^{-31}$ kg and $e = 1.6 \times 10^{-19}$ coulombs, we estimate the collision frequency of an electron in copper to be

$$v_c = \frac{\eta n_0 e^2}{m_e} = \frac{1.7 \times 10^{-8} \times 10^{29} \times (1.6 \times 10^{-19})^2}{9.1 \times 10^{-31}} = 4.8 \times 10^{13} \; \text{sec}^{-1}$$

In the limit of no time variation ($\partial/\partial t = 0$, or $\omega = 0$) in Eq. (9.54), we indeed recover Ohm's law $E = \eta J$ as required by Eq. (9.53). Thus our equations for the three field quantities, J, E, and B, are Eq. (9.54) and

$$-\frac{\partial E}{\partial x} = \frac{\partial B}{\partial t} \frac{\partial^2 \Omega}{\partial u^2} \qquad (9.44)$$

$$-\frac{\partial B}{\partial x} = \mu_0 \left(\varepsilon_0 \frac{\partial E}{\partial t} + J \right) \qquad (9.39)$$

As before, to derive a wave equation for one field quantity, say the current density J, we have to eliminate E and B from these three equations.

The magnetic field B can easily be eliminated as before:

$$\frac{\partial^2 E}{\partial x^2} = \mu_0 \left(\varepsilon_0 \frac{\partial^2 E}{\partial t^2} + \frac{\partial J}{\partial t} \right) \tag{9.55}$$

From Eq. (9.54), we have

$$E = \frac{m_e}{n_0 e^2} \frac{\partial J}{\partial t} + \eta J \tag{9.56}$$

Substituting Eq. (9.56) into (9.55), we obtain

$$\frac{\partial^2}{\partial x^2} \left(\frac{m_e}{n_0 e^2} \frac{\partial J}{\partial t} + \eta J \right) = \varepsilon_0 \mu_0 \frac{\partial^2}{\partial t^2} \left(\frac{m_e}{n_0 e^2} \frac{\partial J}{\partial t} + \eta J \right) + \mu_0 J \tag{9.57}$$

This looks terrible, but don't worry. The term $(m_e/n_0 e^2)(\partial J/\partial t)$ can safely be neglected compared with the term ηJ under most practical conditions. Assuming a sinusoidal current form, $J(x, t) = J_0 \sin(kx - \omega t)$, we see that the amplitude of $m_e/n_0 e^2(\partial J/\partial t)$ is $(m_e \omega/n_0 e^2)J_0$, which is to be compared with ηJ_0. For copper, we have $n_0 = 1 \times 10^{29}$ m^{-3} and $\eta = 1.7 \times 10^{-8}$ $\Omega \cdot$m. Even in the frequency range of microwaves, $\omega = 10^{10} - 10^{12}$ rad/sec, $m_e \omega/n_0 e^2$ is

$$\frac{9.1 \times 10^{-31} \times 10^{12}}{1 \times 10^{29} \times (1.6 \times 10^{-19})^2} = 3.6 \times 10^{-10} \Omega \cdot \text{m}$$

which is much less than the resistivity of copper. Thus the term $(m_e/n_0 e^2)\partial J/\partial t$ can be neglected in comparison with ηJ in Eq. (9.57), and we have

$$\frac{\partial^2 J}{\partial x^2} = \varepsilon_0 \mu_0 \frac{\partial^2 J}{\partial t^2} + \frac{\mu_0}{\eta} \frac{\partial J}{\partial t} \tag{9.58}$$

Next compare the two terms on the right-hand side of this equation. The amplitude of the first term is $\varepsilon_0 \mu_0 \omega^2 J_0$ and that of the second term is $\mu_0 \omega J_0/\eta$. We see that the first term can safely be neglected even in the microwave frequency range, $\omega = 10^{12}$ rad/sec. Thus the original differential equation, Eq. (9.57), has now been greatly simplified to be

$$\frac{\partial^2 J}{\partial x^2} = \frac{\mu_0}{\eta} \frac{\partial J}{\partial t} \tag{9.59}$$

Of course, the other fields, E and B, should be described by the same differential equation,

$$\frac{\partial^2 E}{\partial x^2} = \frac{\mu_0}{\eta} \frac{\partial E}{\partial t} \quad \text{and} \quad \frac{\partial^2 B}{\partial x^2} = \frac{\mu_0}{\eta} \frac{\partial B}{\partial t} \tag{9.60}$$

A differential equation of the preceding form is called *a diffusion equation*. It describes, for example, how rapidly an ink drop placed in water spreads (or diffuses) and is one of the most important equations we have in physics.

The solutions of Eqs. (9.59) and (9.60) can be found for a sinusoidal "wave" without too much difficulty. The term *wave* is used here in a somewhat broader sense than before. The equations contain the first-order time derivatives only and do not have the second-order time derivatives present in the usual wave equations

$$\frac{\partial^2 E}{\partial t^2} = c_w^2 \frac{\partial^2 E}{\partial x^2}$$

Let us assume a solution of the form

$$E(x, t) = A(x)\sin(kx - \omega t) \tag{9.61}$$

for the differential equation for the electric field,

$$\frac{\partial^2 E}{\partial x^2} = \frac{\mu_0}{\eta}\frac{\partial E}{\partial t} \tag{9.62}$$

Since

$$\frac{\partial E}{\partial x} = \frac{dA}{dx}\sin(kx - \omega t) + kA\cos(kx - \omega t)$$

$$\frac{\partial^2 E}{\partial x^2} = \frac{d^2 A}{dx^2}\sin(kx - \omega t) + 2k\frac{dA}{dx}\cos(kx - \omega t) - k^2 A\sin(kx - \omega t)$$

and

$$\frac{\partial E}{\partial t} = -\omega A\cos(kx - \omega t)$$

Eq. (9.62) becomes

$$\left(\frac{d^2 A}{dx^2} - k^2 A\right)\sin(kx - \omega t) + \left(2k\frac{dA}{dx} + \frac{\omega\mu_0}{\eta}A\right)\cos(kx - \omega t) = 0 \tag{9.63}$$

Since Eq. (9.63) must hold for any values of x and t, we must have

$$\frac{d^2 A}{dx^2} - k^2 A = 0 \tag{9.64}$$

and

$$2k\frac{dA}{dx} + \frac{\omega\mu_0}{\eta}A = 0 \tag{9.65}$$

These are mere ordinary differential equations. We assume $A(x) = A_0 e^{-\gamma x}$. Then

$$\frac{dA}{dx} = -\gamma A_0 e^{-\gamma x}$$

and

$$\frac{d^2 A}{dx^2} = \gamma^2 A_0 e^{-\gamma x}$$

Substituting these expressions into Eqs. (9.64) and (9.75), we find

$$\gamma^2 = k^2 \tag{9.66}$$

and

$$-2\gamma k + \frac{\omega \mu_0}{\eta} = 0 \tag{9.67}$$

Then

$$\gamma = k = \sqrt{\frac{\omega \mu_0}{2\eta}}$$

and the electric field becomes

$$E(x, t) = A_0 e^{-kx} \sin(kx - \omega t) \tag{9.68}$$

where

$$k = \sqrt{\frac{\omega \mu_0}{2\eta}} \tag{9.69}$$

A sketch of the preceding solution at a certain time (snapshot) is shown in Figure 9–34. We see that the electric field in the metal spatially damps in an

FIGURE 9–34

Skin effect at a metal surface.

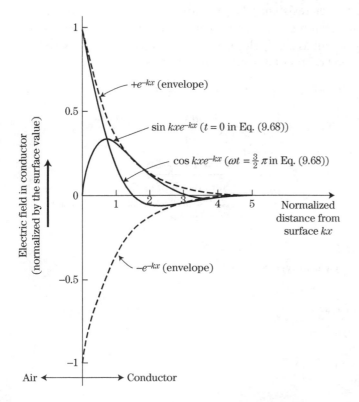

exponential manner. The envelope becomes $1/e = 1/2.7182 \simeq 0.37 = 37\%$ of the value at the metal surface ($x = 0$) at a distance

$$\delta = \sqrt{\frac{2\eta}{\omega\mu_0}} \tag{9.70}$$

This quantity δ is called the *skin depth* and indicates how deep an electromagnetic wave can penetrate into a metal. In the limit of $\omega \to 0$ (dc), δ becomes infinitely large. This is consistent with our experience. A dc field does not have any difficulty in penetrating into a metal although it takes a finite time (skin time) for a dc field to penetrate. As ω increases, the skin depth becomes finite, being inversely proportional to the square root of the frequency. Even at 60 Hz, the skin depth of copper is only 8.5 mm. Thus it is meaningless to make the diameter of high-voltage transmission lines much more than the skin depth since the current is practically limited to exist within a skin depth. As the frequency becomes higher, the skin depth becomes smaller (Figure 9–35). At 1 MHz (AM radio waves), the skin depth is 0.07 mm; that is, the current can flow practically only at the surface of copper.

You must wonder why a higher resistivity results in less spatial damping, as indicated by Eq. (9.70). This is quite opposite to the case of an *RLC* circuit, in which a higher resistance (R) makes the oscillation damp faster (Figure 9–36). Let us see if we can construct an equation similar to Eq. (9.62) using the preceding circuit as a unit element of a transmission line. As before, Kirchhoff's voltage and current laws yield (Figure 9–37)

$$V(x) = V(x + \Delta x) + L\frac{\partial I}{\partial t} + RI$$

$$I(x) = I(x + \Delta x) + C\frac{\partial V}{\partial t}$$

FIGURE 9–35

Skin depth of copper at room temperature as a function of wave frequency.

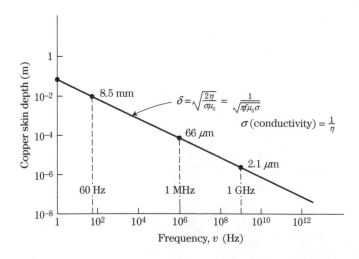

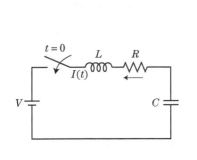

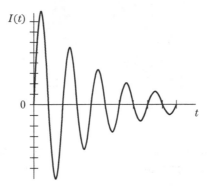

FIGURE 9–36

Damped oscillation of an *RLC* circuit.

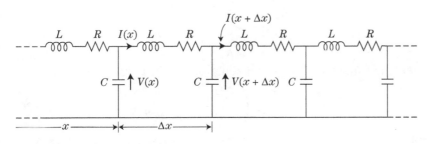

FIGURE 9–37

An incorrect model of a transmission line for simulating the skin effect.

or

$$-\Delta x \frac{\partial V}{\partial x} = L\frac{\partial I}{\partial t} + RI \tag{9.71}$$

$$-\Delta x \frac{\partial I}{\partial x} = C\frac{\partial V}{\partial t} \tag{9.72}$$

From these, we find

$$\frac{\partial^2 V}{\partial x^2} = \frac{C}{(\Delta x)^2}\left(L\frac{\partial^2 V}{\partial t^2} + R\frac{\partial V}{\partial t}\right) \tag{9.73}$$

which is clearly of the form of Eq. (9.20) in which we neglected the term

$$\varepsilon_0 \mu_0 \frac{\partial^2 J}{\partial t^2}$$

compared with $\mu_0/\eta(\partial J/\partial t)$ In other words, we neglected the term associated with the capacitance ε_0 (F/m). In Eq. (9.73), we cannot do this because if we neglect the capacitance ($C \to 0$), we lose all the terms on the right-hand side and end up with a trivial equation

$$\frac{\partial^2 V}{\partial x^2} = 0$$

Thus our model is obviously wrong.

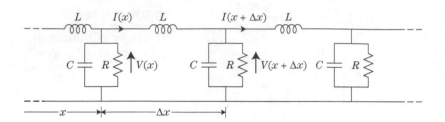

FIGURE 9–38

Correct model: The resistance is in parallel with the capacitance.

A correct model is the one that has a resistance in parallel with the capacitance. For the transmission line (see Figure 9–38), we have

$$-\Delta x \frac{\partial V}{\partial x} = L \frac{\partial I}{\partial t} \tag{9.74}$$

$$-\Delta x \frac{\partial I}{\partial x} = \frac{V}{R} + C \frac{\partial V}{\partial t} \tag{9.75}$$

These yield the following differential equation for the current I

$$\frac{\partial^2 I}{\partial x^2} = \frac{L}{(\Delta x)^2} \left(C \frac{\partial^2 I}{\partial t^2} + \frac{1}{R} \frac{\partial I}{\partial t} \right) \tag{9.76}$$

This looks better, since we now have the term involving the capacitance separated from the term involving the resistance. In fact, if we define $L/\Delta x$ (inductance per unit length), $C/\Delta x$ (capacitance per unit length), and $1/R\Delta x$ (conductance per unit length), and replace $L/\Delta x$ by μ_0, $C/\Delta x$ by ε_0, and $1/R\Delta x$ by $1/\eta$, Eq. (9.38) exactly corresponds to Eq. (9.20). In the transmission line shown (Figure 9–38), it is obvious that as the resistance increases, the line dissipates less energy (since the resistance is in parallel) and the energy can be transferred further spatially, which indicates less spatial damping!

Problems

1. An LC transmission line has the following parameters:

$$\frac{L}{\Delta x} = 1.0 \times 10^{-4} \text{H/m}, \quad \frac{C}{\Delta x} = 20 \times 10^{-12} \text{F/m}$$

Find the velocity of electromagnetic waves on the transmission line. What is the impedance?

2. It is required that the velocity of electromagnetic waves in a coaxial cable is one-half of the velocity of light in vacuum. What dielectric material should be used? Assume $\mu = \mu_0$.

3. Consider the inductance of an LC transmission line having a leakage (or stray) capacitance C_2 (see Figure 9–39).

(a) Show that the differential equation for the voltage is given by

$$\frac{\partial^2 V}{\partial t^2} = \frac{(\Delta x)^2}{LC_1} \frac{\partial^2 V}{\partial x^2} + \frac{C_2}{C_1} (\Delta x)^2 \frac{\partial^4 V}{\partial t^2 \partial x^2}$$

(b) Assuming a sinusoidal voltage wave $V(x, t) = V_0 \sin(kx - \omega t)$, show that the dispersion relation is

$$\omega^2 = \frac{1}{1 + (C_1/C_2)(\Delta x)^2 k^2} \frac{(\Delta x)^2}{LC_1} k^2$$

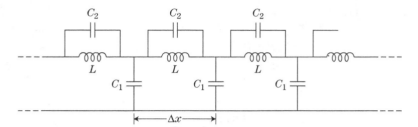

FIGURE 9–39

Problem 3.

Note: This indicates that if there is a stray capacitance across the inductance in each section, the waves are no longer dispersionless. The phase velocity ω/k is now different from the group velocity $d\omega/dk$, as can easily be checked. The dispersion relation can model interesting waves we encounter in several branches of physics. For example, shallow water waves can be well approximated with this dispersion relation. Another example is the sound wave in an ionized gas (plasma), which is called the ion acoustic wave.

4. A coaxial cable has an inner radius of 2×10^{-4} m and an outer radius of 3×10^{-3} m and is filled with a dielectric material of $\varepsilon_r = 2.0$. Find
   (a) The velocity of electromagnetic waves in the cable.
   (b) The characteristic impedance of the cable.

5. A coaxial cable is desired to have 50-Ω characteristic impedance. If the radius of the outer conductor is 2 mm, and the propagation velocity is to be $0.7c$, what dielectric material should be used? What is the inner radius?

6. From the dimensions for E and B, show that the Poynting flux S indeed has the dimensions of watts per square meter. (Introduce a vector $\mathbf{H} = \mathbf{B}/\mu_0$ having a dimension of amperes per meter.)

7. Parallel-wire transmission lines are most commonly used for low-frequency (including dc) power transfer. In the schematic diagram of Figure 9–40, sketch roughly the electric and magnetic field profiles, and show that the Poynting vector is directed from the battery to the load everywhere. Would you expect that the 2 parallel wires will be attracted to each other

or repelled from each other because of the magnetic force? What about the electric force?

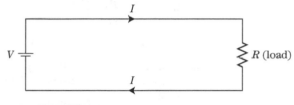

FIGURE 9–40

Problem 7.

8. A long densely wound solenoid is storing magnetic energy. Discuss the mechanism of energy flow in terms of the Poynting vector.

9. In Eq. (9.31), the electric field is in the y direction and the magnetic field is in the z direction. Show that Eq. (9.31) is equivalent to

$$\begin{vmatrix} \mathbf{u_x} & \mathbf{u_y} & \mathbf{u_z} \\ \partial/\partial x & \partial/\partial y & \partial/\partial z \\ 0 & E & 0 \end{vmatrix} = -\frac{\partial B}{\partial t}\mathbf{u_z}$$

if we assume $\partial E/\partial z = 0$. Similarly, Eq. (9.32) may be written as

$$\begin{vmatrix} \mathbf{u_x} & \mathbf{u_y} & \mathbf{u_z} \\ \partial/\partial x & \partial/\partial y & \partial/\partial z \\ 0 & 0 & B \end{vmatrix} = \varepsilon_0\mu_0\frac{\partial E}{\partial t}\mathbf{u_y}$$

with $\partial B/\partial y = 0$. In general,

$$\begin{vmatrix} \mathbf{u_x} & \mathbf{u_y} & \mathbf{u_z} \\ \partial/\partial x & \partial/\partial y & \partial/\partial z \\ A_x & A_y & A_z \end{vmatrix} = \left(\frac{\partial A_z}{\partial y} - \frac{\partial A_y}{\partial z}\right)\mathbf{u_x}$$

$$+ \left(\frac{\partial A_x}{\partial z} - \frac{\partial A_z}{\partial x}\right)\mathbf{u_y} + \left(\frac{\partial A_y}{\partial x} - \frac{\partial A_x}{\partial y}\right)\mathbf{u_z}$$

and this vector differentiation is written as

$$\nabla \times \mathbf{A} \text{ (or curl } \mathbf{A})$$

Using ∇, we may generalize Eq. (9.31) as

$$\nabla \times \mathbf{E} = -\frac{\partial \mathbf{B}}{\partial t}$$

and Eq. (9.32) as

$$\nabla \times \mathbf{B} = \varepsilon_0 \mu_0 \frac{\partial \mathbf{E}}{\partial t}$$

These are called Maxwell's equations in free space in which no conduction currents can exist. In the presence of a conduction current, the second equation should be generalized as

$$\nabla \times \mathbf{B} = \mu_0 \left(\varepsilon_0 \frac{\partial \mathbf{E}}{\partial t} + \mathbf{j} \right)$$

where $\mathbf{j}$ is the conduction current density (A/m^2).

10. A giant laser pulse has a power density 10^{20} W/m^2. Calculate the rms value of the electric field associated with the laser pulse.

11. A radio station is emitting 50 kW radio waves spherically.* Find the rms value of the electric field 1 mi from the station.

12. A 1-g target completely absorbs the energy of a laser pulse (500 MW, 10-nsec duration). Find the momentum to be gained by the target and the velocity.

13. Discuss how the current pulse develops after the switch is closed for $R = 25\Omega$, 50Ω, and 100Ω (Figure 9–41). Note that there is no reflection at the battery end. Does the final current reduce to what you expect?

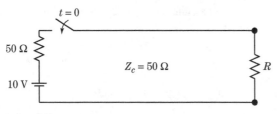

FIGURE 9–41

Problem 13.

14. Repeat Problem 13 for the case of Figure 9–42. The current should eventually approach 1 A. Does it?

*As we will see in Chapter 10, a radio station cannot radiate spherically or in every direction.

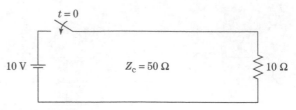

FIGURE 9–42

Problem 14.

15. Derive Eq. (9.38).

16. A coaxial cable has the same characteristic impedance as free space, $Z_0 = 377\Omega$. Can we conclude that electromagnetic waves reaching an open end of the cable should suffer no reflection since the impedances are the same? (The answer is no. Why?)

17. Derive Eq. (9.50).

18. The dispersion relation of electromagnetic waves in a plasma Eq. (9.50) does not allow a solution for ω below the plasma frequency ω_p. Suppose an electromagnetic wave of a frequency much less than ω_p is incident on plasma. Neglecting the term $\partial^2 E / \partial t^2$ compared with $\omega_p^2 E$ in Eq. (9.46), show that the solution for E can be written as

$$E = E_0 e^{-\gamma x} \sin \omega t$$

where $\gamma = \omega_p / c$. The quantity c/ω_{pe} is called the skin depth of a plasma with no resistivity. Evaluate this quantity for the ionospheric plasma assuming $n_0 = 10^{12} \text{m}^{-3}$.

19. Discuss the tunnel effect of electromagnetic waves through a plasma (Figure 9–43). If the plasma is semi-infinite, or if the thickness d is much larger than the skin depth c/ω_p found in Problem 18, the incident wave is one hundred percent reflected. However, if the thickness d is shorter than the skin depth, some energy can go through the plasma, and the reflection is no longer one hundred percent. Can you see why?

20. Show that the energy density of a plane electromagnetic wave in a plasma with amplitude of electric field E_0 given by

$$\frac{1}{4}\varepsilon_0 E_0^2 \left(1 + \frac{\omega_p^2}{\omega^2} + \frac{\omega^2 - \omega_p^2}{\omega^2} \right) = \frac{1}{2}\varepsilon_0 E_0^2 \qquad \text{(A)}$$

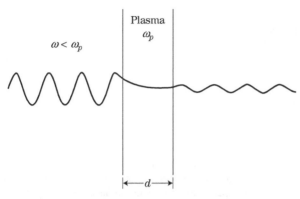

$\omega < \omega_p$

FIGURE 9–43

Problem 19.

Equation (A) indicates the energy partition among the electric field energy, kinetic energy of electrons, and magnetic field energy.

21. Waves tend to be bent toward a region of lower phase velocity. Usually the electron density in the ionospheric plasma increases with the altitude. Consider an oblique incident wave. See why a wave with a frequency higher than the plasma frequency can be reflected back toward the earth by the ionosphere (Figure 9–44).
Note: This bending phenomenon is known as wave refraction. In fact, a shortwave radio with frequencies higher than the plasma frequency of the ionosphere can be effectively reflected. The cutoff phenomenon at the plasma frequency can be well-defined only for normal incidence. Of course, the refraction depends on the wave frequency, and for waves with frequencies much higher than ω_p, the

refraction becomes ineffective. Visible light can go through the ionosphere at any angle of incidence.

A mirage is caused by the same mechanism. (On a hot summer day the highway surface often becomes a perfect mirror, as you must have experienced.) The speed of light in air is slightly less than in vacuum, or the speed of light in air decreases as the air density increases. Any gas becomes less dense as it is heated. Thus the air density at the surface of a hot highway is less than that above where the air temperature is lower. Then the speed of light decreases with the height and light is refracted toward the region of slower velocity.

22. Derive Eqs. (9.73) and (9.76).

23. There is a simple reason for the RLC line in Figure 9–36 to be incorrect and that in Figure 9–37 to be correct. Explain why in terms of the transverse nature of electromagnetic waves.

24. Show that the electromagnetic waves in a plasma can be modeled by the following transmission line if L, C_0, and C_1 are chosen (see Figure 9–45) such that
$$\frac{(\Delta x)^2}{LC_0} = c^2 \quad \text{and} \quad \omega_p^2 = \frac{1}{LC_1}\frac{\partial E}{\partial t}$$

25. Find the wavelength of electromagnetic waves at which copper becomes transparent for the waves.

26. A beam of electromagnetic waves is incident on a plasma lens. Does the lens act as a converging lens or diverging lens?

27. Using the diffusion equation for a resistive medium
$$\frac{\partial^2 E}{\partial x^2} = \frac{\mu_0}{\eta}\frac{\partial E}{\partial t}$$

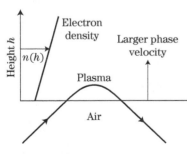

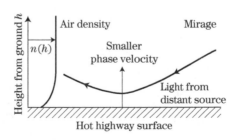

FIGURE 9–44

Problem 21.

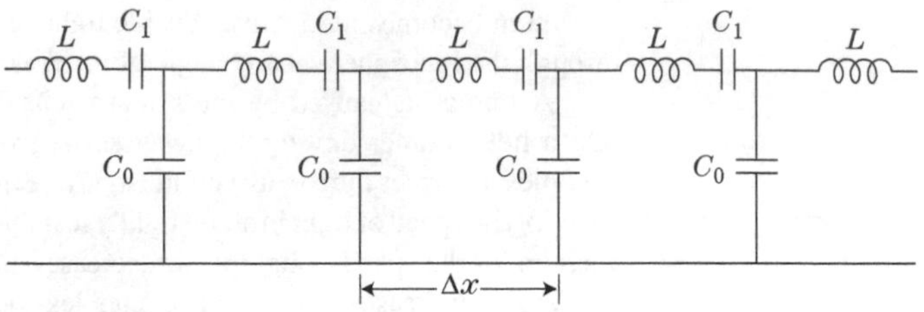

FIGURE 9–45

Problem 24.

estimate the time required for a dc electric field to penetrate fully into a copper slab 2 cm thick. Copper resistivity is $1.7 \times 10^{-8} \Omega \cdot$ m.

28. The resistivity of the earth falls in the range $10^{-2} - 10^{2} \Omega \cdot$ m. Assuming $\eta = 1 \Omega \cdot$ m, evaluate the earth skin depth as a function of frequency. (If the skin depth is much shorter than the wavelength in free space, the earth can be regarded as a good conductor and becomes an effective reflector for electromagnetic waves. For example, an antenna erected from the ground with a height h is effectively $2h$ long because of wave reflection, which is responsible for creation of an image of a conductor above the ground.)

Radiation of Electromagnetic Waves

Introduction

So far we have been discussing how electromagnetic waves behave in some media once they are created without investigating how they can be created. Radio and television waves are transmitted by antennas and are also received by antennas. In this chapter we describe the physical mechanisms behind the radiation of electromagnetic waves.

Fields Associated with a Stationary Charge and a Charge Moving with a Constant Velocity

Suppose we have a point source radiating electromagnetic waves spherically or isotropically in every direction. We saw in the section on the Poynting vector that for the point source, the Poynting flux should be proportional to r^{-2} where r is the distance between the source and the point of observation. The total power that is radiated from the source is conserved. This is written as

$$4\pi r^2 S = \text{power radiated by the source (constant)} \qquad (10.1)$$

unless the medium is absorbing energy. Since the Poynting flux is given by

$$S = c\varepsilon_0 E^2 \qquad (10.2)$$

197

we find that the amplitude of the electric field associated with a spherical wave must be proportional to $1/r$,

$$E \propto \frac{1}{r} \tag{10.3}$$

This dependence contrasts with the electric field due to a stationary point charge

$$E_{\text{stationary}} = \frac{1}{4\pi\varepsilon_0} \frac{q}{r^2} \tag{10.4}$$

which has r^{-2} dependence (Coulomb's law). We thus conclude that a stationary charge cannot radiate electromagnetic waves. We can also draw this conclusion from an alternative point of view. The Poynting vector is the product between the electric field E and the magnetic field B/μ_0. If a charge is not moving, we have no current and thus no magnetic field. The stationary electric field cannot induce the magnetic field either. Therefore the Poynting vector associated with a stationary charge must be zero and we have no energy flow or radiation.

What about a charge in motion with a constant velocity? We now have both an electric field and a magnetic field and the Poynting vector is expected to have a finite nonzero value. Suppose a positive charge q is moving in $+x$ direction with a constant velocity v that is nonrelativistic

$$v = \frac{dx}{dt} = \text{constant} \ll c$$

We consider a circular disk with radius ρ located at $x = 0$ on the x axis and facing perpendicular to the moving charge as shown in Figure 10–1. At a certain instant, the electric field at the edge of the disk is given by Coulomb's law,

$$E = \frac{q}{4\pi\varepsilon_0} \frac{1}{\rho^2 + x^2} \tag{10.5}$$

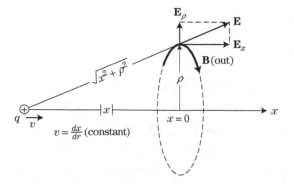

FIGURE 10–1

Electric and magnetic fields due to a charge moving with a constant velocity v.

and its x component is

$$E_x = \frac{q}{4\pi\varepsilon_0} \frac{-x}{\left(\rho^2 + x^2\right)^{3/2}} \tag{10.6}$$

The minus sign is needed because when the charge is to the left of the disk, x is negative and E_x is positive as is shown in the figure. Then the total electric flux *through* the disk is

$$\Phi_E = \int_0^\rho E_x 2\pi\rho \, d\rho = \frac{qx}{2\varepsilon_0} \int_0^\rho \frac{\rho}{\left(\rho^2 + x^2\right)^{3/2}} d\rho$$

$$= -\frac{qx}{2\varepsilon_0}\left(\frac{1}{|x|} - \frac{1}{\sqrt{\rho^2 + x^2}}\right)$$

which is time-varying since x is varying with time. Now we can apply the Ampere–Maxwell law for this time-varying electric flux to find the magnetic field at the edge of the disk.

$$2\pi\rho B = \varepsilon_0\mu_0 \frac{d\Phi_E}{dt} = \varepsilon_0\mu_0 \frac{d\Phi_E}{dx}\frac{dx}{dt} = \frac{\mu_0 q}{2}\frac{\rho^2}{\left(\rho^2 + x^2\right)^{3/2}} v$$

or

$$B = \frac{\mu_0 q v}{4\pi} \frac{\rho}{\left(\rho^2 + x^2\right)^{3/2}} \tag{10.7}$$

A quicker way to find this magnetic field is to use the Biot–Savart law,

$$d\mathbf{B} = \frac{\mu_0}{4\pi} \frac{I \, d\mathbf{l} \times \mathbf{r}}{r^3}$$

in which we replace $I d\mathbf{l}$ by $q\mathbf{v}$, and take $r = \sqrt{\rho^2 + x^2}$. This is a legitimate procedure and you can always replace $I d\mathbf{l}$ by $q\mathbf{v}$ in the Biot–Savart law whenever you want to find the magnetic field due to a charge moving with a constant velocity.

Since both the electric and magnetic fields are now found, we can calculate the Poynting vector at any point around the moving charge. Consider a sphere centered at the charge at a certain instant in time as shown in Figure 10–2. The electric field is normal to the sphere everywhere and the magnetic field is tangent to the sphere and is normal to the electric field. Therefore, the Poynting vector is tangent to the sphere and has no radial component. In other words, the Poynting vector never penetrates through the surface. We conclude that electromagnetic energy cannot be radiated by a charge that is moving with a constant velocity.

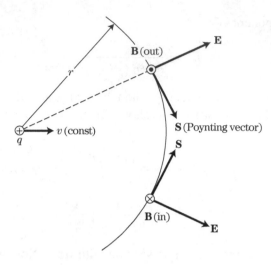

FIGURE 10–2

The Poynting vector associated with a charge moving with a constant velocity cannot go through a spherical surface centered at the charge. The charge cannot radiate.

EXAMPLE 10.1

Using the electric field Eq. (10.5) and the magnetic field Eq. (10.7), calculate the x component of the Poynting vector and then the total energy flow rate through an infinitely large plane placed at the location x that is normal to the x axis. Discuss the result.

Solution

The x component of the Poynting vector (see Figure 10–1) is given by

$$S_x = E_r \frac{B_0}{\mu_0} = \left[\frac{q}{4\pi\varepsilon_0} \frac{\rho}{\left(\rho^2 + x^2\right)^{3/2}} \right] \left[\frac{qv}{4\pi} \frac{\rho}{\left(\rho^2 + x^2\right)^{3/2}} \right] = \frac{q^2 v}{16\pi^2\varepsilon_0} \frac{\rho^2}{\left(\rho^2 + x^2\right)^3} \ (\text{W/m}^2)$$

Then the total energy flow rate becomes

$$\text{Power} = \int_0^\infty S_x 2\pi\rho \, d\rho = \frac{q^2 v}{8\pi\varepsilon_0} \int_0^\infty \frac{\rho^3}{\left(\rho^2 + x^2\right)^3} d\rho$$

The integration can be performed by successive integration by parts and the power is found to be

$$\frac{q^2 v}{32\pi\varepsilon_0} \frac{1}{x^2} \ (\text{W})$$

where $|x|$ is the instantaneous location of the charge as measured from the plane. Noting that

$$v = \frac{dx}{dt}$$

we may rewrite the power as

$$-\frac{q^2}{32\pi\varepsilon_0} \frac{d}{dt}\left(\frac{1}{x}\right) = -\frac{d}{dt}\left[\frac{q^2}{32\pi\varepsilon_0 x}\right]$$

However, the quantity $q^2/(32\pi\varepsilon_0 x_0)$ where $(x_0 > 0)$ is the electric energy that is stored in the space in the region $x > x_0$. Therefore the power calculated from the Poynting flux can be interpreted as the flow rate of the electrostatic energy that is stored in space and has nothing to do with radiation energy. The magnetic energy is of the order of $(v/c)^2 \times$ electric energy, and for a nonrelativistic velocity $v \ll c$, it is negligibly small.

Thus we have seen that (1) a *stationary* charge cannot radiate electromagnetic waves and (2) a charge in motion with a *constant velocity* also cannot either. A charge moving with a constant speed could describe the case in which a charge is going along a circular orbit that can radiate energy. What other situation can we have? We have not considered the more general case in which charged particles are subject to an acceleration or deceleration. This is exactly the case in which we can have radiation of electromagnetic waves. In radio antennas, electrons are forced to go back and forth in time harmonic motion. Electrons are accelerated back and forth with a signal generator that is connected to the antenna and we will find that it can radiate electromagnetic waves. Charged particles trapped in a magnetic field radiate electromagnetic waves at the cyclotron frequency or synchrotron frequency if the particle energy is relativistic. In this case, the centripetal acceleration associated with the circular motion of the charged particles is responsible for the radiation.

Radiation Fields Due to an Accelerated or a Decelerated Charge

Suppose a positive charge q originally at rest at point A is accelerated in the x direction as shown in Figure 10–3. The acceleration lasts for Δt seconds only until the charge reaches point B, after which the charge moves with a constant velocity $\mathbf{v} = \mathbf{a}\Delta t$. We assume nonrelativistic velocities in that the velocity is much less than the speed of light. Let us examine how the electric lines of force associated with the charge will look. The electric field lines of a charge that is stationary (at A) or moving with a constant velocity (at C) are just directed radially outward and are the Coulomb fields. The charge is accelerated from point A to point B. Thus the electric field lines when the charge was at A and at a point after the acceleration occurs, say at C, are all radially outward although they are not concentric. Since the lines of force must be continuous, these nonconcentric lines must be connected somehow. Therefore the effect of the acceleration appears as kinked electric lines of force as shown. *The kinks, which are disturbances in the electric field lines caused*

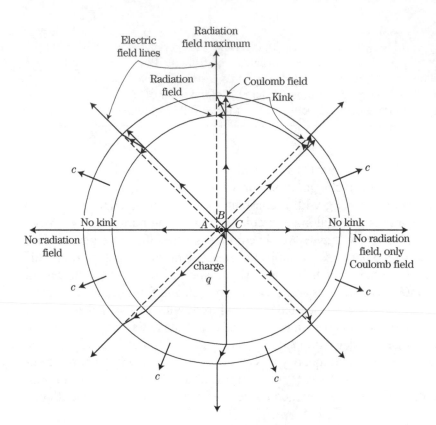

FIGURE 10–3

A charge under acceleration does radiate electromagnetic waves. Notice the kinks in the electric field lines.

by the acceleration of the charge, propagate with the speed of light. It takes Δt sec for the accelerated charge to move from A to B. The separation between the larger circle and the smaller circle is approximately $c\Delta t \simeq$ constant; if the charge is accelerated and moves sufficiently slowly, relativistic effects can be neglected. In the kinks, we obviously have electric field components that are perpendicular to the Coulomb fields. These transverse components are responsible for the radiation of electromagnetic waves.

Consider a point Q in Figure 10–4 that is normal to the velocity of the charge at a certain instant. Let t be the time after the charge passes point B (after the charge is accelerated). At Q, we have two electric fields. The first is the radially directed Coulomb field E_0 that is given by

$$E_0 = \frac{q}{4\pi\varepsilon_0}\frac{1}{r^2} = \frac{q}{4\pi\varepsilon_0}\frac{1}{(ct)^2} \tag{10.8}$$

The second component is the radiation field E_t, which is perpendicular to the Coulomb field. (The subscript t here indicates "transverse.") From the triangle OPQ, we identify the edge $OP = c\Delta t$ (constant) as being the distance in which

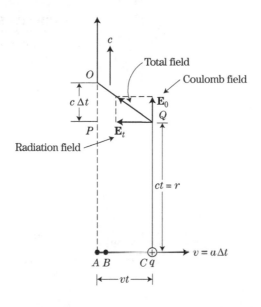

FIGURE 10–4

Coulomb field E_0 and the radiation field E_t that is perpendicular to the motion of the charge.

we have radiation. Since

$$\frac{OP}{PQ} = \frac{c\,\Delta t}{vt} = -\frac{E_0}{E_t} \tag{10.9}$$

we find

$$E_t = -\frac{q}{4\pi\varepsilon_0 c^2}\frac{v}{\Delta t}\frac{1}{r} \tag{10.10}$$

Recall that the charge was initially at rest and it was accelerated to a final velocity v. This acceleration is

$$a = \frac{dv}{dt} \simeq \frac{v_{\text{final}} - v_{\text{initial}}}{\Delta t} = \frac{v}{\Delta t}$$

This is exactly what we anticipated. The transverse (or radiation) electric field is proportional to the acceleration $a = v/\Delta t$ and $1/r$! (The minus sign in Eq. (10.10) is due to the direction of E_t which is opposite to the direction of the acceleration.)

At an arbitrary point, we have the electric fields at an arbitrary angle θ

$$E_t = -\frac{q}{4\pi\varepsilon_0}\frac{v\sin\theta}{c^2\Delta t r} \tag{10.11}$$

as is clear from Figure 10–5. In general, the radiation electric field due to a charge q under acceleration $\mathbf{a}$ (vector) is given by

$$\mathbf{E}_t = \frac{q}{4\pi\varepsilon_0 c^2}\frac{\mathbf{u_r} \times (\mathbf{u_r} \times \mathbf{a})}{r} \tag{10.12}$$

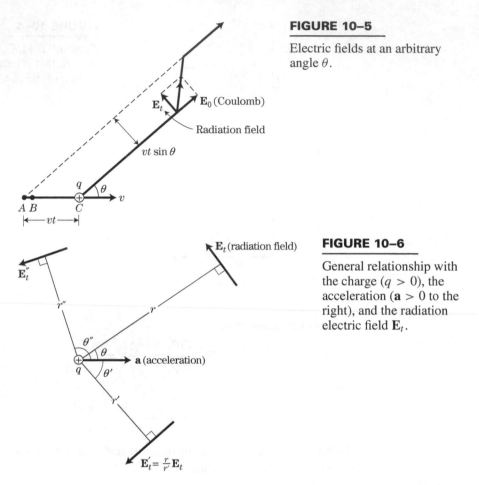

FIGURE 10–5

Electric fields at an arbitrary angle θ.

FIGURE 10–6

General relationship with the charge ($q > 0$), the acceleration ($\mathbf{a} > 0$ to the right), and the radiation electric field $\mathbf{E}_t$.

where the direction of $\mathbf{E}_t$ is normal to the radial vector $\mathbf{r}$, $\mathbf{u_t} = \mathbf{r}/r$ is the unit vector in the radial direction, and θ is the angle between the radial vector $\mathbf{r}$ and the acceleration $\mathbf{a}$. Note that the electric field depends on the sign of the charge q. If q is negative, the direction of the electric field must be reversed. Figure 10–6 shows the case of positive charge with acceleration in the positive x direction.

One point must be considered here with some caution. In general, the acceleration $\mathbf{a}$ is a function of time. For example, if the charge moves back and forth sinusoidally, the acceleration must also vary sinusoidally. Since the radiation reaches the point P after a time r/c, what an observer sees at the point P is the electric field due to the acceleration that occurred r/c sec before! Thus the correct expression for the electric field is

$$E_t\,(r, \theta, t) = -\frac{q}{4\pi\,\varepsilon_0 c^2}\,\frac{\sin\theta}{r}\,a_{t-r/c} \tag{10.13}$$

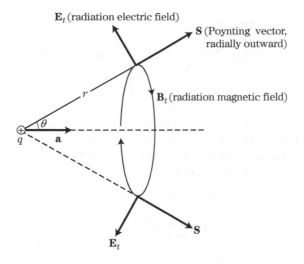

FIGURE 10–7

Radiation electric and magnetic fields, $\mathbf{E}_t$, $\mathbf{B}_t$, and the Poynting vector $\mathbf{S}$ due to an accelerated charge.

This electric field is called the *retarded electric field*. The term $a_{t-r/c}$ indicates the acceleration that occurred r/c seconds before relative to the electric field that is observed at the time t. It properly takes into account the effect of the finite propagation time of the electromagnetic radiation.

The radiation magnetic field associated with the radiation electric field can easily be calculated from (see Sections 9.2 and 9.3)

$$\mathbf{B}_t = \frac{1}{c}\mathbf{u_n} \times \mathbf{E}_t \qquad (10.14)$$

Its magnitude is

$$B_t\,(r, \theta, t) = \frac{q}{4\pi\varepsilon_0 c^2}\frac{\sin\theta}{r}a_{t-r/c} \qquad (10.15)$$

which is normal to the electric field and the radial vector $\mathbf{r}$. Figure 10–7 shows the radiation electric field, magnetic field, and the Poynting vector due to acceleration $\mathbf{a}$. The Poynting vector $\mathbf{S} = \frac{1}{\mu_0}\mathbf{E} \times \mathbf{B}$ is radially outward, and its magnitude is given by

$$S = \frac{q^2}{16\pi^2\varepsilon_0 c^3}\frac{\sin^2\theta}{r^2}a^2_{t-r/c}, \quad (\text{W/m}^2) \qquad (10.16)$$

The total power emitted by a charge q that is subject to an acceleration a can be found to be

$$P = \frac{1}{4\pi\varepsilon_0}\frac{2q^2a^2}{3c^3}, \quad (\text{W}) \qquad (10.17)$$

The details are shown in the following example.

EXAMPLE 10.2

Show that the instantaneous radiation power emitted by a charge q subject to an acceleration a is Eq. (10.17).

Solution

The Poynting vector is the local power density (W/m^2). Then if we integrate the Poynting vector over the entire spherical surface having a radius r, we should obtain the total power. To carry out the surface integration, we note that the area of the thin circular belt having a radius $r \sin \theta$ and a width $rd\theta$ (see Figure 10–8) is

$$dA = 2\pi r \sin \theta \, rd\theta = 2\pi r^2 \sin \theta \, d\theta$$

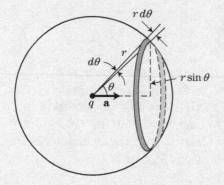

FIGURE 10–8

Then the power through this differential area is

$$dP = S \, dA = S \, 2\pi r^2 \sin \theta \, d\theta \quad \text{(W)}$$

Substituting the expression for S [Eq. (10.16)] and integrating, we find,

$$P = \int dP = \frac{q^2 a^2}{8\pi \varepsilon_0 c^3} \int_0^\pi \sin^3 \theta \, d\theta$$

However,

$$\int_0^\pi \sin^3 \theta \, d\theta = \int_{-1}^1 \sin^2 \theta \, d(\cos \theta) = \int_{-1}^1 (1 - x^2) dx = 2 - \frac{2}{3} = \frac{4}{3}$$

Then the total power becomes

$$P = \frac{q^2 a^2}{8\pi \varepsilon_0 c^3} \times \frac{4}{3} = \frac{1}{4\pi \varepsilon_0} \frac{2q^2 a^2}{3c^3} \quad \text{(W)}$$

Note that the acceleration appears as a^2 and the formula is equally applicable for the case in which the charge is subject to deceleration ($a < 0$). This is known as Larmor's radiation formula. It is valid only if the charge velocity is nonrelativistic, $v \ll c$.

The Poynting vector **S** has a strong angular dependence $\sin^2\theta$, which becomes a maximum at an angle that is perpendicular to the acceleration. This effect is called the *directivity* of the radiation intensity. All antennas have directivity and it is in fact impossible to have an antenna radiate isotropically or equally in every angular direction.

Radiation from an Oscillating Dipole Charge and a Dipole Antenna

Let us apply these basic results to more practical situations. The first example we choose is an oscillating dipole charge, in which two equal but opposite charges undergo a sinusoidal oscillation. Let the amplitude of the oscillations be x_0 (m) and assume the oscillations for the positive charge (Figure 10–9),

$$x_+(t) = x_0 \sin \omega t$$

and for the negative charge,

$$x_-(t) = -x_0 \sin \omega t$$

Then the acceleration for the positive charge is

$$a_+ = \frac{d^2 x_+}{dt^2} = -x_0 \omega^2 \sin \omega t \qquad (10.18a)$$

and the acceleration of the negative charge is

$$a_- = +x_0 \omega^2 \sin \omega t \qquad (10.18b)$$

The radiation electric field is given by

$$E = \frac{1}{4\pi \varepsilon_0 c^2} \frac{\sin \theta}{r}(-qa_+ + qa_-) = \frac{1}{4\pi \varepsilon_0 c^2} \frac{\sin \theta}{r} \times 2qx_0\omega^2 \sin \omega t$$

$$= \frac{qx_0\omega^2 \sin \omega t}{2\pi \varepsilon_0 c^2} \frac{\sin \theta}{r}$$

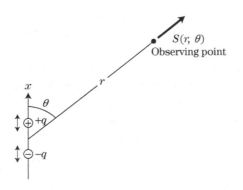

FIGURE 10–9

Oscillating dipole charge radiates electromagnetic waves.

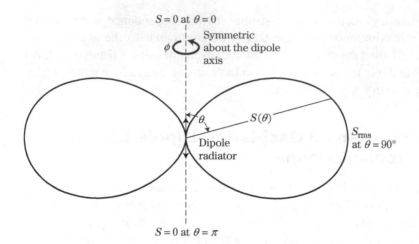

FIGURE 10–10

Angular dependence of the Poynting vector due to an oscillating dipole charge. The Poynting vector **S** is maximum at $\theta = 90°$ and it is symmetric about the dipole axis.

The Poynting flux at a point distance r (m) away from the dipole charge is

$$S = \frac{q^2 x_0^2}{4\pi^2 \varepsilon_0 c^3} \frac{\sin^2 \theta}{r^2} \omega^4 \sin^2 \omega t \qquad (10.19)$$

and its rms value is

$$S_{\text{rms}} = \frac{q^2 x_0^2 \omega^4}{8\pi^2 \varepsilon_0 c^3} \frac{\sin^2 \theta}{r^2}, \quad (\text{W/m}^2) \qquad (10.20)$$

Note in Figure 10–10 that the power is most effectively radiated in the direction that is perpendicular to the dipole charge ($\theta = \pi/2$) as in the case of the radiation intensity due to a single charge. In fact, any antenna will have a strong angular dependence which is desirable for commercial radio or television stations.

Next, we consider a short radio antenna. We assume that the *length of the antenna is much shorter than the wavelength of the radio wave*. This assumption means that the current on the antenna can be assumed to have the same phase everywhere. The entire antenna is oscillating with a certain frequency ω. Of course, the conduction electrons in the antenna are actually doing the oscillating.

Since there are many conduction electrons, we have to add up the radiation electric fields due to the individual electrons. But the story is much simpler, since we already have the Poynting flux due to an oscillating charge [see Eq. (10.16)]. Suppose that the antenna is l (m) long, has a cross section of A (m^2), and has a conduction electron density of n (m^{-3}). The total number of conduction electrons is Aln and the charge is $q = eAln$ (coulombs). If we assume that a current $I_0 \sin \omega t$ flows through the antenna, then the velocity

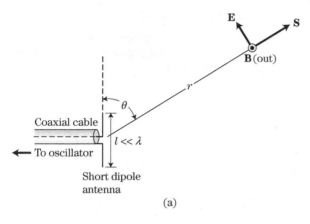

(a)

FIGURE 10–11

(a) Radiation fields due to a short ($l \ll \lambda$) dipole antenna. (b) Formation of a dipole antenna from an open-ended transmission line.

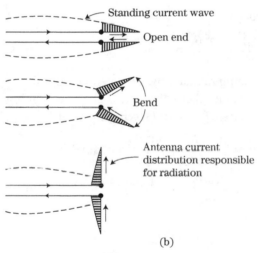

(b)

of each electron is

$$v = \frac{l}{q} I_0 \sin \omega t \tag{10.21}$$

since $I = enAv$. Therefore the acceleration of each electron is

$$a = \frac{dv}{dt} = \frac{l}{q} I_0 \omega \cos \omega t \tag{10.22}$$

or

$$qa = l\omega I_0 \cos \omega t \tag{10.23}$$

Substituting Eq. (10.23) into Eqs. (10.13) and (10.16), we find

$$E_t = \frac{l\omega I_0}{4\pi \varepsilon_0 c^2} \frac{\sin \theta}{r} \cos \omega t \tag{10.24}$$

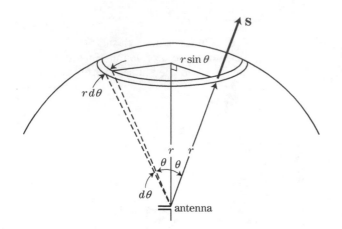

FIGURE 10–12

Integration of Poynting vector **S** over the entire sphere yields the total power radiated from the antenna.

and the corresponding Poynting flux is

$$S = \frac{(l\omega I_0)^2}{16\pi^2\varepsilon_0 c^3} \frac{\sin^2\theta}{r^2} \cos^2\omega t \qquad (10.25)$$

Let us evaluate the total power being radiated by the antenna. For this we have to integrate the Poynting flux over a spherical surface. The power going through the thin ring with an area $2\pi r \sin\theta \times r d\theta$ (see Figure 10–12) is

$$dP = S_{\text{ave}} \times 2\pi r^2 \sin\theta d\theta = \frac{(I\omega I_0)^2}{32\pi^2\varepsilon_0 c^3} \frac{\sin^2\theta}{r^2} 2\pi r^2 \sin\theta d\theta$$

$$= \frac{(l\omega I_0)^2}{16\pi\varepsilon_0 c^3} \sin^3\theta d\theta \qquad (10.26)$$

Then the total radiated power is

$$P = \int_0^\pi \frac{(l\omega I_0)^2}{16\pi\varepsilon_0 c^3} \sin^3\theta d\theta = \frac{(l\omega I_0)^2}{16\pi\varepsilon_0 c^3} \int_0^\pi \sin^3\theta d\theta \qquad (10.27)$$

The integration was performed in Example 10.2 and its value was found to be 4/3. Therefore the total power radiated by the short antenna is

$$P = \frac{(l\omega I_0)^2}{16\pi\varepsilon_0 c^3} \times \frac{4}{3} = \frac{(l\omega I_0)^2}{12\varepsilon_0 c^3}, \; (\text{W}) \qquad (10.28)$$

This antenna is connected to a source of the electromagnetic energy, typically through a transmission line, and the antenna absorbs the energy. This is very similar to a load resistor that is connected to a voltage source. Therefore, we can consider the antenna as being a resistor that absorbs the energy and the resistance is called the *radiation resistance*. The value of this radiation resistance can be calculated from

$$R_{\text{rad}}I_{\text{rms}}^2 = \frac{1}{2}R_{\text{rad}}I_0^2 = P \qquad (10.29)$$

or

$$R_{rad} = \frac{(l\omega)^2}{6\pi \varepsilon_0 c^3}, \quad (\Omega) \tag{10.30}$$

Since

$$c = \frac{1}{\sqrt{\varepsilon_0 \mu_0}}, \quad \omega = 2\pi \nu = \frac{2\pi c}{\lambda}$$

we can rewrite the preceding expression as

$$R_{rad} = \frac{2\pi}{3} \sqrt{\frac{\mu_0}{\varepsilon_0}} \left(\frac{l}{\lambda}\right)^2 = 790 \left(\frac{l}{\lambda}\right)^2, \quad (\Omega) \tag{10.31}$$

where $\sqrt{\mu_0/\varepsilon_0} = 377 \ \Omega$ is the characteristic impedance of free space. Remember that the expressions for power and radiation resistance are all subject to the restriction, $l \ll \lambda$ (short antenna). The generator connected to the antenna has to supply the power that is radiated into space in order to maintain a steady state.

The radiation power in Eq. (10.31) is subject to the condition that $l \ll \lambda$ and the resultant radiation resistance is small, which indicates short antennas are rather inefficient. Practical antennas used in communication have a length that is comparable with the wavelength. For example, a center-fed half dipole $(l = \lambda/2)$ has a radiation resistance of

$$R_{rad} = \frac{1.22}{2\pi} \sqrt{\frac{\mu_0}{\epsilon_0}} = 73 \ \Omega$$

where the factor 1.22 arises from the evaluation of the integral

$$\int_0^\pi \frac{\cos^2\left(\frac{\pi}{2} \cos\theta\right)}{\sin\theta} d\theta = 1.22$$

(This will be discussed in Problem 11.18. A long antenna can be analyzed by superposing contributions from short antennas.)

The radiation power due to the oscillating electric dipole in Eq. (10.20) is proportional to ω^4. This is known as Rayleigh's law. It is valid if the dipole radiator size is much smaller than the wavelength $ka \ll 1$ where $k = \frac{2\pi}{\lambda}$ and a being a characteristic dimension of the radiator such as the radius of molecules. Air molecules radiate or re-radiate light when exposed to solar radiation through a dipole radiation mechanism since the molecule size is much smaller than the wavelength of visible light. Because of the sensitive dependence of the radiation power on the frequency ($\propto \omega^4$) or wavelength ($\propto 1/\lambda^4$), short wavelength blue light is scattered much more efficiently than red light. The blueness of the sky can thus be explained in terms of *Rayleigh scattering*. If the radiator size is comparable with or longer than the wavelength, such as scattering by water droplets in clouds, the scattering power becomes insensitive to the frequency and scattered light appears white. Scattering in this regime is called *Mie scattering*.

Cyclotron and Synchrotron Radiation

Charged particles in a magnetic field undergo a circular motion as shown in Figure 10–13. The magnetic force $\mathbf{F}_B = q\mathbf{v} \times \mathbf{B}$ is balanced with the mechanical centrifugal force $\mathbf{F}_c = \dfrac{mv^2}{\rho}\mathbf{u}_\rho$ where $\mathbf{u}_\rho$ is the unit vector in the radial direction. The charged particle is under continuous acceleration due to the magnetic force which is radially inward and radiation of the electromagnetic wave is thus expected. The magnitude of the acceleration is

$$a = \frac{F_B}{m} = \frac{F_c}{m} = \frac{v^2}{\rho}$$

Then the radiation power [see Eq. (10.17)] is

$$P = \frac{q^2 a^2}{6\pi \varepsilon_0 c^3} = \frac{q^4 v^2 B^2}{6\pi \varepsilon_0 m c^3} = \frac{q^2 v^4}{6\pi \varepsilon_0 c^3 \rho^2}, \quad \text{(W)} \tag{10.32}$$

that is valid only for a nonrelativistic velocity $v \ll c$. Relativistic effects enhance the radiation power to

$$P = \frac{q^2 a^2 \gamma_{\text{rel}}^4}{6\pi \varepsilon_0 c^3} \tag{10.33}$$

where γ_{rel} is the relativity factor given by

$$\gamma_{\text{rel}} = \frac{1}{\sqrt{1 - \left(\dfrac{v}{c}\right)^2}}$$

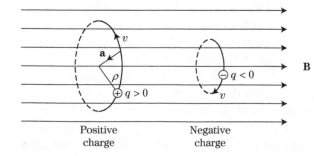

Positive charge   Negative charge

FIGURE 10–13

A charged particle in a magnetic field undergoes a cyclotron motion and radiates since the particle is under centripetal acceleration.

For a charge undergoing circular motion at a radius ρ, the acceleration is $a = v^2/\rho$, since

$$qvB = \frac{m\gamma_{\text{rel}}v^2}{\rho}$$

and thus

$$a = \frac{qvB}{m\gamma_{\text{rel}}} = \frac{v^2}{\rho}$$

Note the relativistic change in the mass, $\gamma_{\text{rel}}m$. In the nonrelativistic case, the radiated electromagnetic fields all have the single frequency that is the cyclotron frequency $\omega = \omega_c = qB/m$. In the relativistic case, harmonics of the fundamental frequency $n\omega_c$ appear and in the highly relativistic case $\gamma_{\text{rel}} \gg 1$, the radiation frequency spectrum becomes essentially continuous.

In summary, we have seen that if charged particles are subject to acceleration, radiation of electromagnetic waves will occur. Radiation from antennas is due to the acceleration of conduction electrons. Radiation carries energy away. Therefore to maintain a steady state, there must be a source of energy that is fed into the radiation system.

There are radiation mechanisms that do not require an acceleration of the charges. A typical example is Cherenkov radiation, which occurs when a charge is moving in a material medium with a velocity that is larger than the velocity of light in the medium,

$$v > \frac{1}{\sqrt{\varepsilon\mu_0}}$$

This can happen when a relativistic charge with a velocity close to the velocity of light enters a material medium. Cherenkov radiation is similar to a shock wave caused by a supersonic object with $v > c_s$.

Problems

1. In the electron gun of an oscilloscope, a 20-kV potential is applied between anode and cathode, which are 5 cm apart. Estimate the maximum radiation electric field at a point 1 m away from the gun assuming there are about 4×10^7 electrons in the gun.

2. We have seen that for the creation of electromagnetic waves, we must have charge acceleration perpendicular to the direction of the Poynting vector. This is due to the *transverse* nature of electromagnetic waves. Discuss how we can create sound waves in air that are *longitudinal*.

3. An AM radio station of 1 MHz frequency uses an antenna 20 m long placed well above the ground.
   (a) What is the radiation resistance of the antenna?
   (b) If the station is to be operated at 50 kW power, determine the value of the rms current that should be supplied to the antenna.
   Note: If an antenna l(m) high is erected above the ground, its "effective" length is $2l$(m). The reason is that the earth is a mirror (reflector) for electromagnetic waves. Frequently, conducting wires are placed radially away from the centerline of the antenna in

order to increase the conductivity of the ground in the region of the antenna. This "method of images" will be studied in more advanced electromagnetic classes.

4. X-rays (short wavelength electromagnetic waves, $\lambda \simeq 10^{-10} - 10^{-9}$ m) can be created when energetic electrons hit a surface of a hard metal such as tungsten. Explain qualitatively the radiation mechanism.

5. An electron having an energy of 10 keV ($= 10^4 \times 1.6 \times 10^{-19}$ J) undergoes cyclotron motion in a magnetic field of 1 T.

(a) What is the acceleration of the electron?

(b) Evaluate the initial rate of electron energy loss caused by the cyclotron radiation.

Interference and Diffraction

Introduction

A harmonic wave in the typical form used in previous chapters is given by

$$A \sin(kx - \omega t)$$

which is characterized by the amplitude A, the angular frequency ω and the wavelength $\lambda = 2\pi/k$. If we have only one wave source, the preceding expression is, in general, sufficient. However, as soon as we have more than one wave source that all have the same frequency, the total amplitude at a given observation point must be the sum of contributions from each wave source. We assume that all of the waves have their amplitudes polarized in the same direction. What is important here is the phase difference between the waves, which we have not considered in detail so far. For example, if two waves of equal amplitude and frequency having no phase difference (or a phase difference of an integer multiple of 2π) are added, the amplitude is simply doubled, and the intensity is quadrupled. But if the two waves are out of phase or have a phase difference of $\pm(2m + 1)\pi$ where m is an integer, the total amplitude becomes zero. Therefore, depending on the phase difference, the total amplitude and thus the wave intensity can vary. Interference is the most fundamental nature of all wave phenomena. If one physical quantity exhibits interference, that quantity should have a wave nature. In this chapter we study the interaction among more than one wave, sometimes infinitely many waves.

Interference Between Two Harmonic Waves

Suppose we have two point wave sources that are separated by a distance d as shown in Figure 11–1. We assume that the two sources radiate waves at exactly the identical frequencies. Such a situation can easily be realized if, for example, two identical antennas are connected to a common signal generator.

Consider a point P at which we place a wave detector. The distances between P and the wave sources are x_1 and x_2, respectively. At P we simultaneously detect two different waves since

1. The field amplitudes, which are proportional to $1/x_1$ and $1/x_2$, respectively, are different.

2. The phases $2\pi x_1/\lambda$ and $2\pi x_2/\lambda$, which are proportional to x_1 and x_2, are different.

The amplitude difference may not be important if the point P is far away such that $x_1, x_2 \gg d$.

The phase difference given by

$$\phi = \frac{2\pi}{\lambda}(x_2 - x_1), \quad (\lambda = \text{wavelength}) \tag{11.1}$$

is, however, important and plays major roles in interference and diffraction. The quantity $x_2 - x_1$ is called the *path difference*. In Figure 11–1, the path

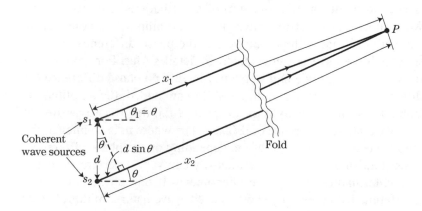

FIGURE 11–1

Waves created by two coherent sources can interfere with each other. Here we assume $x_1, x_2 \gg d, \lambda$ (wavelength). In this case, the path difference can be approximated by $d \sin \theta$.

difference is given approximately by

$$d \sin \theta \tag{11.2}$$

if $x_1, x_2 \gg d$.

Consider two harmonic waves

$$E_1 = E_0 \sin(kx_1 - \omega t), \quad k = \frac{2\pi}{\lambda} \tag{11.3}$$

$$E_2 = E_0 \sin(kx_2 - \omega t) \tag{11.4}$$

created by the sources S_1 and S_2, respectively. Using the phase difference given by Eq. (11.1), we may rewrite Eq. (11.4) as

$$E_2 = E_0 \sin(kx_1 - \omega t + \phi) \tag{11.5}$$

Since we're assuming that the two waves are propagating in a linear medium, these two fields at the point P is just the sum of the two fields. The total field becomes

$$E = E_1 + E_2 = 2E_0 \sin\left(kx_1 - \omega t + \frac{\phi}{2}\right) \cos\left(\frac{\phi}{2}\right) \tag{11.6}$$

which is still a propagating wave. Recall that

$$\sin A + \sin B = 2 \sin\left(\frac{A+B}{2}\right) \cos\left(\frac{A-B}{2}\right)$$

The amplitude, however, strongly depends on the phase difference ϕ. The effective amplitude is

$$E_m = 2E_0 \left|\cos\left(\frac{\phi}{2}\right)\right| \tag{11.7}$$

and it can vary between 0 and $2E_0$ depending on the phase difference. The maximum amplitude $E_m = 2E_0$ is obtained when $|\cos(\phi/2)| = 1$ or

$$\phi = 0, \pm 2\pi, \pm 4\pi, \ldots = 2m\pi \quad (m = \text{integer}) \tag{11.8}$$

and the minimum amplitude $E_m = 0$ when $\cos(\phi/2) = 0$ or

$$\phi = \pm \pi, \pm 3\pi, \ldots = (2m+1)\pi \tag{11.9}$$

In terms of $d \sin \theta$, we have

$$d \sin \theta = m\lambda \quad \text{for maxima} \tag{11.10}$$

and

$$d \sin \theta = \left(m + \frac{1}{2}\right)\lambda \quad \text{for minima} \tag{11.11}$$

These results can be found intuitively if we graphically superpose two waves as shown in Figure 11–2.

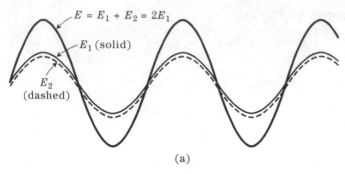

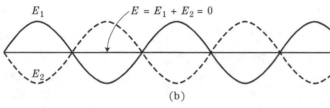

FIGURE 11–2

Graphical superposition of two waves: (a) In phase—two waves add up and the amplitude is a maximum. (b) Out of phase—two waves cancel each other and the amplitude is zero.

FIGURE 11–3

Interference of sound waves propagating through two different paths. Depending on the distance d, the received intensity varies.

Consider the setup for demonstrating sound wave interference (Figure 11–3). The path difference in this case is $2d$. If

$$2d = m\lambda$$

the sound intensity should be maximum at the receiving end and if

$$2d = \left(m + \frac{1}{2} \right) \lambda$$

the sound intensity should be minimum. This value will be 0 under ideal conditions.

The recipe for various interference and diffraction phenomena that we are going to study is nothing more than Eq. (11.6), that is, superposition of two or more sinusoidal waves with equal amplitudes but with different phases. If we have many wave sources, we have to add up more waves to find the total field but the mathematics involved is no more than adding up sinusoidal functions.

The distinction between interference and diffraction is not clear. Both are caused by the interaction among more than one wave. We use the following convention to distinguish the two, if we have relatively few wave sources, we use the term *interference*. If we have to add up many (sometimes infinitely many) waves, *diffraction* is used.

Young's Experiment

We know that light is an electromagnetic *wave*. The wavelengths of the visible light spectrum range from 4000 Å (or 4×10^{-7} m) to 7000 Å (7×10^{-7} m). (1 Å (angstrom) = 10^{-10} m.) This seemingly obvious fact was not so obvious before Young did the famous experiment in 1801–1803, now known as Young's double-slit experiment. The experiment clearly demonstrated the wave nature of light. (We will learn that light also behaves as particles, known as photons. In fact, both a wave and a particle nature coexist in light or in any moving physical object. For example, energetic electrons also have both a particle and a wave nature. As we will see later, electron microscopes can "see" better than optical microscopes, since "the wavelengths of electron waves" are much shorter than those of visible light. Quantum mechanics has been able to unify the wave and particle nature.)

The principle of Young's experiment is shown in Figure 11–1. Instead of two point sources, as in Figure 11–1, Young used two narrow parallel slits that were illuminated by a monochromatic light source as shown in Figure 11–4.

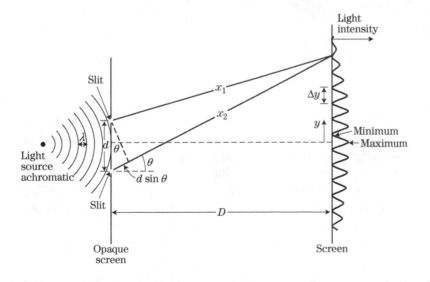

FIGURE 11–4a

Arrangement of Young's double slit experiment. In practice, the screen distance D is much larger than the slit separation d.

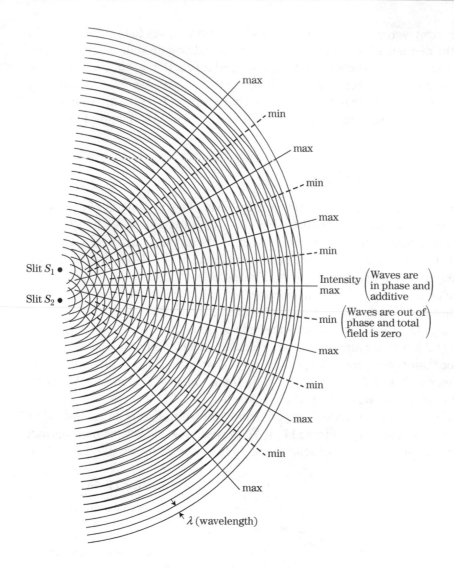

FIGURE 11–4b

Qualitative illustration of the interference mechanism. Two families of concentric circles indicate radiation from each slit and are drawn with the same interval corresponding to the wavelength λ. Where the two families of circles intersect, the wave from each slit are in phase and the wave intensity becomes a maximum. Intensity minima occur between neighboring maxima.

If the slit opening is narrow enough, an interesting thing happens. The slits act as if they were new light sources. If the light source is placed on the bisector line of the two slits, these slits become light sources of equal phase since the slits are equal distances away from the light source (no path difference). The slits act as line light sources rather than point sources. Therefore light emitted from the slits consists of cylindrical rather than spherical waves. However, we do not have to worry about this since all we need is the path difference. Of course, the slits can be replaced with two pinholes but this only complicates the analysis.

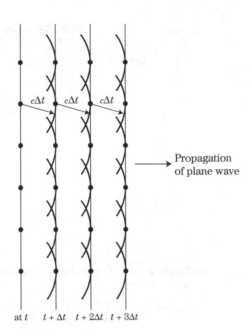

FIGURE 11–5

Huygens' principle applied to a plane wave.

That narrow openings (such as the slits) act as new light sources can be understood from Huygens' principle, which may be stated as follows: All points on a wavefront act as new point sources. To illustrate, let us consider a trivial case of a plane wave. As shown in Figure 11–5, the new wavefront is formed as a tangent plane to all the waves created by the points on the old wavefront. The new wavefront then further creates another wavefront. This process continues and is observed as wave propagation.

Suppose the plane wave encounters an obstacle with a small opening (Figure 11–6). In the opening, we can have only a few Huygens' points (only one is shown in the figure) and beyond the opening, the wave is not a plane wave anymore. If the opening is very small, it acts as a point wave source. Beyond the opening, we essentially have a spherical wave. The image formed on a screen is then widely spread. Later, we will calculate this intensity profile for the case of a narrow slit in the section on diffraction.

Now we return to Young's experiment. Since the light intensity is proportional to E^2, where E is the total electric field, we obtain from Eq. (11.7)

$$I(\theta) = I_0 \cos^2\left(\frac{\phi}{2}\right) \tag{11.12}$$

where

$$\phi = \frac{2\pi d}{\lambda}\sin\theta$$

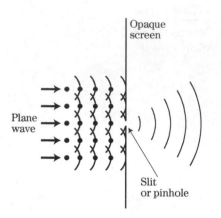

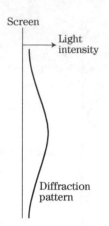

FIGURE 11–6

A plane wave is converted into a spherical wave when a light wave goes through a pinhole or into a cylindrical wave when the light wave goes through a long narrow slit.

If a screen is placed at a distance $D \, (\gg d)$ parallel to the slits, we find

$$I(y) = I_0 \cos^2 \left(\frac{\pi d}{\lambda} \frac{y}{D} \right) \qquad (11.13)$$

where

$$\sin \theta \simeq \tan \theta = \frac{y}{D} \, (\ll 1)$$

is used assuming a small angle θ. Then the maxima are located at

$$y = 0, \quad \pm \frac{\lambda D}{d}, \quad \pm 2 \frac{\lambda D}{d}, \dots$$

and the minima at

$$y = \pm \frac{1}{2} \frac{\lambda D}{d}, \quad \pm \left(1 + \frac{1}{2} \right) \frac{\lambda D}{d}, \quad \pm \left(2 + \frac{1}{2} \right) \frac{\lambda D}{d}, \dots$$

The separation between neighboring maxima (or minima) is

$$\Delta y = \frac{\lambda D}{d} \qquad (11.14)$$

Therefore by measuring Δy, D, and d, the wavelength λ can be determined.

The experiment works better for narrower slit openings. Larger slit openings complicate the intensity pattern formed on the screen because of diffraction effects, which we will study later. If the opening is increased further, the interference pattern disappears and we simply have two slit images, although somewhat blurred. This is because for a large opening, the slits do not behave as line sources.

EXAMPLE 11.1

In Young's double-slit arrangement, assume the slit separation d is 0.1 mm and slit-screen distance D is 50 cm. If a separation between neighboring maxima (or minima) of 2.5 mm is observed, what is the wavelength of light illuminating the slits?

Solution

From Eq. (11.14), we find

$$\lambda = \frac{\Delta y d}{D} = \frac{0.25 \times 0.01}{50} = 5.0 \times 10^{-5} \text{ cm} = 5000 \text{ Å}$$

EXAMPLE 11.2

Two loudspeakers connected to a common audio amplifier are 5 m apart (Figure 11–7). As one walks along a straight path 100 m away from the speakers, find the separation distance Δy between two maxima in the observed intensity. Assume a wavelength of $\lambda = 30$ cm.

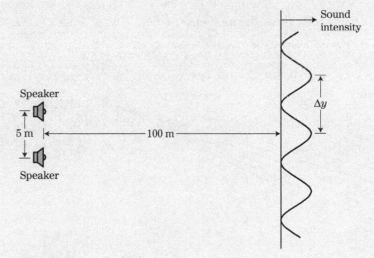

FIGURE 11–7

Two loudspeakers are connected to a common audio amplifier.

Solution

From Eq. (11.14), we write

$$\Delta y = \frac{\lambda D}{d} = \frac{0.3 \times 100}{5.0} = 6.0 \text{ m}$$

Multi-slit Structure

Let us see what would happen if the number of slits is increased. We assume that all slits are equally spaced and illuminated by a common, monochromatic light source. The case of six slits is shown in Figure 11–8. The phase difference between two neighboring waves is

$$\delta = \frac{2\pi d}{\lambda} \sin\theta \qquad (11.15)$$

Thus the total electric field is given by

$$E = E_0 \left[\sin(kx - \omega t) + \sin(kx - \omega t + \delta) + \sin(kx - \omega t + 2\delta)\right.$$
$$\left. + \sin(kx - \omega t + 3\delta) + \sin(kx - \omega t + 4\delta) + \sin(kx - \omega t + 5\delta)\right] \qquad (11.16)$$

We could add these six terms one by one, but there is a more elegant way to do this. In alternating current circuit theory, we find that two oscillating voltages V_1 and V_2 with a phase difference ϕ can be added graphically. The terminology used there is to model the voltages as "phasors" (see Figure 11–9). If we use this technique for the case of a double slit, we immediately find that the amplitude of the total electric field E is given by

$$E = 2E_0 \cos\left(\frac{\phi}{2}\right)$$

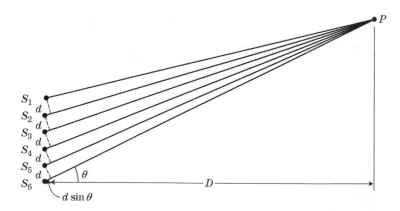

FIGURE 11–8

Six coherent light sources are equally spaced along a straight line.

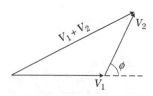

FIGURE 11–9

In alternating current circuit theory, two voltages can be added vectorially.

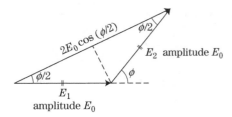

FIGURE 11–10

Vectorial addition of two electric fields in the case of a double-slit arrangement.

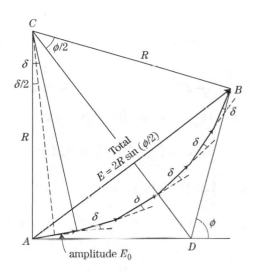

FIGURE 11–11

Addition of six electric fields. Note that the phase difference between any two neighboring fields is δ, which has a constant value.

which is consistent with Eq. (11.7) and shown in Figure 11–10. Note that the vectors are introduced here for mathematical convenience and have nothing to do with the electric field vector associated with the light. The true electric field vectors are all in the same direction if the observation point is sufficiently far from the light sources.

Now we add up the six fields using the alternating current circuit theory. Since the phase difference δ between two neighboring fields is the same everywhere, the six vectors form an arc with a radius R as shown in Figure 11–11. We write

$$\frac{E_0/2}{R} = \sin\frac{\delta}{2} \tag{11.17}$$

$$\phi = 6\delta \tag{11.18}$$

and

$$E = 2R\sin\left(\frac{\phi}{2}\right) \tag{11.19}$$

Eliminating R and ϕ, we find

$$E = E_0 \frac{\sin(6\delta/2)}{\sin(\delta/2)} \tag{11.20}$$

A more elegant way to derive this is to use complex variables. Using the Euler formula, $\sin A = \text{Im}\,(e^{iA})$ where Im indicates the imaginary part, Eq. (11.16) can be written as

$$E = E_0\,\text{Im}[e^{iX}(1 + e^{i\delta} + \cdots + e^{i5\delta})]$$

where $X = kx - \omega t$. Since the magnitude of e^{iX} is 1, the amplitude of E is given by

$$E_0|1 + e^{i\delta} + e^{i2\delta} + e^{i3\delta} + e^{i4\delta} + e^{i5\delta}| = E_0|(1 + e^{i\delta})e^{i2\delta}(e^{-i2\delta} + 1 + e^{i2\delta})|$$

Noting $e^{iA} + e^{-iA} = 2\cos A$, the amplitude becomes

$$2E_0 \cos\left(\frac{\delta}{2}\right)[1 + 2\cos(2\delta)]$$

which is identical to Eq. (11.20). [Use $\sin 3A = 3\sin A - 4\sin^3 A$ in Eq. (11.20).]

We can easily generalize this to the case of N light sources as

$$E = E_0 \frac{\sin(N\delta/2)}{\sin(\delta/2)} \tag{11.21}$$

The case of the double slit corresponds to $N = 2$, and we indeed recover Eq. (11.7),

$$E\,(N = 2) = E_0 \frac{\sin(\delta)}{\sin(\delta/2)} = E_0 \frac{2\sin(\delta/2)\cos(\delta/2)}{\sin(\delta/2)} = 2E_0 \cos\left(\frac{\delta}{2}\right)$$

The light intensity corresponding to the electric field given by Eq. (11.21) becomes

$$I = I_0 \frac{\sin^2(N\delta/2)}{\sin^2(\delta/2)} \tag{11.22}$$

Here we need some mathematics. We want to know what would happen to the function

$$f(x) = \frac{\sin Nx}{\sin x}, \quad (N = \text{integer})$$

if we let $\sin x$ approach zero. The function $\sin x = 0$ occurs when

$$x = m\pi \quad (m = \text{integer})$$

Then $\sin Nx = \sin(Nm\pi)$ also becomes zero and we end up with $0/0$, which is undefined. Let $x = m\pi + \varepsilon$, with ε a small value. Since

$$\sin(m\pi + \varepsilon) = \sin m\pi \cos \varepsilon + \cos m\pi \sin \varepsilon. = \pm \sin \varepsilon$$

and $\sin N(m\pi + \varepsilon) = \pm \sin N\varepsilon$, we find

$$\lim_{x \to m\pi} f(x) = \lim_{\varepsilon \to 0} \frac{\sin N\varepsilon}{\sin \varepsilon} = \frac{N\varepsilon}{\varepsilon} = N$$

which is finite.

Then the light intensity I becomes a maximum whenever $\sin(\delta/2) = 0$ and its peak value is proportional to $N^2 \cdot \sin(\delta/2) = 0$ yields

$$\frac{\delta}{2} = 0, \quad \pm\pi, \quad \pm 2\pi, \ldots$$

or

$$\alpha \sin \theta = m\lambda \quad (m = \text{integer}) \tag{11.23}$$

The function

$$f(\delta) = \frac{1}{N^2} \frac{\sin^2(N\delta/2)}{\sin^2(\delta/2)} \tag{11.24}$$

indicates the relative intensity (maximum chosen as 1.0) and is plotted in Figure 11–12 for $N = 2, 5$, and 10, as a function of $a/\lambda \sin \theta$. It can be seen that as N increases, the interference pattern becomes sharper and sharper. You may imagine what profile will result for, say $N = 100$. The profile will be extremely sharp and will simply look like vertical lines located at

$$\frac{d}{\lambda} \sin \theta = 0, \pm 1, \pm 2, \ldots$$

There is an optical device called a *spectrometer* from which one can determine the intensity at a particular wavelength. In other words, a spectrometer can Fourier-analyze a light wave and may be called an optical spectrum analyzer. The principle is the one we have just studied. The number of light sources is usually in the tens of thousands or more. You may imagine how sharp the interference pattern should be for a large numerical value of the integer N. Spectrometers have so-called gratings, a structure similar to the fine grooves on an LP record, a CD, or a DVD. (Actually, LP records can be considered to be a rough spectrometer. You can easily see the color spectrum on the record surface.)

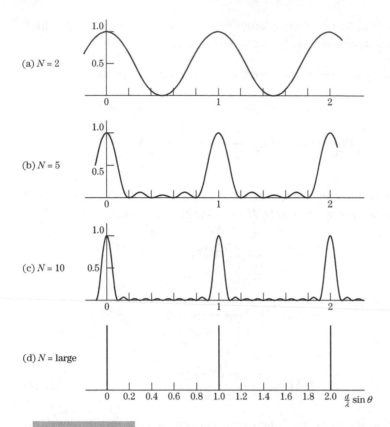

(a) N = 2

(b) N = 5

(c) N = 10

(d) N = large

$$0 \quad 0.2 \quad 0.4 \quad 0.6 \quad 0.8 \quad 1.0 \quad 1.2 \quad 1.4 \quad 1.6 \quad 1.8 \quad 2.0 \quad \frac{d}{\lambda}\sin\theta$$

FIGURE 11–12

(a) Equation (11.24) plotted as a function of $\delta/2\pi = d\sin\theta/\lambda$ for the case of $N = 2$. (This corresponds to the intensity profile in Young's experiment.) (b) $N = 5$. (c) $N = 10$. (d) $N = $ large.

EXAMPLE 11.3

Consider a grating having 5000 grooves per centimeter. There the separation between grooves (corresponding to the slits) is

$$d = \frac{1 \text{ cm}}{5000} = 2 \times 10^{-4} \text{ cm}$$

Determine the groove angles for red and violet light.

Solution

For red light of $\lambda = 7000$ Å $= 7 \times 10^{-5}$ cm, the sharp peaks appear at

$$\sin\theta_{\text{red}} = 0, \pm\frac{\lambda}{a}, \pm2\frac{\lambda}{a}, \ldots = 0, \pm0.35, \pm0.70, (\pm1.05\ldots)$$

where the final term may be neglected. This corresponds to

$$\theta_{\text{red}} = 0°, \pm20.5°, \pm44.5°$$

For violet light of $\lambda = 4000$ Å $= 4 \times 10^{-5}$ cm, the angles become

$$\theta_{\text{violet}} = 0°, \pm11.5°, \pm23.6°, \pm36.9°, \pm53.1°$$

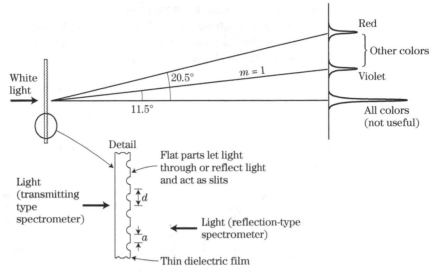

FIGURE 11–13

Grating spectrometer. Light falls normal to the grating and $d = 2 \times 10^{-4}$ cm.

All colors fall on $\theta = 0°$ and this angular location is not useful. The next peak in Figure 11–13 (called order 1) can be used for spectrum analysis. Colors in the wavelength range 4000–7000 Å fall between $\theta = 11.5°$ and $\theta = 20.5°$. For the second order, the angular range is between 23.6° and 44.4°, which is also useful. In this example, the red light cannot produce third-order (and higher) order peaks and the violet light cannot produce a fifth- (and higher) order peaks.

The result that we obtained in Eq. (11.22) can also be applied to the interference pattern formed by an antenna array. In radio communication, it is often desired that the antennas have a sharp directivity. As we saw in Chapter 10, a single dipole antenna erected vertically radiates uniformly in all horizontal directions or it has no horizontal directivity. For commercial radio stations, this is a desired feature. However, for example, antenna arrays used for the instrumental landing of airplanes require a strong horizontal directivity. (As the wavelengths become shorter, the directivity can be realized using a principle completely different from interference. In microwaves, parabolic antennas are used. For such short wavelengths, the concept of geometric optics can be applied and parabolic antennas can be regarded as concave mirrors in geometrical optics.)

The wavelengths of electromagnetic waves used in radio communication are of course much longer than optical wavelengths and the spacing between wave sources (d) can be chosen to be of the order of the wavelength. In fact, the spacing d is chosen at $\lambda/2$ in many applications. Consider an antenna array consisting of four coherent dipole antennas as shown in Figure 11–14. In Eq. (11.22) we choose $N = 4$, and $d = \lambda/2$. Then the intensity becomes

$$I = I_0 \frac{\sin^2(2\pi \sin\theta)}{\sin^2\left(\dfrac{\pi}{2}\sin\theta\right)} \qquad (11.25)$$

which can be calculated easily with either a programmable calculator or a computer (see Figure 1l–15).

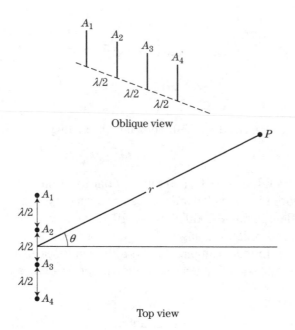

Oblique view

Top view

FIGURE 11–14

Antenna array for producing a strong horizontal directivity. The antennas are $\lambda/2$ apart.

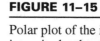

FIGURE 11–15

Polar plot of the interference intensity by the antenna arrangement of Figure 11–14.

We can intuitively see that the array does not radiate waves along the array itself ($\theta = 90°$) since in this direction, A_1 and A_2 are out of phase ($\theta = \pi$), and A_3 and A_4 are also out of phase. Thus, the electric field at $\theta = 90°$ is zero.

Of course each antenna radiates radio waves with a power determined by the current supplied from the source and the radiation resistance of the antenna. By making an array, the angular distribution of the power can be made so that the power can be radiated in a preferential direction. It is like making a high hill by collecting rocks that are widely distributed. By increasing the number of antennas, the directivity can be made sharper and sharper. This is similar to a grating spectrometer.

Similar array arrangement can be used for receiving antennas when a sharp directivity (or high resolving power) is required. In radio astronomy, an antenna array consisting of tens of antennas is not unusual.

Optical Interference in Thin Films

We frequently observe that a motor oil film on a water surface appears to be colored. Also, good cameras all have lenses that are coated with an antireflecting material (such as MgO_2) to minimize the light reflection from the lenses. Here we analyze the mechanism behind these phenomena. The reflection properties do depend upon the wavelength of the incident wave.

In Section 9.6, we learned that whenever electromagnetic waves in air try to penetrate into a medium having a characteristic impedance that is less than that of air $\sqrt{\mu_0/\varepsilon_0} = 377\Omega$, the reflected electric field (or voltage) suffers a phase change of $\pi (\theta = 180°)$. Such a medium is called a "hard" medium. For example, glass has a relative permittivity of about 2.3 in the frequency range of visible light. Then the velocity of light in glass is approximately $c/\sqrt{2.3} = 0.67c$ and the characteristic impedance of glass for visible light is approximately 250 Ω. This yields the electric field reflection coefficient,

$$\Gamma = \frac{250 - 377}{250 + 377} = -0.20$$

and the power reflection coefficient

$$\Gamma^2 = 0.04$$

for light incident normal upon a flat glass surface. This indicates that almost 4% of the incident light energy is reflected at the glass surface, and 96% can penetrate into the glass. Γ itself is negative, and the reflected electric field is subject to a phase change of π upon reflection. For oblique incidence, the analysis becomes complicated and we do not consider it here.

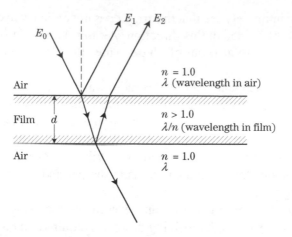

FIGURE 11–16

Reflections of light at both
film surfaces cause
interference. The electric field
E_1 changes its sign relative to
E_0, but E_2 does not.

With this knowledge, we can now analyze the interference caused by a thin
dielectric film (such as water, soap, oil, etc.) placed in air (see Figure 11–16).
Consider monochromatic light with a wavelength λ in air propagating almost
normal to the film. The electric field E_1 is the one reflected at the upper film
surface and is almost out of phase with respect to the incident field E_0. The
film is a medium that is harder than air. The electric field E_2 is due to the
reflection at the lower surface and it will be in phase with respect to E_0, since
air is softer than the film. But E_2 travels an additional distance $2d$ than E_1.
We will have to take the path difference into account in order to calculate the
total phase difference between E_1, and E_2. Since the velocity of light in the
film is smaller than that in air by a factor of $\sqrt{\varepsilon/\varepsilon_0} = \sqrt{\varepsilon_r}$ with ε_r the relative
dielectric constant, the wavelength in the film becomes shorter by the same
factor. We define the *index of refraction* by

$$n = \sqrt{\varepsilon_r} = \frac{c}{c_{\text{film}}} \tag{11.26}$$

which is the ratio between the velocities of light in the two media.

The phase difference between E_1 and E_2 resulting from the path difference alone is

$$\frac{2\pi \cdot 2d}{\lambda/n} \quad \text{(rad)}$$

However, since E_1 has a phase difference of π relative to E_0, the net phase
difference between E_1 and E_2 is

$$\phi = n\frac{4\pi d}{\lambda} - \pi$$

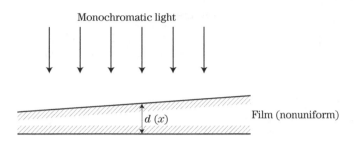

Monochromatic light

$d(x)$

Film (nonuniform)

Reflected
light
intensity

x

FIGURE 11–17

Dielectric film of nonuniform
thickness can cause
interference stripes.

If this total phase difference is an integer multiple of 2π or

$$n\frac{4\pi d}{\lambda} - \pi = m \times 2\pi$$

we have constructive interference or the reflected light will be intensified. We
may rewrite the preceding equation as

$$2d = \left(m + \frac{1}{2}\right)\frac{\lambda}{n} \quad (m = 0, 1, 2, 3, \ldots) \tag{11.27}$$

Destructive interference occurs if

$$2d = \frac{m\lambda}{n} \tag{11.28}$$

If the thickness gradually varies as shown in Figure 11–17, many stripes
will appear at the locations where

$$d = \left(m + \frac{1}{2}\right)\frac{\lambda}{2n} \quad (m = 0, 1, 2, 3, \ldots) \tag{11.29}$$

is satisfied. This is related to Newton's rings. (See Problem 11.7.)

Coatings on camera lenses work on the same principle except that we now
have three media: air, film, and glass (see Figure 11–18). We assume that
$n_g > n_f > n_a = 1$ holds where n_g is the index of refraction of the glass lens
and n_f is that of the film. The reflected light now suffers a phase change of π
at *both* surfaces and we have

$$2d = \frac{m\lambda}{n_f} \quad \text{for constructive interference} \tag{11.30}$$

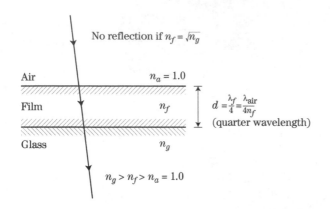

FIGURE 11–18

A dielectric film with a thickness of $\lambda/4$ coating on glass can prevent light reflection.

and

$$2d = \left(m + \frac{1}{2} \right) \frac{\lambda}{n_f} \quad \text{for destructive interference} \tag{11.31}$$

A case of particular importance is destructive interference which is indicated by a minimum reflected light intensity. The minimum thickness corresponds to $m = 0$ and we have

$$d = \frac{\lambda}{4n_f} = \frac{\text{wavelength in film}}{4} \tag{11.32}$$

This is the well-known *quarter wavelength coating* that is routinely applied to high-quality optical devices. It is possible to reduce the reflection to be less than 0.1%, depending on the uniformity of the coating and the glass surface.

The $\lambda/4$ coating alone, however, cannot completely eliminate reflection. It gives us only a necessary condition. Another condition to be imposed is that the amplitudes of both reflected waves E_1 and E_2 be the same. Then the waves that are reflected at both surfaces exactly cancel and we can have complete destructive interference. Let Γ_1 be the electric field reflection coefficient at the air–film boundary and Γ_2 that of the film–glass boundary. The impedances of air, film, and glass are, $Z_a = 377\,\Omega$, $Z_f = 377/n_f\,\Omega$, and $Z_g = 377/n_g\,\Omega$, respectively. Then the reflection coefficient at the air–film boundary is

$$\Gamma_1 = \frac{1 - n_f}{1 + n_f} \tag{11.33}$$

and that at the film–glass boundary is

$$\Gamma_2 = \frac{n_f - n_g}{n_f + n_g} \tag{11.34}$$

Thus the electric field E_1 of the wave reflected at the air–film boundary is

$$E_1 = \Gamma_1 E_0 \tag{11.35}$$

and the electric field E_2 of the wave reflected at the film–glass boundary is

$$E_2 = -\Gamma_2 E_0 \tag{11.36}$$

where the minus sign here is due to the path difference of

$$2 \times \frac{\lambda_f}{4} = \frac{\lambda_f}{2}$$

between the two reflected waves. (Recall that a path difference of one-half wavelength corresponds to a phase difference of π.) For complete destructive interference, $E_1 + E_2 = 0$ is required. Then the condition for no reflection is

$$\Gamma_1 = \Gamma_2$$

or, in terms of n's, we have

$$(1 - n_f)(n_f + n_g) = (1 + n_f)(n_f - n_g)$$

which yields

$$n_f = \sqrt{n_a n_g} = \sqrt{n_g} \tag{11.37}$$

This can be rewritten in terms of the characteristic impedances as

$$Z_f = \sqrt{Z_a Z_g} \tag{11.38}$$

You may ask why we can neglect the reflection at the first surface in calculating the second reflection. The reason is that we are seeking the condition for complete transmission and, in this case, we may assume the incident wave E_0 in calculating the second reflected wave. An analysis based on multiple reflections yields exactly the same result.

EXAMPLE 11.4

Find the thickness of the coating and its index of refraction to minimize light reflection on the surface of glass having $n_g = 1.5$. Assume $\lambda = 5000$ Å in air.

Solution

From $n_f^2 = n_g$, we find $n_f = \sqrt{1.5} = 1.22$. The coating should be $\lambda_f/4$, where λ_f is the wavelength in the film. Then

$$\frac{1}{4}\lambda_f = \frac{1}{4n_f}\lambda_{\text{air}} = \frac{1}{4 \times 1.22} \times 5000 = 1000 \text{ Å}$$

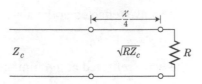

FIGURE 11–19

Quarter wavelength impedance transformer.

The concept of impedance matching using a quarter wavelength medium can also be applied to transmission line problems. When a load resistance $R(\neq Z_c)$ is to be matched to a transmission line having a characteristic impedance Z_c, one should insert another transmission line that has a characteristic impedance $\sqrt{RZ_c}$ and is a quarter wavelength long between the load and the transmission line. The impedance seen by the transmission line is then Z_c or matching is achieved. Note that the wavelength λ' is that of waves in the inserted transmission line and it may not be equal to the wavelength in the transmission line that is to be matched (see Figure 11–19).

11.6   **Diffraction I (Fraunhofer Diffraction)**

We have seen that a narrow opening, such as a slit in a flat surface, appears to act as a light source that radiates waves in the region beyond the narrow opening. Even though a plane wave is incident upon the narrow slit, a cylindrical wave will be emitted in the region behind it (Figure 11–20). In other words, light does not always travel along a straight line. Another example is an AM radio wave that is received even behind a high mountain. The radio waves can

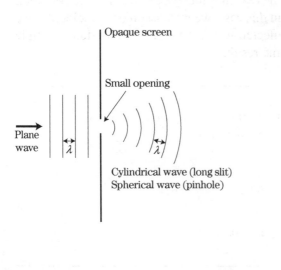

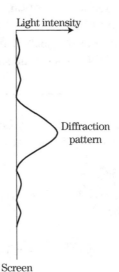

FIGURE 11–20

A small opening (pinhole or a narrow slit) "diffracts" light. Light does not travel in a straight line.

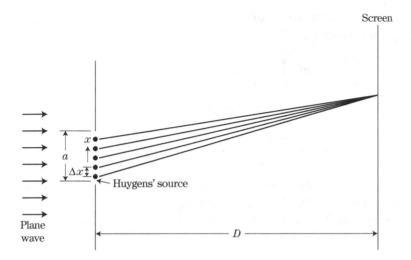

FIGURE 11–21

A large number of coherent light sources to replace the opening.

propagate around the mountain without too much difficulty. Television waves, on the other hand, are difficult to receive in similar circumstances. You may intuitively see that waves with short wavelengths tend to travel along straight lines and those with long wavelengths suffer a stronger bending. This bending process is called *refraction* or *diffraction.*

The analysis that we are going to do for diffraction is very similar to what we did for the multi-slit (grating) structure. As Huygens' principle assures us, we may assume a large number of infinitesimal light sources that are equally spaced at the slit opening that has a width a (Figure 11–21). The difference between this case and the multi-slit case is that here we have no well-defined spacing between two neighboring light sources. Rather, we consider an infinitely large number of light sources and eventually let $\Delta x \to 0$.

If D is sufficiently large, the phase difference between two neighboring waves is given by

$$\delta = \frac{2\pi}{\lambda} \Delta x \sin \theta \tag{11.39}$$

that is a constant along x. The diffraction in this case is called *Fraunhofer diffraction* and is the easiest to analyze. We may substitute the preceding phase difference δ into Eq. (11.22) for the multi-slit case,

$$I = I_0 \frac{\sin^2 \left(\frac{\pi}{\lambda} N \Delta x \sin \theta \right)}{\sin^2 \left(\frac{\pi}{\lambda} \Delta x \sin \theta \right)} \tag{11.40}$$

However, $N\Delta x = a$ and if we make N very large, Δx becomes very small. Then we may approximate

$$\sin\left(\frac{\pi}{\lambda}\Delta x \sin\theta\right) \simeq \frac{\pi}{\lambda}\Delta x \sin\theta = \frac{1}{N}\frac{\pi}{\lambda}a \sin\theta$$

Therefore the intensity becomes proportional to a function $(\sin\alpha/\alpha)^2$,

$$I = I_0\left(\frac{\sin\alpha}{\alpha}\right)^2, \quad \alpha = \frac{\pi}{\lambda}a \sin\theta \qquad (11.41)$$

which gives the Fraunhofer diffraction pattern (as a function of the angular location θ) caused by a single slit with an opening width a.

The function $(\sin\alpha/\alpha)^2$ is plotted in Figure 11–22. The angular spread is approximately given by $\Delta\alpha \simeq \pi$ or

$$\sin\theta = \frac{\lambda}{a} \qquad (11.42)$$

and on a screen a distance D away, the vertical spread Δy is given approximately by

$$\Delta y \simeq D\frac{\lambda}{a} \qquad (11.43)$$

Therefore, the ideal square-shaped image that would be observed if there were no diffraction appears as a blurred image without clear-cut edges as shown in Figure 11–23. As the wavelength λ increases, the image becomes more widely spread out; that is, long wavelengths suffer stronger diffraction.

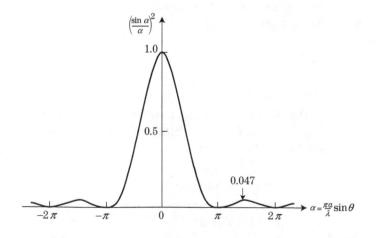

FIGURE 11–22

The function $(\sin\alpha/\alpha)^2$ plotted versus α. It does not diverge at $\alpha = 0$.

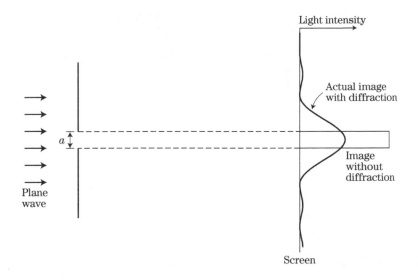

FIGURE 11–23

Diffraction pattern caused by a small opening.

Resolution of Optical Devices

As long as light has a wave nature, we cannot avoid diffraction. Diffraction imposes a serious limitation on optical devices such as telescopes, microscopes, the human eye, and so on. Most optical devices have circular lenses, or apertures, and we would have to analyze the diffraction caused by a circular aperture rather than a one-dimensional slit. The analysis, however, is complicated and we do not attempt to do it here. It only introduces a numerical factor of 1.22 in Eq. (11.42) and the angular spread for a circular aperture is given by

$$\sin\theta = 1.22\frac{\lambda}{a} \qquad (11.44)$$

where a is the diameter of the aperture.

Consider two light sources that are separated by an angle β as seen from a lens (Figure 11–24). Images formed by the lens are inevitably blurred because of diffraction. Their angular spread is given by Eq. (11.44) about each image. If β becomes small, the two images are superposed and we cannot tell which is which anymore. It is obvious that this critical angle is approximately given by θ itself.

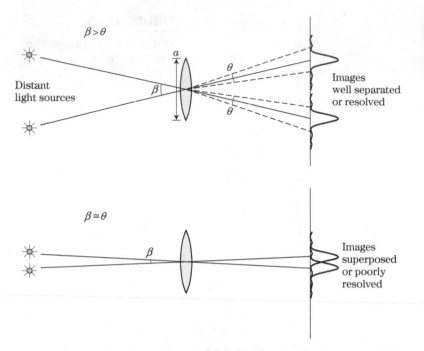

FIGURE 11–24

Two diffraction patterns (or images) get superposed on each other as the angle β becomes smaller. Light intensities are additive when the sources are incoherent.

EXAMPLE 11.5

Let us take a human eye as an example. We assume it has a 5-mm aperture and wish to find the resolution for green light, $\lambda = 5500 \text{ Å}$. The human eye is most sensitive to the color green.

Solution

From Eq. (11.44) (notice $\sin\theta \simeq \theta$ for small θ),

$$\theta \simeq 1.22 \times \frac{5.5 \times 10^{-5}}{0.5} = 1.3 \times 10^{-4} \text{ rad}$$

Thus if the two headlights of a car are 1.5 m apart, a human eye cannot resolve them as two separate light sources at a distance more than

$$D = \frac{1.5}{1.3 \times 10^{-4}} = 11.5 \text{ km}$$

This sounds too good. Of course, we have assumed that the resolution is limited only by diffraction and have neglected other effects such as the finite size of the photoreceptors of the retina, aging, and so on.

Equation (11.44) also explains why an electron microscope can "see" better than an optical microscope. The diffraction becomes smaller as the wavelength decreases and therefore the resolving power improves as the wavelength decreases. According to de Broglie, any object having a momentum p has a wave nature. Its wavelength is

$$\lambda_{\text{deBroglie}} = \frac{h}{p} \tag{11.45}$$

where $h = 6.63 \times 10^{-34}$ J · sec is *Planck's constant*, which we will encounter when studying the photoelectric effect. Consider an energetic electron having an energy of $100\,\text{keV} = 10^5 \times 1.6 \times 10^{-19}$ J. From

$$\frac{1}{2}mv^2 = 10^5 \times 1.6 \times 10^{-19} \text{ J}$$

we find $mv = 1.7 \times 10^{-22}$ kg m/sec. Then the wavelength associated with the electron is

$$\lambda = 3.9 \times 10^{-12} \text{ m} = 3.9 \times 10^{-2} \text{ Å}$$

which is roughly 10^5 times shorter than the wavelength of visible light (4000–7000 Å). Thus the diffraction is expected to be extremely small and the resolution of electron microscopes is expected to be much better than that of optical microscopes.

Another example is an astronomical telescope. The larger the aperture diameter a is, the more light is collected and the brighter the image becomes. However, a more important benefit is the higher resolving power of the telescope.

Diffraction II (Fresnel Diffraction)

In Fraunhofer diffraction, we assumed that the distance between the slit and the screen is large enough so that the phase varies linearly with x for $0 \leq x \leq a$. If the screen is brought closer to the slit, however, this linear phase variation breaks down.

We first consider an obstacle blocking a light beam. If there were no diffraction, the image on the screen would be a sharply defined step function. The actual image, however, is blurred with a wavy structure as shown qualitatively in Figure 11–25. Light can even go around the obstacle.

In order to find the light intensity at the point P on the screen at a distance y above the edge of the obstacle, consider the phase difference between the waves emitted from points A and B. The path difference is approximately

$$\sqrt{D^2 + h^2} - D \simeq D\left(1 + \frac{h^2}{2D^2}\right) - D = \frac{h^2}{2D} \quad (D \gg h)$$

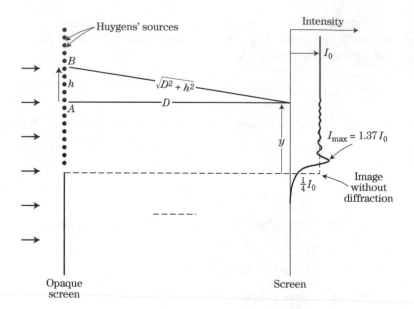

FIGURE 11–25

Fresnel diffraction pattern caused by a semi-infinite opaque screen.

which is proportional to h^2. This is in contrast with the previous cases of interference and the Fraunhofer diffraction. In those cases, we assumed that the distance D was almost infinite and the phase difference was proportional to h.

The phase difference corresponding to the path difference $h^2/2D$ is

$$\phi(h) = \frac{2\pi}{\lambda} \frac{h^2}{2D} = \frac{\pi h^2}{\lambda D} \tag{11.46}$$

in contrast with the case of multi-slit interference and Fraunhofer diffraction in which we found

$$\phi(h) = \frac{2\pi}{\lambda} h \sin\theta$$

that was independent of the separation distance D as noted in Figure 11–26.

To derive this difference more carefully, consider two sources located at h_1 and h_2 as indicated in Figure 11–27. The path difference is given by

$$\sqrt{D^2 + h_2^2} - \sqrt{D^2 + h_1^2} \simeq D\left(1 + \frac{h_2^2}{2D^2}\right) - D\left(1 + \frac{h_1^2}{2D^2}\right) = \frac{1}{2D}(h_2^2 - h_1^2)$$

Letting $h_2 = h + h_1$, we find

$$\text{Path difference} = \frac{1}{2D}\left(2hh_1 + h^2\right)$$

If $D \gg h_1, h_2$, we may approximate

$$\sin\theta \simeq \frac{h_1}{D}$$

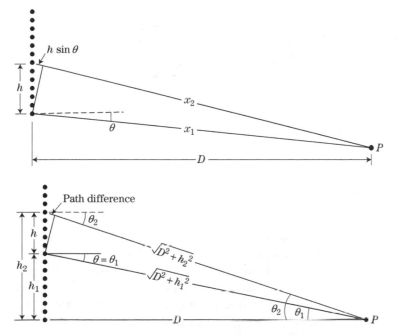

FIGURE 11–26

If the point P is very far away from the light sources, the path difference is $h \sin \theta$ and it is independent of the distance D.

FIGURE 11–27

As D becomes smaller, the difference between the angles θ_1 and θ_2 is not negligible.

Then the path difference becomes

$$h \sin \theta + \frac{h^2}{2D}$$

In the case of multi-slit interference and Fraunhofer diffraction, we retained only the first term by assuming $D \rightarrow \infty$. If the angle θ is small, the path difference is dominated by the quadratic term $h^2/2D$, which characterizes Fresnel diffraction.

What about the amplitude of the electric fields? As we have seen before, the amplitude of cylindrical waves is inversely proportional to the square root of the distance. In the present case, the amplitude of the electric field emitted at A in Figure 11–25 is

$$E_A \propto \frac{1}{\sqrt{D}}$$

and that of the field emitted at B is

$$E_B \propto \frac{1}{(D^2 + h^2)^{1/4}}$$

If $D^2 \gg h^2$, the amplitude difference is still negligible and we may assume that the electric fields all have the same amplitude. Otherwise, the analysis would be insurmountably complicated.

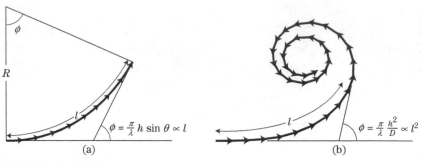

FIGURE 11–28

Phase diagrams for (a) Fraunhofer and (b) Fresnel diffraction.

Now we are ready to draw a phase vector diagram for the Fresnel diffraction in Figure 11–28. For comparison, the phase diagram for the Fraunhofer diffraction is also shown. This latter diagram forms a circular arc since the phase difference is proportional to the distance h. In contrast, for the case of Fresnel diffraction, the phase difference increases more rapidly since it is proportional to h^2 and the phase diagram becomes a spiral. Unfortunately, there are no simple mathematical equations to describe the spiral. The only thing we can do is describe it in terms of a parameter h, noting that the length l along the spiral curve is related to the phase angle ϕ through

$$\phi = \frac{\pi}{\lambda}\frac{h^2}{D}, \quad l = \text{const} \times h$$

where l is the length along the spiral. The spiral shown in Figure 11–29 is known as the Cornu spiral, which can be parametrically described by the

FIGURE 11–29

Cornu spiral.

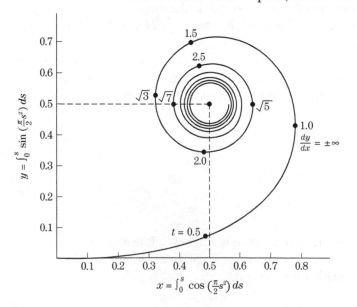

so-called Fresnel integrals

$$C(s) = \int_0^s \cos\left(\frac{\pi}{2}s^2\right) ds \tag{11.47}$$

$$S(s) = \int_0^s \sin\left(\frac{\pi}{2}s^2\right) ds \tag{11.48}$$

Using the spiral, we can qualitatively discuss how a Fresnel diffraction pattern appears when formed by an obstacle with a sharp edge such as a knife. We choose the reference electric field E_R at the origin O in Figure 11–25 as the one emitted by the source located at the same height as the observing point on the screen $h = 0$. Consider an observing point on the screen above the edge $y = 0$. The Huygens' light sources below the point A in Figure 11–25 are in the third quadrant in Figure 11–30 and those above the point A are in the first quadrant. The amplitude of the electric field at the point P on the screen is then given by the length AP in Figure 11–30. As y increases from zero, the length AP first increases and then oscillates about the constant length AA', which corresponds to the field amplitude on the screen well above the edge. It finally assumes a constant value corresponding to the length AA'. Note that the field amplitude at $y = 0$ is just one half of the unperturbed field at $y = \infty$ and thus the light intensity at $y = 0$ is one-quarter of the unperturbed intensity.

If the observing point P' on the screen is below the edge or behind the obstacle ($y < 0$), the Huygens' light sources start at $h = -y$ in the first quadrant

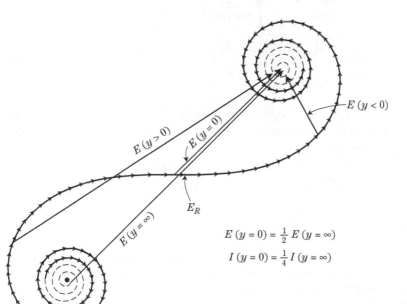

FIGURE 11–30

Method to find the total electric field at various positions y on the screen. Note:

$$E(y = 0) = \frac{1}{2}E(y = \infty).$$

$E(y < 0)$

$E(y > 0)$ $E(y = 0)$

E_R

$E(y = \infty)$

$$E(y = 0) = \tfrac{1}{2}E(y = \infty)$$
$$I(y = 0) = \tfrac{1}{4}I(y = \infty)$$

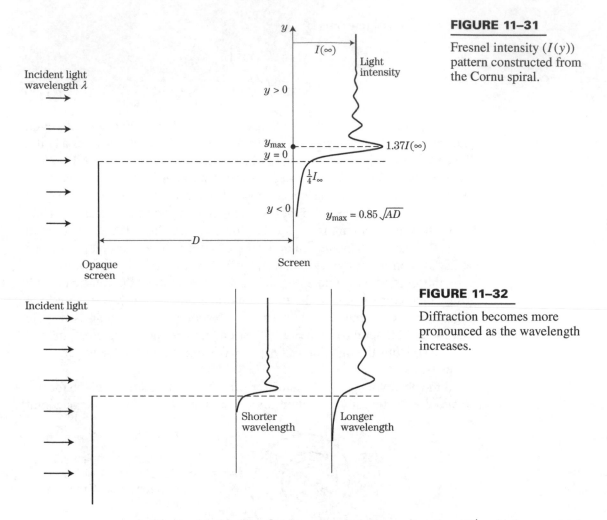

FIGURE 11–31

Fresnel intensity ($I(y)$) pattern constructed from the Cornu spiral.

FIGURE 11–32

Diffraction becomes more pronounced as the wavelength increases.

in Figure 11–30. The field amplitude is now given by AP', which monotonically decreases as the observing point is lowered, and finally becomes zero.

The light intensity $I(y) \propto E^2(y)$ is qualitatively shown in Figure 11–31. The Fresnel diffraction along with the Fraunhofer diffraction becomes more pronounced as the wavelength increases (Figure 11–32).

11.9 Problems

1. In Young's double-slit experiment, a fringe spacing of $\Delta y = 5$ mm is observed. Assuming the slit separation $d = 0.1$ mm and the slit-screen distance $D = 1$ m, find the wavelength.

2. In Young's double-slit experiment, one slit is covered with a thin mica film. Discuss what changes should result.

3. What would happen if the whole apparatus of Young's double-slit experiment is immersed in water having an index of refraction of $n = 1.3$?

4. In the diagram (Figure 11–33) S_1 and S_2 are two coherent wave sources, which radiate waves spherically in every direction. The field amplitude thus has $1/r$ dependence. Assuming that the field vector is normal to the page, find the wave intensity along the x axis as a function of x. (*Hint:* At P, the amplitudes of the fields emitted by S_1 and S_2 are not equal.)

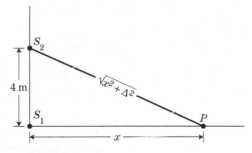

FIGURE 11–33

Problem 4.

5. A dielectric film is coated on glass surface to prevent reflection of light having $\lambda = 700$ nm. If the glass has an index of refraction of 1.5, find the index of refraction of the film and its thickness.

6. *Directional coupler.* In microwave waveguide circuits, a device called a directional coupler is frequently used, when it is desired that microwave energy be branched off into another waveguide system. It consists of two waveguides joined together (Figure 11–34). Through the wall, two holes

are $\lambda/4$ apart drilled. Explain why no microwaves can exist in the region III.

7. *Newton's rings.* A planoconvex lens rests on a flat glass surface. Light of wavelength λ falls normal to the plane surface (Figure 11–35).
   (a) Find the spacing d as a function r, the radial position. Assume R (curvature radius) $R \gg r$.
   (b) Show that bright interference fringes are located at the positions given by

$$r = \left[\left(m + \frac{1}{2} \right) R\lambda \right]^{1/2}$$

where m is an integer.

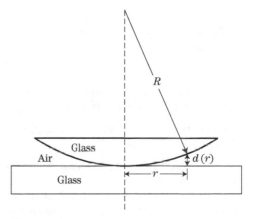

FIGURE 11–35

Problem 7.

8. A spy satellite is claimed to be able to resolve two points on the earth that are separated by a distance

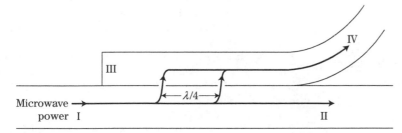

FIGURE 11–34

Problem 6.

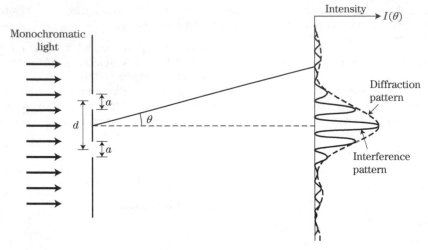

FIGURE 11–36

Problem 10.

of 50 cm. Assuming the satellite is 200 km high and $\lambda = 500\,\text{nm} = 5000\,\text{Å}$, find the minimum diameter of the telescope carried by the satellite.

9. In single-slit diffraction, what would happen if the slit opening is doubled?

10. *Diffraction in double-slit experiment.* Assume each slit has an opening a in Young's double-slit experiment (Figure 11–36). Then, in addition to the interference, we expect diffraction due to the finite aperture. Show that the intensity on the screen is given by

$$I = I_0 \cos^2 \beta \left(\frac{\sin \alpha}{\alpha} \right)^2$$

where

$$\alpha = \frac{\pi a}{\lambda} \sin \theta$$

$$\beta = \frac{\pi d}{\lambda} \sin \theta$$

11. In Problem 10, assuming $d = 2a$, plot the light intensity as a function of θ.

12. Explain why an AM radio can be received better than an FM radio in mountain areas.

    (*Note:* You may wonder why the FM radio or television stations do not use lower frequencies since waves with lower frequencies can be diffracted more

and received better in the presence of obstacles. The reason is simply that it is impossible. When we say an AM radio wave has a frequency, say, of 540 kHz, what we actually mean is that the radio station uses a frequency *band* 520 kHz $< f <$ 560 kHz where ± 20 kHz corresponds to the maximum audio frequency. For FM signals the required frequency *band* is of the order of 1 MHz. The carrier frequency must be much higher than this and in fact all FM radio stations have carrier frequencies around 100 MHz.)

13. The schematic diagram of a spectrometer is shown in Figure 11–37. Light enters the entrance slit, is reflected by a spherical mirror ($M1$), hits a grating (G), is reflected by another mirror ($M2$), and finally emerges from an exit slit. The grating is rotatable and a desired wavelength can thus be chosen. The distance between the grating and the slit corresponds to the slit-screen distance D in the text.

    (a) It is desired that the *dispersion* ($\Delta\lambda/\Delta y$) be 10 Å/mm at the exit slit. Assuming $\lambda = 500\,\text{nm} = 5000\,\text{Å}$, $D = 1.3$ m, and m (order) $= 1$, find the number of grooves (in 1 cm) of the grating.

    (b) Assuming that the grating has a width of 10 cm, estimate the half-width of the intensity profile at $\lambda = 5000\,\text{Å}$.

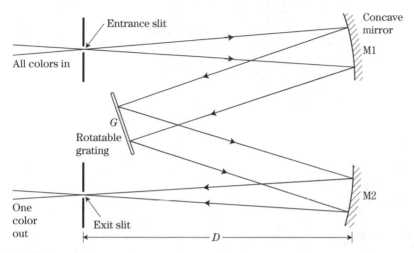

FIGURE 11–37

Problem 13.

14. Hold horizontally an LP record (if you can find one!) close to your eyes and let the record reflect light coming from a distant light source. You will see a color spectrum. Explain. (The color spectrum seen on a compact disk is also caused by diffraction.)

15. Look at a distant point light source (a flashlight will do) through a nylon stocking. You will find several colored rings around the light source. Can you explain? (Rings around the moon on a foggy night are caused by the same principle. They are due to diffraction by small water drops.)

16. Consider a glass surface coated with a quarter-wavelength-thick dielectric film as in Figure 11–18. Show that the effective impedance for light incident on the coated glass is given by Z_f^2/Z_g, where Z_f is the characteristic impedance of film and Z_g is that of glass.

17. Light of wavelength 600 nm = 6000 Å (in air) is incident normally on a plastic film that has a permittivity $\varepsilon = 4\varepsilon_0$ (Figure 11–38).
   (a) If the film is to be a quarter wavelength thick, what should the thickness be?
   (b) Calculate how much (in %) energy is reflected and transmitted.

(*Hint:* Use the result of Problem 16.)

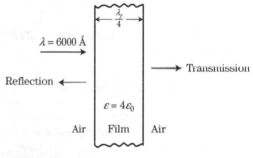

FIGURE 11–38

Problem 17.

18. Radiation from an antenna having a size comparable with the wavelength can be analyzed as an interference problem.
   (a) Calculate the radiation electric field from a half-wavelength dipole antenna (Figure 11–39). Assume that the current is distributed on the antenna as $I(z, t) = I_0 \cos(kz) \sin(\omega t)$.
   (b) Sketch the radiation field pattern as a function of θ.

Note: The power radiated by a half-wavelength antenna is given by

$$P = 73.2(\Omega)I_0^2 \text{ (W)}$$

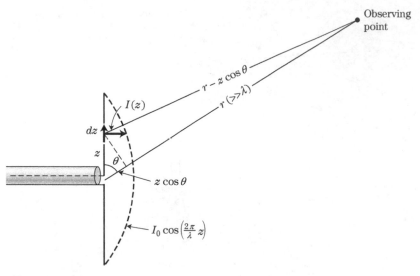

FIGURE 11–39

Problem 18.

where 73.2 (Ω) is the radiation resistance of the half-wave dipole antenna.

19. **(a)** Using the expression for the radiation electric field due to an oscillating dipole, Eq. (10.24), show that the amplitude of the radiation electric field due to a small oscillating current loop (Figure 11–40) is azimuthal (directed in ϕ direction) and given by

$$E_\phi(r, \theta) = \frac{\mu_0 Mc}{4\pi r} \frac{1}{4} \left(\frac{\omega}{c}\right)^2 \sin\theta$$

where $M = I_0 \pi a^2$ is the magnetic dipole moment.

(b) Calculate the total power radiated by the magnetic dipole.

20. **(a)** Determine the orbit radius of a geostationary satellite.

(b) If a microwave parabola antenna carried by the satellite is to cover a circular area with a radius of 300 km on the earth, what should the diameter of the antenna be? Assume $f = 4$ GHz.

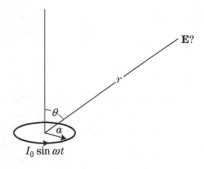

FIGURE 11–40

Problem 19.

21. Two incandescent light bulbs of equal size (2 cm in radius), one with opaque glass and the other with transparent glass, are placed side by side. Assuming the pupil diameter of 3 mm and $\lambda = 550$ nm, estimate the observing distance at which the bulbs appear to be of equal size.

Geometrical Optics

12.1 Introduction

In Chapter 11, we learned that the diffraction pattern depends strongly on the size of the aperture a of the optical instruments. A small circular aperture essentially acts as a point light source or a pinhole and no clear image can be formed on the screen behind the hole because of diffraction. As the aperture size increases, the diffraction becomes less significant and the image becomes clearer.

Geometrical optics is one branch of optics in which we are able to neglect interference and diffraction. We assume that a light beam propagates along a straight line in a uniform medium, neglecting any increase in the angular spread that is inevitable as the beam propagates. On encountering a foreign medium (from air to glass, for example), the light beam changes its direction, being reflected and refracted, but both the reflected and refracted beams again travel along straight lines. Geometrical optics can greatly simplify the analysis of conventional optical devices such as mirrors and lenses as long as the medium (e.g., lens glass) is uniform. Another requirement is that the boundary between two different media be smooth. For example, this is usually satisfied for the transition between air and glass in optical devices. Otherwise, the reflection and refraction of the optical wave may become random and the light beam may be randomly directed in space.

Reflection and Refraction

Suppose a beam of light is incident obliquely on a flat surface of glass with an index of refraction $n_g = \sqrt{\varepsilon_g/\varepsilon_0}$ as shown In Figure 12–1. The speed of light in the glass is c/n_g, and the wavelength in glass is shorter than that in air by a factor of n_g/n_a, where n_a is the index of refraction of air which is close to unity. (The permittivity of air at one atmospheric pressure, $0°$ C is approximately 1.0003.)

You may know that the angle of reflection θ_1' is equal to the angle of incidence θ_1 just like the case of an elastic ball hitting a heavy wall or the dribbling of a basketball on a court. Here we prove this and also find the relationship between the angle of incidence θ_1 and the angle of refraction θ_2. Before doing this, we need some mathematics. As we have seen, a plane wave (Figure 12–2) propagating in the x direction is described by

$$E(x, t) = E_0 \sin(kx - \omega t) \qquad (12.1)$$

where $k = 2\pi/\lambda$ is the wavenumber. How can we describe the wave propagating in the direction that is not on the axis as shown in Figure 12–3? It can be written using vector notation as

$$E(x, y, t) = E_0 \sin(ks - \omega t) \qquad (12.2)$$

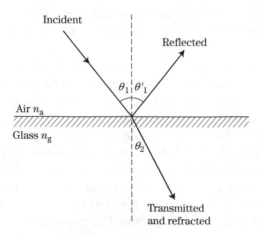

FIGURE 12–1

Light falling on a glass surface at an angle θ_1.

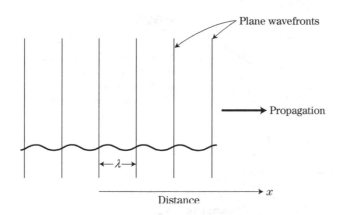

Plane wavefronts

Propagation

$\leftarrow \lambda \rightarrow$

Distance

x

FIGURE 12–2

Plane wave propagating in the x direction.

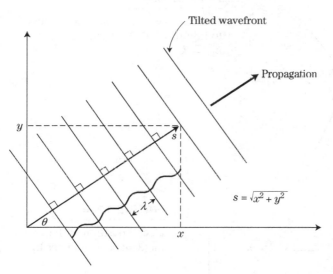

Tilted wavefront

Propagation

y

s

λ

$s = \sqrt{x^2 + y^2}$

θ

x

FIGURE 12–3

Plane wave propagating in a direction that is oblique with respect to the coordinates.

where $s = \sqrt{x^2 + y^2}$ is the distance along the straight line OS. The wave number k is still defined in terms of the wavelength $k = 2\pi/\lambda$. Now we rewrite ks as (note that $\cos^2 \theta + \sin^2 \theta = 1$)

$$ks = k \cos \theta\, s \cos \theta + k \sin \theta\, s \sin \theta$$

From Figure 12–4, we note that $s \cos \theta = x$, and $s \sin \theta = y$. Thus

$$E(x, y, t) = E_0 \sin(ks - \omega t) \tag{12.2}$$

$$ks = k \cos \theta\, x + k \sin \theta\, y \tag{12.3}$$

It is now clear that k is actually a vector directed in the direction of the wave propagation. Its magnitude is still $2\pi/\lambda$ and its x component is $k \cos \theta$ and its

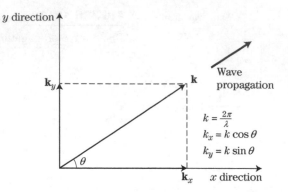

FIGURE 12–4

The wave number k is actually a vector directed in the direction of the propagation of the wave. The magnitude of the wave vector is $2\pi/\lambda$.

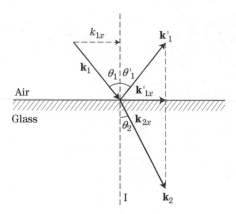

FIGURE 12–5

Wavenumber $\mathbf{k}$ diagram corresponding to Figure 12–4. Note that $k_{1x} = k'_{1x} = k_{2x}$.

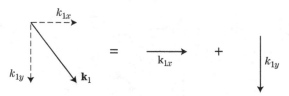

FIGURE 12–6

Decomposition of the incident wave number $\mathbf{k_1}$.

y component is $k \sin \theta$. We can write ks using the vector notation as

$$ks = \mathbf{k} \cdot \mathbf{r}, \quad \mathbf{r} = \mathbf{x} + \mathbf{y} \tag{12.4}$$

and a wave propagating in the direction $\mathbf{r}$ is conveniently written as

$$E = E_0 \sin(\mathbf{k} \cdot \mathbf{r} - \omega t) \tag{12.5}$$

For the problem of reflection and refraction, we assign $\mathbf{k_1}$, $\mathbf{k'_1}$, and $\mathbf{k_2}$ for the incident, reflected, and refracted waves, respectively (Figure 12–5). The incident wave is actually composed of two waves, one propagating in the negative y direction and another propagating in the positive x direction (Figure 12–6). The component propagating along the x direction never touches the boundary and therefore the x component of k should not change. Thus

$$k'_{1x} = k_{1x}$$

The vertical component k_{1y} should simply change its sign upon reflection. Thus

$$k'_{1y} = -k_{1y}$$

Therefore, the angles θ_1 and θ'_1 must be equal to each other.

For refraction, the x components k_{1x} and k_{2x} are the same as shown in Figure 12–7. The number of waves contained in a given distance along the x

(a)

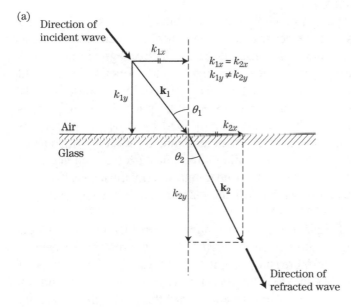

FIGURE 12–7

(a) Change in the vector **k** due to refraction. The component parallel to the boundary (k_x) remains unchanged. (b) Illustration of refraction by wave planes. Note that the wavelength along x (λ_x) does not change.

(b)

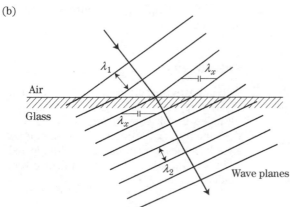

axis is the same in both air and glass. Thus

$$k_1 \sin \theta_1 = k_2 \sin \theta_2 \qquad (12.6)$$

However, the light velocity in glass is c/n_g, or

$$\frac{\omega}{k_2} = \frac{c}{n_g}$$

Thus, we find after recalling $\omega/k_1 = c/n_a$

$$\frac{\sin \theta_1}{\sin \theta_2} = \frac{k_2}{k_1} = \frac{n_g}{n_a} \qquad (12.7)$$

This is known as *Snell's law*.

There is a more rigorous procedure to derive Snell's law using the proper boundary conditions for the electric and magnetic fields associated with light. In other words, Snell's law is a consequence of electromagnetic properties formulated by Maxwell. The physical meaning behind refraction is that light (or any electromagnetic wave) is bent toward a region of lower phase velocity as we briefly saw before in Chapter 11. Since the light velocity in glass is smaller than that in air, light appears to be pulled by the glass and k_{2y} is indeed larger than k_{1y}. Reflection, on the other hand, is caused by impedance mismatching, which we also have studied. Glass has a lower characteristic impedance than does air and some electromagnetic energy must be reflected at the air glass interface (about 4% for ordinary glass if light is normally incident upon the glass).

As an example of light refraction in a nonuniform medium, let us take a look at the bending of sunlight in the earth's atmosphere. The air density gradually decreases with increasing height (Figure 12–8). The velocity of light in air is not exactly equal to c, but it is slightly less, and this depends on the air density. Thus the light beam is bent toward the lower phase velocity region where the air density is greater. This example is in contrast with the mirage discussed in Chapter 9 (see problem 9.21).

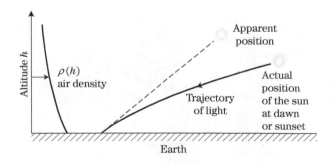

FIGURE 12–8

The index of refraction of air (1 atmosphere, 0° C) is approximately 1.0003. At higher altitudes, the air density decreases and so does the index of refraction. The velocity of light becomes a function of the altitude and light is refracted or bent toward the slower velocity region.

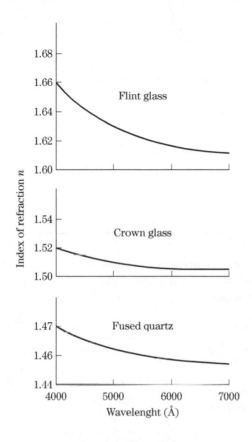

FIGURE 12–9

The index of refraction of typical glass and quartz as functions of wavelength.

The index of refraction n of a given material (say, glass) is not a constant but depends on the wavelength λ. The index of refraction of flint glass, crown glass, and quartz as a function of the wavelength in the visible range 4000–7000 Å is shown in Figure 12–9. It can be seen that blue light is refracted more than red light and this property can be used to explain the operation of a prism spectrograph and color splitting in a rainbow. This wavelength dependence of the index of refraction is undesirable for lenses because it causes chromatic aberration as we will see later.

The approximate dependence of the index of refraction on the wavelength λ is

$$n(\lambda) \simeq A + \frac{B}{\lambda^2} \tag{12.7}$$

where A and B are constants that are specific to a given material. This is known as Cauchy's formula and can be derived from the general form of the

permittivity,

$$\varepsilon(\omega) = \varepsilon_0 \left(1 - \frac{\omega_p^2}{\omega^2 - \omega_0^2} \right)$$

where ω_p is the "plasma frequency" $\omega_p = \sqrt{n_e e^2 / \varepsilon_0 m_e}$ with n_e being the electron density of a particular state of atom and ω_0 is the frequency of the bound harmonic motion of electron in the state.

EXAMPLE 12.1

Monochromatic (single color or wavelength) light is incident upon the side surface of a 45° prism at an angle of 40°. If the index of refraction of glass is $n = 1.5$ at this wavelength, find the total deflection angle α (see Figure 12–10).

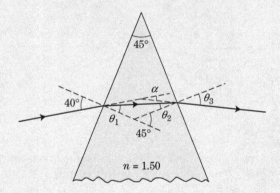

FIGURE 12–10

A single color of light entering a prism.

Solution

For the refraction at the first air–glass surface, we write using Snell's law that

$$\frac{\sin 40°}{\sin \theta_1} = 1.5$$

Then

$$\theta_1 = \sin^{-1}\left(\frac{\sin 40°}{1.5} \right) = 25.4°$$

Since

$$\theta_1 + \theta_2 = 45°, \text{ we find } \theta_2 = 19.6°$$

Applying again Snell's law at the exit glass-air surface,

$$\frac{\sin \theta_3}{\sin \theta_2} = 1.5$$

we find

$$\theta_3 = \sin^{-1}(\sin \theta_2 \times 1.5) = 30.2°$$

Then the total deflection angle is

$$\alpha = (40° - \theta_1) + (\theta_3 - \theta_2) = 25.2°$$

EXAMPLE 12.2

Repeat Example 12.1 by assuming another wavelength for which $n = 1.46$ (see Figure 12–11).

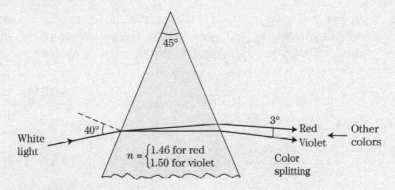

FIGURE 12–11

White light incident upon a prism.

Solution

$$\theta_1 = \sin^{-1} \frac{\sin 40°}{1.46} = 26.1°$$
$$\theta_2 = 45° - \theta_1 = 18.9°$$
$$\theta_3 = \sin^{-1}(\sin \theta_2 \times 1.46) = 28.2°$$

Then

$$\alpha = 40° - \theta_1 + \theta_3 - \theta_2 = 23.2°$$

Total Reflection

Consider a light beam emerging from the water surface into air. Water has an index of refraction of approximately 1.3. From Snell's law (see Figure 12–12), we write:

$$\frac{\sin \theta_2}{\sin \theta_1} = \frac{1}{n_w}$$

As the incident angle θ_2 increases, the refraction angle also θ_1 increases, and it finally reaches 90°. The incident angle at which this occurs is

$$\sin \theta_2 = \frac{1}{n_w} = \frac{1}{1.3}$$

or $\theta_2 = 50.3°$ for water. At angles larger than this, light is completely reflected at the boundary and no light will come out of water. The angle is called the *critical angle for total reflection.*

The optical fiber uses this principle of total reflection as shown in Figure 12–13. The light beam can be effectively guided along the fiber. Again, for light not to be scattered at the surface, the fiber surface must be extremely smooth without any perturbing lumps. Recent developments in the manufacturing of high-quality optical fibers have led to the use of laser beams for

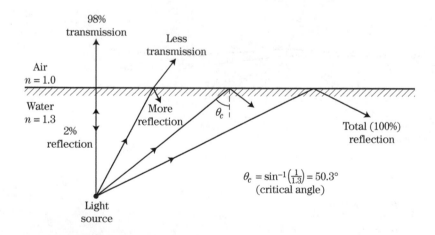

FIGURE 12–12

Total reflection occurs when the incident angle is larger than the critical angle θ_c.

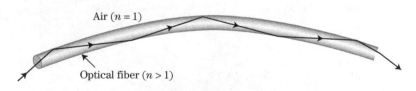

FIGURE 12–13

A hair-thin optical fiber can guide the light that is totally reflected at the surface. The fiber is a dielectric waveguide for light.

communication. The important aspects of optical fibers for communication are:

1. Small light attenuation and scattering.
2. Low dispersion.

Otherwise, the signals may be strongly damped and/or dispersed and the launched signal from the transmitter would become severely deformed. In practical fibers used in optical communication, the index of refraction is tailored to be nonuniform across the fiber's cross-section to minimize dispersion.

EXAMPLE 12.3

For what incident angles does total reflection occur at the vertical surface? Assume an index of refraction $n = 1.3$ (see Figure 12–14).

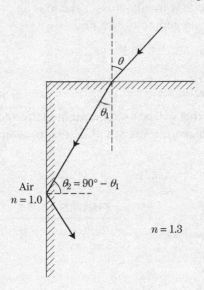

FIGURE 12–14

A wave incident upon a corner.

Solution

From Snell's law,

$$\frac{\sin \theta}{\sin \theta_1} = n = 1.3$$

Since $\theta_2 = 90° - \theta_1$ and the critical angle $\theta_c = \theta_2$ is given by

$$\sin \theta_2 = \cos \theta_1 = \frac{1}{n} = \frac{1}{1.3}$$

we find

$$\sin\theta = 1.3 \times \sin\theta_1 = 1.3\sqrt{1 - \cos^2\theta} = 1.3 \times \sqrt{1 - (1/1.3)^2} = 0.83$$

or

$$\theta = 56°$$

Total reflection at the vertical surface occurs if the angle θ is less than 56°.

Reflection at Spherical Surfaces (Mirrors)

Light reflection at a spherical surface works in the same way as that at the flat surface. The incident angle θ is now with respect to the radius of curvature R of the surface as shown in Figure 12–15. Let us place a point light source at a distance o from the mirror. We want to determine where a light beam leaving the source at an angle α to the axis finally crosses the axis again. Let this distance from the mirror be i. Using geometry, we find

$$\alpha + 2\theta = \beta, \quad \alpha + \theta = \gamma$$

or eliminating θ,

$$\alpha + \beta = 2\gamma \tag{12.8}$$

Here we make a major assumption that will be used throughout this section. That is, we assume that all the angles α, β, θ are small. When measured in radians, this means

$$\alpha, \beta, \gamma \ll 1 \quad \text{(rad)}$$

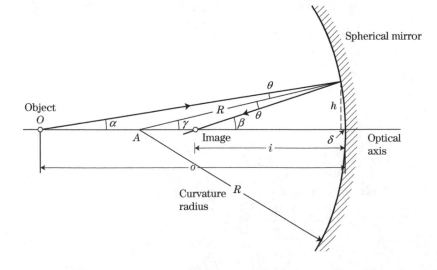

FIGURE 12–15

Reflection at a spherical surface.

This assumption can alternatively be put as $h, \delta \ll o, i, R$ where h is the height of the point where the light beam hits the mirror and δ is the deviation from the mirror center. This assumption is required to avoid spherical aberration and is equivalent to the assumption that the apertures of mirrors (and lenses) are small.

If α, β, γ are small, we may approximate

$$\alpha \simeq \tan\alpha, \; \beta \simeq \tan\beta, \; \gamma \simeq \tan\gamma$$

The Taylor series expansion for $\tan x$ (Chapter 3) is given by

$$\tan x = x + \frac{1}{3}x^3 + \frac{2}{15}x^5, \ldots$$

with $\tan\alpha = h/(o - \delta) \simeq h/o \; (\delta \ll o)$, $\tan\beta = h/i$ and $\tan\gamma = h/R$. Substituting these into Eq. (12.8), we find

$$\frac{1}{o} + \frac{1}{i} = \frac{2}{R} \tag{12.9}$$

which is the mirror formula. We define the focal length of the mirror as the image distance formed when the light source is placed at large distances from the mirror $o \to \infty$ and obtain $f \equiv R/2$. Using this *focal length*, the mirror formula (Figure 12–16) can be written as

$$\frac{1}{o} + \frac{1}{i} = \frac{1}{f} \tag{12.10}$$

Equation (12.10) can be generalized if we properly interpret the sign ($+$ or $-$) of all the quantities, o, i, and f. The sign convention we adopt is:

$o, i > 0$ for object and image in front of the mirror

$o, i < 0$ for object and image behind the mirror

$f > 0$ for concave mirrors ($R > 0$)

$f < 0$ for convex mirrors ($R < 0$).

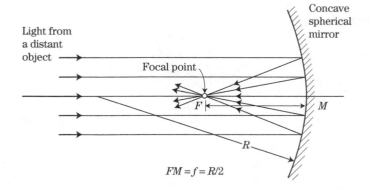

Concave spherical mirror

Light from a distant object

Focal point

$FM = f = R/2$

FIGURE 12–16

Rays parallel to the axis all converge (or are focused) at the focal point.

Let us work on the example in Figure 12–17. If an object is placed at $o = +30$ cm the image is formed at $i = +15$ cm as we easily find from Eq. (12.10). We write from Figure 12–18 that

$$i = \frac{R}{2} + \frac{h'}{\tan 2\theta}, \quad h' = \frac{i}{o}h$$

and

$$\tan \theta = \frac{h}{R}$$

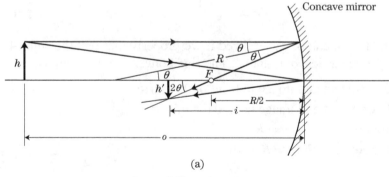

FIGURE 12–17

An example in which $o = 30$ cm, $f = 10$ cm, and $i = 15$ cm.

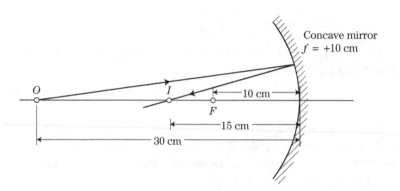

FIGURE 12–18

Finding the image location by ray tracing. (a) Concave mirror. (b) Convex mirror.

(a)

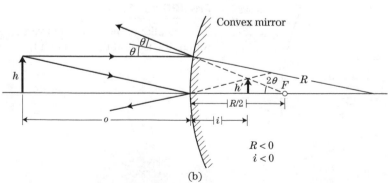

(b)

Again assuming a small angle θ so that $\tan \theta \simeq \theta$, we readily find

$$\frac{1}{o} + \frac{1}{i} = \frac{2}{R}$$

The ratio between the heights h' and h is called the magnification m. We define

$$m = -\frac{i}{o} \tag{12.11}$$

so that m becomes positive for an *erect image* (arrows in the same direction) and negative for an *inverted image* (arrows in the opposite direction). In the preceding example,

$$m = -\frac{15}{30} = -0.5$$

which tells us that the image size is one-half of the object size and it is inverted.

There is one more complication. Consider now $o = +5$ cm in the preceding example. If we directly use the formula

$$\frac{1}{5} + \frac{1}{i} = \frac{1}{10}$$

we find $i = -10$ cm ($m = +2.0$), which tells us that the image is formed behind the mirror ($i < 0$) (Figure 12–19). But light can never go into the region behind the mirror, and the image in this case is not formed by a real light beam intersecting with the axis. Rather, the image is formed by the beam extending into the region behind the mirror (broken line in the figure). Such an image is called a *virtual image*. The image formed by real light rays is called a *real image*.

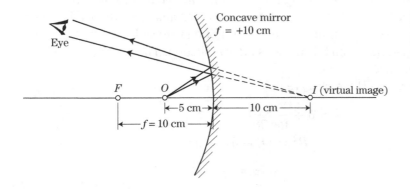

FIGURE 12–19

Example of virtual image. $i < 0$.

EXAMPLE 12.4

Virtual object. Consider a light beam falling on a concave mirror as shown in Figure 12–20. To find the location of the image, using Eq. (12.10), we insert the distance $o = -30$ cm since the beam intersects with the axis behind the mirror.

Solution

From

$$\frac{1}{-30} - \frac{1}{i} = \frac{1}{20}$$

we find $i = 12$ cm and m $= +0.4$.

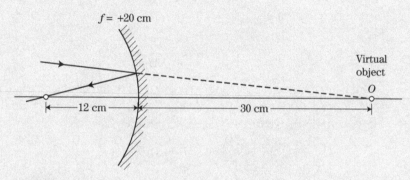

FIGURE 12–20

Creation of a virtual image.

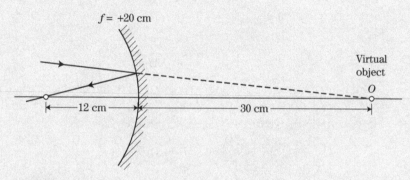

◆12.5◆ ## Spherical Aberration of Mirrors

In deriving the mirror formula, we assumed that h, the vertical distance between P and the axis, is small in comparison with the radius of curvature R. Under this condition, the light beams that are parallel to the axis are all focused at a point called the focal point $i = R/2$ and this distance is called the focal distance. Let us remove this assumption here and actually determine where the parallel beam hits the axis after reflection. The maximum value of h is obviously R.

In Figure 12–21 we observe that

$$i = \delta + \frac{h}{\tan 2\theta} = \delta + h\frac{\cos 2\theta}{\sin 2\theta}$$
$$R^2 = (R - \delta)^2 + h^2$$
$$\sin \theta = \frac{h}{R}$$

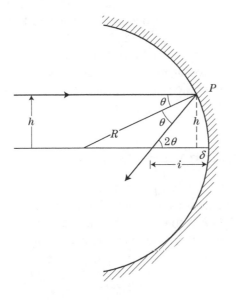

FIGURE 12–21

Ray tracing without assuming $h \ll R$.

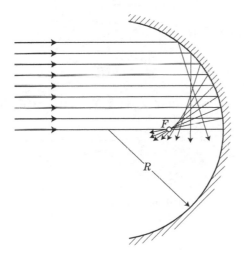

FIGURE 12–22

Incident rays parallel to the axis are not focused if the mirror aperture is large (spherical aberration of mirrors).

Eliminating δ and θ, we find

$$i = R \left(1 - \frac{1}{2\sqrt{1 - (h/R)^2}} \right) \qquad (12.12)$$

If $h \ll R$, we indeed recover $i = R/2 = f$. As h increases, the image distance i becomes smaller and at $h = (\sqrt{3}R/2)$, i becomes zero. The variation of the focal distance with the height h is called *spherical aberration* and it cannot be avoided as long as the reflector surface is spherical (Figure 12–22).

However, if h is small or if the opening aperture is kept small, spherical aberration can practically be neglected.

Spherical aberration can be avoided for light beams that are parallel to the axis by using a parabolic mirror. The proof will be left as a problem. Microwave antennas and reflecting astronomical telescopes usually have this parabolic shape to eliminate spherical aberration.

Refraction at Spherical Surfaces

Refraction at spherical surfaces provides us with a basis for understanding how lenses work. Here again we assume that all angles α, β, γ, θ are small implying that the height h is much less than the radius of curvature R. This is necessary if the spherical aberration is to be negligible. In Figure 12–23, we have

$$\gamma = \beta + \theta_2$$
$$\theta_1 = \alpha + \gamma$$

However, Snell's law requires

$$\frac{\sin \theta_1}{\sin \theta_2} = n$$

where n is the index of refraction of glass. Then for small θ_1 and θ_2, we have

$$\frac{\sin \theta_1}{\sin \theta_2} \simeq \frac{\theta_1}{\theta_2} = \frac{\alpha + \gamma}{\gamma - \beta} = n$$

or

$$\alpha + n\beta = (n - 1)\gamma$$

Using

$$\alpha \simeq \frac{h}{o}, \quad \beta \simeq \frac{h}{i}, \quad \text{and} \quad \gamma \simeq \frac{h}{R}$$

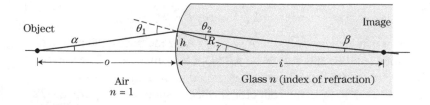

FIGURE 12–23

Refraction at a spherical surface. R is the curvature radius of the surface.

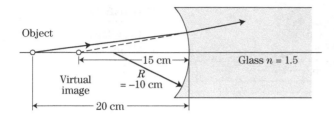

FIGURE 12–24

Example of refraction at a concave spherical surface.

we obtain

$$\frac{1}{o} + \frac{n}{i} = (n-1)\frac{1}{R}$$

(12.13)

The sign convention we adopt here is as follows:

$i > 0$ for an image in the glass or behind the surface,

$i < 0$ for an image in air or in front of the surface,

in contrast with the case of mirrors. For the radius of curvature R,

$R > 0$ for a convex surface as seen by the incident light,

$R < 0$ for a concave surface as seen by the incident light.

As an example, consider a concave glass surface shown in Figure 12–24. Since R is negative, we have

$$\frac{1}{20} + \frac{1.5}{i} = (1.5 - 1)\frac{1}{-10}$$

Solving for i,

$$i = -15 \text{ cm}$$

which indicates that a virtual image is formed in front of the surface.

Lenses

The formula that we have just obtained enables us to analyze lenses that have two spherical surfaces with two different radii of curvatures R_1 and R_2. These surfaces must have a common optical axis. Otherwise the lens cannot form clear images. For the lens shown in Figure 12–25, $R_1 > 0$ and $R_2 < 0$ according to the sign convention for the radius of curvature.

At the first surface, we have

$$\frac{1}{o} + \frac{n}{i'} = (n-1)\frac{1}{R_1}, \quad (R_1 > 0)$$

(12.14)

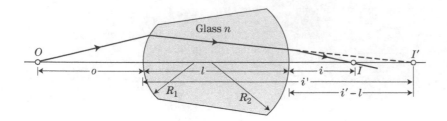

FIGURE 12–25

A thick lens used for analysis.

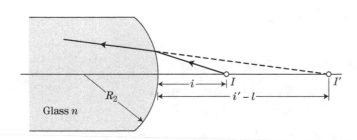

FIGURE 12–26

The rays can be reversed as explained in Eq. (12.15).

where i' is the location of the image in the glass medium if the second surface is absent. Refraction at the second surface obeys

$$-\frac{n}{i'-l} + \frac{1}{i} = (1-n)\frac{1}{R_2}, \quad (R_2 < 0) \tag{12.15}$$

Here, $i' - l$ is the virtual object distance, and the order in $1 - n$ indicates refraction of light entering the air from the glass. In Figure 12–25, the radius of the second surface is negative (concave for the incoming light). If the light propagation direction is reversed as shown in Figure 12–26, the curvature radius is now positive and refraction is described by

$$\frac{1}{i} - \frac{n}{i'-l} = (n-1)\frac{1}{-R_2}$$

Now we neglect l compared with the distance i', that is, we assume that the lens is very thin, $l \to 0$. Adding the two equations, Eq. (12.14) and Eq. (12.15), we find

$$\frac{1}{o} + \frac{1}{i} = (n-1)\left(\frac{1}{R_1} - \frac{1}{R_2}\right) = \frac{1}{f} \tag{12.16}$$

where the focal length f is defined by

$$\frac{1}{f} = (n-1)\left(\frac{1}{R_1} - \frac{1}{R_2}\right) \tag{12.17}$$

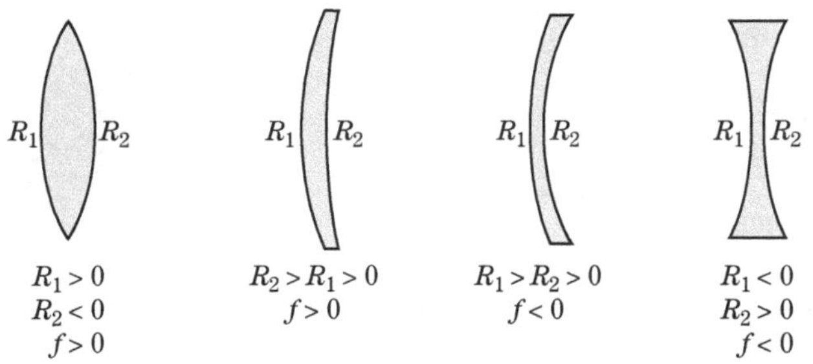

R_1 R_2 R_1 R_2 R_1 R_2 R_1 R_2

$R_1 > 0$ $R_2 < 0$ $f > 0$ $R_2 > R_1 > 0$ $f > 0$ $R_1 > R_2 > 0$ $f < 0$ $R_1 < 0$ $R_2 > 0$ $f < 0$

FIGURE 12–27

Four fundamental lenses.

If $f > 0$, the lens is a *converging lens*, and if $f < 0$, the lens is a *diverging lens*. Four fundamental lenses are illustrated in Figure 12–27. The lens formula

$$\frac{1}{o} + \frac{1}{i} = \frac{1}{f}$$ (12.18)

is identical in form with the mirror formula except for the sign convention for the image distance i. The magnification m is still defined by

$$m = -\frac{i}{o}$$

and the magnification for a lens can be interpreted in exactly the same way as the magnification for mirrors.

EXAMPLE 12.5

Find the location and nature (real or virtual, erect or inverted) of the final image formed in Figure 12–28.

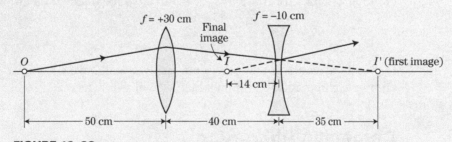

FIGURE 12–28

Solution

For the first convergent lens, the lens formula gives

$$\frac{1}{50} + \frac{1}{i} = \frac{1}{30}$$

or

$$i = 75 \text{ cm}$$

The magnification due to the first lens is

$$m_1 = -\frac{75}{50} = -1.5$$

For the second lens, we have a virtual object distance of -35 cm. Then

$$-\frac{1}{35} + \frac{1}{i'} = -\frac{1}{10}$$

or $i' = -14.0$ cm, and $m_2 = -0.4$. Thus the final image is formed at 17.5 cm to the left of the divergent lens and the final magnification is

$$m = m_1 m_2 = 0.6$$

Since $m > 0$, the image is erect. The image is virtual since the ray does not intersect with the axis after going through the second lens.

In the preceding example the separation between the lenses approaches zero. Then the virtual object distance for the divergent lens becomes -75 cm, and the final image is formed at i' that is determined from

$$\frac{1}{50} + \frac{1}{i'} = \frac{1}{30} - \frac{1}{10}$$

In general, if two lenses of focal lengths f_1 and f_2 are attached together, the compound lenses form a lens with an effective focal length determined from

$$\frac{1}{f_{\text{eff}}} = \frac{1}{f_1} + \frac{1}{f_2} \tag{12.19}$$

High-quality cameras and other optical devices all have compound lenses to eliminate chromatic aberration which we will study next.

Chromatic Aberration

Besides spherical aberration, which is common to both mirrors and lenses, lenses have another deleterious aberration. As we have seen before, the index of refraction n depends on wavelength. Blue light is refracted more than red light and the separation of colors takes place (Figure 12–29). This is extremely undesirable for lenses since we usually want the lenses to form images of blue light and red light exactly at the same location in space. Otherwise images become blurred. This is called *chromatic aberration*.

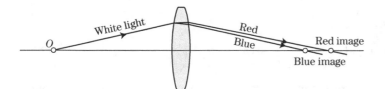

FIGURE 12–29

Prism action of lens causes chromatic aberration.

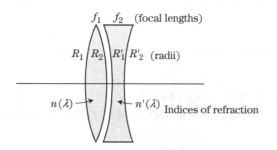

FIGURE 12–30

Compound lens to eliminate chromatic aberration (achromatic lens).

The focal length for blue light is given by

$$\frac{1}{f_B} = (n_B - 1)\left(\frac{1}{R_1} - \frac{1}{R_2}\right) \tag{12.20a}$$

and for red light by

$$\frac{1}{f_R} = (n_R - 1)\left(\frac{1}{R_1} - \frac{1}{R_2}\right) \tag{12.20b}$$

where n_B is the index of refraction at the blue wavelength 5000 Å and n_R is that at the red wavelength 6500 Å. If $n_B = n_R$, we have no problem. However, all glass materials used for lenses have dispersion in that the velocity of light depends on the wavelength.

There is an ingenious way to eliminate the chromatic aberration. Consider a compound lens (Figure 12–30). Since the focal length of a compound lens is given by

$$\frac{1}{f} = \frac{1}{f_1} + \frac{1}{f_2}$$

the focal length for blue light becomes

$$\frac{1}{f_B} = \frac{1}{f_{1B}} + \frac{1}{f_{2B}} = (n_B - 1)\left(\frac{1}{R_1} - \frac{1}{R_2}\right) + (n'_B - 1)\left(\frac{1}{R'_1} - \frac{1}{R'_2}\right)$$

and that for red light is

$$\frac{1}{f_R} = (n_R - 1)\left(\frac{1}{R_1} - \frac{1}{R_2}\right) + (n'_R - 1)\left(\frac{1}{R'_1} - \frac{1}{R'_2}\right)$$

Thus if $f_B = f_R$ or

$$(n_B - n_R)\left(\frac{1}{R_1} - \frac{1}{R_2}\right) = (n'_R - n'_B)\left(\frac{1}{R'_1} - \frac{1}{R'_2}\right) \tag{12.21}$$

the chromatic aberration can be corrected. Since $n_B > n_R$, the compound lens must be composed of one convergent lens and one divergent lens.

EXAMPLE 12.6

Crown glass has $n_B = 1.510$, $n_R = 1.505$ and flint glass has $n_B = 1.630$, $n_R = 1.615$. Design an achromatic compound lens having a focal length of 50 cm. You may assume the compound lens consists of two lenses attached together having a common curvature radius as shown (Figure 12–31).

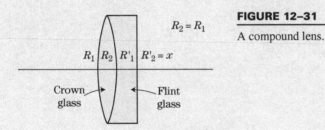

FIGURE 12–31

A compound lens.

Solution

From Eq. (12.21), we have

$$0.005\left(\frac{1}{R_1} - \frac{1}{R_2}\right) = -0.015\left(\frac{1}{R'_1} - \frac{1}{R'_2}\right)$$

where $R_2 = R'_1$ and $R'_2 = \infty$. Then

$$\frac{1}{R_1} - \frac{1}{R_2} = -3\frac{1}{R_2}$$

or

$$R_2 = -2R_1 \tag{i}$$

From the formula for a compound lens, we have

$$\frac{1}{50} = 0.51\left(\frac{1}{R_1} - \frac{1}{R_2}\right) + 0.63\left(\frac{1}{R_2} - \frac{1}{\infty}\right)$$

or

$$\frac{1}{50} = 0.51\frac{1}{R_1} + 0.12\frac{1}{R_2} \tag{ii}$$

Solving (i) and (ii) for R_1 and R_2, we find

$$R_1 = 22.5\,\text{cm} \quad \text{and} \quad R_2 = -45\,\text{cm}$$

You should check that the compound lens indeed has a focal length of 50 cm.

Optical Instruments

Human Eye

The human eye is modeled with a single refraction surface having a curvature radius of 5.7 mm in the relaxed condition as shown in Figure 12–32. The curvature radius can be controlled by the muscles attached to the lens. The index of refraction of the liquid in the eye, called the vitreous humor, is 1.34. Paraxial light entering the eye from the left is focused at the retina 22.5 mm from the surface of the eye, which is consistent with the refraction formula

$$\frac{1}{\infty} + \frac{n}{i} = \frac{n-1}{R} \rightarrow \frac{1.34}{i} = \frac{0.34}{5.7} \rightarrow i = f = 22.5\,\text{mm}$$

For a light beam coming from the right, the focal point is at 16.8 mm in front of the surface.

The focal length of a short-sighted eye (myopia) is shorter than normal (Figure 12–33). This can be corrected by wearing a corrective divergent lens. If the maximum object distance of a myopic eye is d, a divergent lens having a focal length of $-d$ will form an image of the parallel beam at the retina. Let

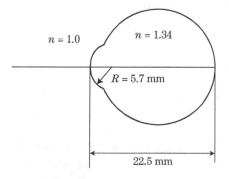

FIGURE 12–32

Simplified model of the human eye. The curvature radius R is adjustable.

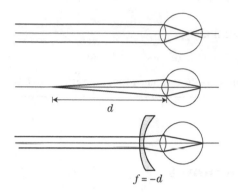

FIGURE 12–33

A myopic eye and its correction with a divergent lens.

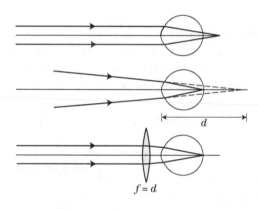

FIGURE 12–34

A hyperopic eye and its correction by a convergent lens. The distance d is the far point of the eye.

the focal length of a myopic eye be f_M. With a divergent lens, the effective focal length

$$\frac{1}{f_{eff}} = \frac{1}{f_M} - \frac{1}{d}$$

can be adjusted to the proper value of $f_{eff} = 22.5$ mm.

In contrast, the focal length of a far-sighted eye (hyperopia) is longer than normal. It can be corrected by wearing a convergent lens having a focal length $f = d$ where d is the far point of the eye. (Figure 12–34).

Magnifying Glass

A convergent lens ($f > 0$) can create an erect virtual image with a magnification larger than unity $m > 1$ if the object is placed just inside the focal point at a distance o that is slightly smaller than the focal distance f (Figure 12–35). If the image is formed at infinity by placing the object at the focal position, the angular magnification in comparison with when it is placed at 25 cm in

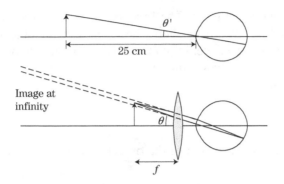

FIGURE 12–35

Magnifying glass. The object is placed at the focal point of a converging lens and the image is formed at infinity. The image size on the retina is $m = \theta/\theta' = 25/f$ times larger than the case without the glass (top figure). If the image is formed at 25 cm, the magnification becomes $m' = 1 + \frac{25}{f}$.

front of the eye for most distinct viewing without the lens is

$$m = \frac{\theta}{\theta'} = \frac{25}{f} \qquad (12.22)$$

If the image is formed at $i = -25$ cm by placing the object just inside the focal point, the lens formula

$$\frac{1}{o} = \frac{1}{f} + \frac{1}{25}$$

yields a magnification

$$m = -\frac{-25}{o} = 1 + \frac{25}{f}$$

which is slightly larger than $25/f$.

Microscope

A microscope consists of two convergent lenses, one called the objective lens and the other called the eyepiece. The standard microscope tube length, defined by the distance between the focal points, is 16 cm, as shown in Figure 12–36. The objective forms a real, inverted image if the object is placed

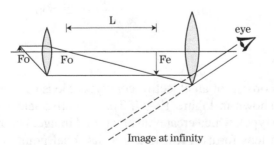

FIGURE 12–36

Principle of a microscope. Fo and Fe are the focal points of the objective and eyepiece lenses, respectively.

just beyond the focal point. The function of the eyepiece is to magnify the image. Therefore, the image should be formed by the objective lens near the focal point of the eyepiece. In practice, the focal lengths of the objective and eyepiece are much smaller than the tube length. The magnification by the objective is therefore

$$m_1 = -\frac{L}{f_1}$$

where f_1 is the focal length of the objective lens. The magnification by the eyepiece is

$$m_2 = \frac{25}{f_2}$$

if the image is formed at infinity. Here f_2 is the focal length of the eyepiece. The total magnification is

$$m = m_1 m_2 = -\frac{L}{f_1}\frac{25}{f_2}$$

If the image is formed at 25 cm, the magnification is modified to be

$$m = -\frac{L}{f_1}\left(1 + \frac{25}{f_2}\right) \tag{12.23}$$

The maximum useful magnification of optical microscopes is about 600. This limitation is due partly to the diffractive nature of light and partly to the physiological property of the retina. The resolution limit of the human eye is about 0.1 mm while the resolution limit of the objective lens is of the order of the wavelength,

$$\Delta x \simeq \frac{\lambda}{2n \sin\theta}$$

where n is the index of refraction of the lens and θ is the angle subtended by the lens. For $\sin\theta \simeq 1$ and $n = 1.5$, we have $\Delta x \simeq 1.7 \times 10^{-7}$ m. Then the maximum magnification is

$$m_{max} \simeq \frac{10^{-4}\text{m}}{1.7 \times 10^{-7}\text{ m}} \simeq 600$$

Telescope

A telescope consists of an objective convergent lens (or concave mirror) and eyepiece as shown in Figure 12–37. The objective lens can be a divergent lens (Galileo type), which creates an erect final image. For an object far away, the objective lens forms a real image at its focal point. The image is then

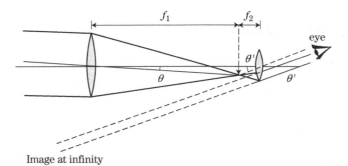

FIGURE 12–37

Principle of a telescope. The eyepiece can be a divergent lens. In this case, the image is erect.

magnified by the eyepiece. The angular magnification is given by

$$m = -\frac{\theta'}{\theta} = -\frac{f_1}{f_2} \tag{12.24}$$

if the final image is formed at infinity. Here, the angle θ is the angle subtended by the object and θ' is the angle subtended by the intermediate image. If the final image is formed at 25 cm, the magnification is modified to be

$$m = -f_1 \left(\frac{1}{f_2} + \frac{1}{25} \right) \tag{12.25}$$

since the object distance for the eyepiece should be chosen so that

$$\frac{1}{o} - \frac{1}{25} = \frac{1}{f_2}$$

and the magnification in this case is

$$m = -\frac{f_1}{o} = -f_1 \left(\frac{1}{f_2} + \frac{1}{25} \right) \tag{12.26}$$

The eyepiece can be a divergent lens (Galileo type). If $f_2 < 0$, the magnification is positive and the image is erect, as desired for opera glasses. However, the magnification is limited to a value of less than 10. Binoculars have convergent eyepieces. The image inversion is performed with two prisms, such as one in the x direction and other in the y direction, in the plane perpendicular to the light propagation.

The viewfinder telescope for rifles has a third convergent lens (called the erector) between the objective and the eyepiece to achieve image inversion. This will be studied in Problem 12.35.

Physical Meaning of Focusing

Spherical surfaces used for reflection and refraction of light are characterized by the focal lengths. Consider a spherical reflector as shown in Figure 12–38. We saw that two rays leaving the object are focused to the same image position, irrespective of the height h, provided that h is much smaller than the radius of curvature R or provided that the spherical aberration is negligible. Here it will be shown that the time required for light to travel along either path is the same. This is the physical meaning of focusing.

In Figure 12–38 the length OA is

$$OA = \sqrt{(o - \delta)^2 + h^2} \simeq o - \delta + \frac{1}{2}\frac{h^2}{o}$$

and the length IA is

$$IA = \sqrt{(i - \delta)^2 + h^2} \simeq i - \delta + \frac{1}{2}\frac{h^2}{i}$$

where we have made use of binomial expansion and noting that $\delta, h \ll o, i$.

The quantity δ can be found from Figure 12–39. For the triangle OAB, we have

$$R = \sqrt{(R - \delta)^2 + h^2}$$

and after neglecting the small contribution of δ^2

$$\delta = \frac{h^2}{2R}$$

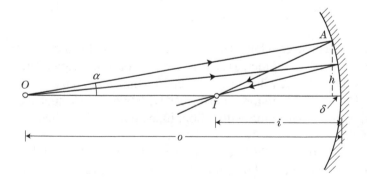

FIGURE 12–38

Time required for light to travel from the object to the image is independent of α or h provided that h is sufficiently small.

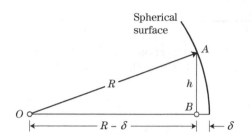

FIGURE 12–39

If $h, \delta \ll R, \delta \sim h^2/2R$.

Then the time required for light to travel from O to I in Figure 12–38 is

$$t = \frac{OA}{c} + \frac{AI}{c} = \frac{1}{c}(o + i - 2\delta) + \frac{h^2}{2c}\left(\frac{1}{o} + \frac{1}{i}\right)$$

$$= \frac{1}{c}(o + i) + \frac{h^2}{2c}\left(\frac{1}{o} + \frac{1}{i} - \frac{2}{R}\right)$$

However, the mirror formula was

$$\frac{1}{o} + \frac{1}{i} = \frac{2}{R}$$

and we see that the time is independent of the height h or the light path.

Similarly, it can be shown that in the case of lenses, the light propagation time is independent of the path as long as spherical aberration is negligible. Consider a convergent lens shown in Figure 12–40. The geometrical distance along path 1 is obviously longer than that along path 2. However, the *optical distance* defined by the sum of

$$n \times \text{geometrical distance}$$

in each medium can be shown to be independent of which path light takes.

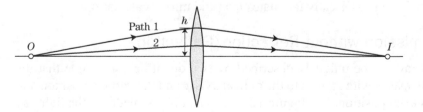

FIGURE 12–40

If h is small, the light propagation times along paths 1 and 2 are the same.

EXAMPLE 12.7

Recalling Eq. (12.13), show that the light propagation time along the path OAI does not depend on the height h and is equal to $(1/c)(o + n \times i)$.

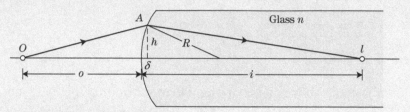

FIGURE 12–41

Solution

The distances from Figure 12–41 are

$$OA = \sqrt{(o + \delta)^2 + h^2} \simeq o + \delta + \frac{h^2}{2o}$$

$$AI = \sqrt{(i - \delta)^2 + h^2} \simeq i - \delta + \frac{h^2}{2i}$$

where $\delta = h^2/2R$ as in the case of the spherical mirror. Then the propagation time along the path OAI is

$$\frac{OA}{c} + \frac{AI}{c/n} = \frac{1}{c}(o + ni) + \frac{h^2}{2c}\left[\frac{1}{o} + \frac{n}{i} + (1 - n)\frac{1}{R}\right]$$

The quantity in the brackets [ ] is zero from Eq. (12.13). Therefore the optical path $OA + nAI$ is independent of the height h as long as spherical aberration is negligible.

Matrix Method in Geometrical Optics

In a sophisticated optical system, several lenses correct various forms of aberrations and a calculation based on ray tracing becomes overly cumbersome. The matrix method greatly facilitates lens and mirror calculations.

Transmission without Refraction/Reflection

A light ray can be uniquely described by two quantities, the angle that the light ray makes with respect to the optical axis ϕ and the vertical position h at a given axial position. See Figure 12–42. In a uniform medium, the light ray travels along a straight line as shown in Figure 12–43. The angle relative to

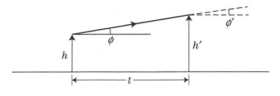

FIGURE 12–42

Transmission of a light ray over a distance t along the axis. The angle with respect to the axis remains unchanged as long as the medium is uniform.

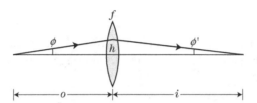

FIGURE 12–43

Ray angles ϕ, ϕ' before and after refraction and the vertical position h (unchanged) in a thin lens. Note that the angle ϕ' is negative in this case.

the axis is constant in this case, $\phi' = \phi$, but the height h changes according to

$$h' = h + t\phi$$

where t is the transmission distance along the axis. We can write $\phi' = \phi$ and $h' = h + t\phi$ in matrix notation as

$$\begin{pmatrix} h' \\ \phi' \end{pmatrix} = \begin{pmatrix} 1 & t \\ 0 & 1 \end{pmatrix} \begin{pmatrix} h \\ \phi \end{pmatrix}$$

Then the matrix for transmission is

$$\mathbf{T} = \begin{pmatrix} 1 & t \\ 0 & 1 \end{pmatrix} \tag{12.27}$$

Thin Lens

Let us consider a thin lens with a focal length f. The relationship between the object distance o, the image distance i, and the focal length f is the familiar one,

$$\frac{1}{o} + \frac{1}{i} = \frac{1}{f}$$

In terms of the ray angles ϕ, ϕ' before and after the refraction along with the height h within the lens which is unchanged, and we can write

$$\phi = \frac{h}{o}, \phi' = -\frac{h}{i} = -\frac{h'}{i}$$

or

$$h' = h = h + 0 \times \phi$$
$$\phi' = -\frac{h}{f} + \phi$$

This can be written as the following matrix

$$\begin{pmatrix} h' \\ \phi' \end{pmatrix} = \begin{pmatrix} 1 & 0 \\ -\frac{1}{f} & 1 \end{pmatrix} \begin{pmatrix} h \\ \phi \end{pmatrix} = \mathbf{L} \begin{pmatrix} h \\ \phi \end{pmatrix}$$

where $\mathbf{L}$ is called the *thin lens matrix*,

$$\mathbf{L} = \begin{pmatrix} 1 & 0 \\ -\frac{1}{f} & 1 \end{pmatrix} \tag{12.28}$$

If the object is at a distance o from the lens and the image is at a distance i from the lens, the system matrix, object, and image is given by the product of the three matrices,

$$\begin{pmatrix} 1 & i \\ 0 & 1 \end{pmatrix} \begin{pmatrix} 1 & 0 \\ -\frac{1}{f} & 1 \end{pmatrix} \begin{pmatrix} 1 & o \\ 0 & 1 \end{pmatrix} = \begin{pmatrix} \frac{f-i}{f} & \frac{if-io+fo}{f} \\ -\frac{1}{f} & \frac{f-o}{f} \end{pmatrix} = \begin{pmatrix} A & B \\ C & D \end{pmatrix} \tag{12.29}$$

Since

$$h' = Ah + B\phi$$

the condition $B = 0$ means that the height h' is independent of the angle of the ray. That is, all rays converge to the same height irrespective of the angle of the rays. This is the condition for focusing and $B = 0$ in this example reproduces the lens formula,

$$if - io + of = 0$$

or

$$\frac{1}{o} + \frac{1}{i} = \frac{1}{f}$$

The focal length of the lens can be identified from

$$f = -\frac{1}{C}$$

This can be seen from the relationship

$$\phi' = Ch + D\phi$$

The focal position can be found by assuming a parallel ray $\phi = 0$. Then the focal position is

$$f = -\frac{h}{\phi'} = -\frac{1}{C}$$

The magnification is given by

$$m = -\frac{i}{o} = A$$

Compound Lens

The matrix of the combination of two thin lenses with focal lengths f_1 and f_2 that are separated by a distance d consists of two refraction matrices and one transmission matrix,

$$\begin{pmatrix} 1 & 0 \\ -\frac{1}{f_2} & 1 \end{pmatrix} \begin{pmatrix} 1 & d \\ 0 & 1 \end{pmatrix} \begin{pmatrix} 1 & 0 \\ -\frac{1}{f_1} & 1 \end{pmatrix} = \begin{pmatrix} -\frac{d-f_1}{f_1} & d \\ -\frac{f_1+f_2-d}{f_1 f_2} & -\frac{d-f_2}{f_2} \end{pmatrix}$$

The effective focal length can be found from

$$f = -\frac{1}{C} = \frac{f_1 f_2}{f_1 + f_2 - d} \tag{12.30}$$

In contrast with the case of a single thin lens, for incident rays parallel to the optical axis, the focal length f is not the distance between the second lens and the focal point. The distance between the second lens and the focal point is given by

$$f_2' = -\frac{A}{C} = \frac{(f_1 - d)f_2}{f_1 + f_2 - d} \tag{12.31}$$

since

$$h' = Ah + B\phi, \, \phi' = Ch + D\phi$$

the incident ray parallel to the axis corresponds to $\phi = 0$, and thus

$$h' = Ah, \phi' = Ch$$

yields

$$\frac{1}{f_2'} = -\frac{\phi'}{h'} = -\frac{A}{C}$$

The difference between f and f_2' is

$$f - f_2' = \frac{fd}{f_1}$$

The physical meanings of f and f_2' are shown in Figure 12–44. The intersection between the extensions of the paraxial ray and the final focusing ray determines the *principal plane* of the second lens H2 as shown in the upper Fig. 12. 44. It is at a distance

$$\frac{fd}{f_1} \tag{12.32}$$

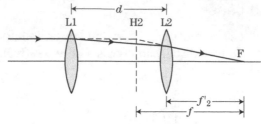

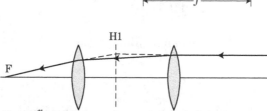

FIGURE 12–44

Compound lens with the separation d. The focal length f is the distance between the focal position F and principal planes H1, H2, where f_1'(f_2') is the distance between the focal point and the respective lenses.

to the left of the second lens. By reversing the ray direction, the principal plane of the first lens can be similarly found and H1 as shown in Figure 12–44. It is located at a distance

$$\frac{fd}{f_2} \tag{12.33}$$

to the right of the first lens. If the object distance o is measured from H1 and the image distance i from H2, the lens formula

$$\frac{1}{o} + \frac{1}{i} = \frac{1}{f} = \frac{1}{f_1} + \frac{1}{f_2} - \frac{d}{f_1 f_2}$$

still holds where f is the effective focal length.

EXAMPLE 12.8

Achromatic compound lens. Two lenses with focal lengths f_1 and f_2 are a distance d apart. Show that if

$$d = \frac{f_1 + f_2}{2}$$

the compound lens becomes achromatic.

Solution

The condition for an achromatic lens is that the focal length be independent of the index of refraction $n(\lambda)$. Let the focal lengths of the lenses be

$$\frac{1}{f_1} = (n-1)\left(\frac{1}{R_1} - \frac{1}{R_2}\right), \frac{1}{f_2} = (n-1)\left(\frac{1}{R_3} - \frac{1}{R_4}\right)$$

Then the effective focal length is

$$\frac{1}{f} = \frac{1}{f_1} + \frac{1}{f_2} - \frac{d}{f_1 f_2}$$

$$= (n-1)\left(\frac{1}{R_1} - \frac{1}{R_2} + \frac{1}{R_3} - \frac{1}{R_4}\right) - (n-1)^2 d\left(\frac{1}{R_1} - \frac{1}{R_2}\right)\left(\frac{1}{R_3} - \frac{1}{R_4}\right)$$

The condition

$$\frac{d}{dn}\left(\frac{1}{f}\right) = 0$$

yields

$$\frac{1}{R_1} - \frac{1}{R_2} + \frac{1}{R_3} - \frac{1}{R_4} - 2(n-1)d\left(\frac{1}{R_1} - \frac{1}{R_2}\right)\left(\frac{1}{R_3} - \frac{1}{R_4}\right) = 0$$

Then

$$d = \frac{1}{2(n-1)}\left(\frac{1}{\dfrac{1}{R_1} - \dfrac{1}{R_2}} + \frac{1}{\dfrac{1}{R_3} - \dfrac{1}{R_4}}\right) = \frac{1}{2}(f_1 + f_2)$$

Refraction at a Spherical Surface: Thick Lens

We next consider refraction at a spherical boundary with a curvature radius R between media with indices n and n' (Figure 12–45). Snell's law yields

$$\phi + \gamma = \frac{n'}{n}(\gamma + \phi')$$

or noting $\gamma = h/R$,

$$\phi' = \left(\frac{n}{n'} - 1\right)\frac{h}{R} + \frac{n}{n'}\phi$$

Then refraction at a spherical surface can be described by

$$\begin{pmatrix} h' \\ \phi' \end{pmatrix} = \begin{pmatrix} 1 & 0 \\ \left(\frac{n}{n'} - 1\right)\frac{1}{R} & \frac{n}{n'} \end{pmatrix} \begin{pmatrix} h \\ \phi \end{pmatrix} = \mathbf{R}\begin{pmatrix} h \\ \phi \end{pmatrix}$$

where $\mathbf{R}$ is the refraction matrix,

$$\mathbf{R} = \begin{pmatrix} 1 & 0 \\ \left(\frac{n}{n'} - 1\right)\frac{1}{R} & \frac{n}{n'} \end{pmatrix} \tag{12.34}$$

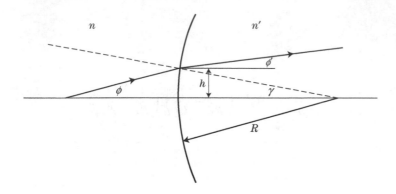

FIGURE 12–45

Refraction at a spherical surface.

At the air–glass boundary $n = 1, n' = n_g$ and we have

$$\mathbf{R} = \begin{pmatrix} 1 & 0 \\ \left(\frac{1}{n_g} - 1\right)\frac{1}{R} & \frac{1}{n_g} \end{pmatrix}$$

For a flat air–glass boundary with $R = \infty$,

$$\mathbf{R} = \begin{pmatrix} 1 & 0 \\ 0 & \frac{1}{n_g} \end{pmatrix}$$

A thick lens shown in Figure 12–46 in air with thickness t, index of refraction n_g, and radii of curvature R_1 and R_2 can be analyzed by combining the refractions at both surfaces and the transmission through the lens. The refraction matrix at the first (entrance) surface is

$$\mathbf{R}_1 = \begin{pmatrix} 1 & 0 \\ \left(\frac{1}{n_g} - 1\right)\frac{1}{R_1} & \frac{1}{n_g} \end{pmatrix} \tag{12.35}$$

The transmission matrix over the thickness t is

$$\mathbf{T} = \begin{pmatrix} 1 & t \\ 0 & 1 \end{pmatrix} \tag{12.36}$$

and the refraction matrix at the second (exit) surface is

$$\mathbf{R}_2 = \begin{pmatrix} 1 & 0 \\ (n_g - 1)\frac{1}{R_2} & n_g \end{pmatrix} \tag{12.37}$$

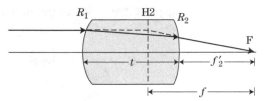

FIGURE 12–46

Focusing in a thick lens. Paraxial ray from the left is focused at F after two refractions. The effective focal length of a thick lens f in Eq. (12.38) is the distance between the second principal plane H2 and F, while f_2' in Eq. (12.39) is the distance between the second vertex and F.

The total matrix (system matrix) of the lens is given by the product of the three matrices,

$$\mathbf{R_2 T R_1} = \begin{pmatrix} 1 & 0 \\ (n_g - 1)\frac{1}{R_2} & n_g \end{pmatrix} \begin{pmatrix} 1 & t \\ 0 & 1 \end{pmatrix} \begin{pmatrix} 1 & 0 \\ \left(\frac{1}{n_g} - 1\right)\frac{1}{R_1} & \frac{1}{n_g} \end{pmatrix}$$

which reduces to

$$\begin{pmatrix} A & B \\ C & D \end{pmatrix} = \begin{pmatrix} 1 - \frac{t}{R_1}\left(1 - \frac{1}{n_g}\right) & \frac{t}{n_g} \\ -(n_g - 1)\left(\frac{1}{R_1} - \frac{1}{R_2} - \frac{(n_g-1)^2 t}{n_g R_1 R_2}\right) & 1 + \frac{t}{n_g R_2}(n_g - 1) \end{pmatrix}$$

The inverse effective focal length of the thick lens can be found from

$$\frac{1}{f} = -C = (n_g - 1)\left(\frac{1}{R_1} - \frac{1}{R_2} - \frac{(n_g - 1)^2 t}{n_g R_1 R_2}\right)$$

which can be simplified to

$$\frac{1}{f} = \frac{1}{f_1} + \frac{1}{f_2} - \frac{t}{n_g f_1 f_2} \tag{12.38}$$

The *focal position* measured from the exit surface is

$$f_2' = -\frac{A}{C} = f\left[1 + \frac{t}{R_2}\left(\frac{1}{n_g} - 1\right)\right] \tag{12.39}$$

The difference between f and f_2' determines the location of the principal plane of the second vertex H2 where

$$\text{H2:} \quad \frac{f t}{n_g f_1} \quad \text{to the left of the second vertex.}$$

Similarly, the principal plane of the first vertex is H1 where

$$H1: \frac{ft}{n_g f_2} \text{ to the right of the first vertex}$$

When the thickness t is negligible, we recover the thin lens result

$$\frac{1}{f} = \frac{1}{f_1} + \frac{1}{f_2} = (n_g - 1)\left(\frac{1}{R_1} - \frac{1}{R_2}\right)$$

EXAMPLE 12.9

A thick converging lens has $R_1 = 10$ cm, $R_2 = -5$ cm, t (thickness) = 5 cm, and n_g (refraction index) = 1.5. Determine the lens matrix. When an object is placed 20 cm in front of the spherical surface with a curvature radius $R_1 = 10$ cm, where is its image formed? Find the magnification.

Solution

The lens matrix is

$$\mathbf{L} = \begin{pmatrix} 1 - \frac{t}{R_1}\left(1 - \frac{1}{n_g}\right) & \frac{t}{n_g} \\ -(n_g - 1)\left(\frac{1}{R_1} - \frac{1}{R_2} - \frac{(n_g-1)^2 t}{n_g R_1 R_2}\right) & 1 + \frac{t}{n_g R_2}(n_g - 1) \end{pmatrix}$$

$$= \begin{pmatrix} 0.8333 & 3.3333 \text{ cm} \\ -0.1333\frac{1}{\text{cm}} & 0.6667 \end{pmatrix}$$

The focal length is

$$f = -\frac{1}{C} = 7.5 \text{ cm}$$

The principal plane of the first surface is at

$$H1: \frac{ft}{f_2 n_g} = \frac{7.5 \times 5}{10 \times 1.5} = 2.5 \text{ cm to the right of the first vertex}$$

and H2 is at

$$H2: \frac{ft}{f_1 n_g} = 1.25 \text{ cm to the left of the second vertex}$$

If an object is placed 20 cm in front of the first vertex, the distance between the object and H1 is $o = 22.5$ cm. From

$$\frac{1}{o} + \frac{1}{i} = \frac{1}{f} \rightarrow \frac{1}{i} = \frac{1}{f} - \frac{1}{o} = \frac{1}{7.5} - \frac{1}{22.5}$$

we find $i = 11.25$ cm from H2 to the right. The image is at 10 cm from the second vertex. The magnification is

$$m = -\frac{i}{o} = -0.5$$

Mirror

The matrix for a spherical mirror can be obtained in a similar manner. The mirror matrix is

$$\mathbf{M} = \begin{pmatrix} 1 & 0 \\ -\frac{2}{R} & 1 \end{pmatrix} = \begin{pmatrix} 1 & 0 \\ -\frac{1}{f} & 1 \end{pmatrix} \tag{12.40}$$

where the sign convention for mirrors should be followed, namely $R > 0$ for a concave mirror and $R < 0$ for a convex mirror. Consider a simple example. If an object is placed $o = 30$ cm in front of a convex mirror with a focal length of -20 cm, its image is formed $i = 12$ cm behind the mirror as calculated from the mirror formula

$$\frac{1}{o} + \frac{1}{i} = \frac{1}{f}$$

The magnification is $i/o = 0.4$. This can be analyzed using the pertinent matrices as follows.

$$\begin{pmatrix} 1 & i \\ 0 & 1 \end{pmatrix} \begin{pmatrix} 1 & 0 \\ +\frac{1}{20} & 1 \end{pmatrix} \begin{pmatrix} 1 & 30 \\ 0 & 1 \end{pmatrix} = \begin{pmatrix} 1 + \frac{i}{20} & 30 + \frac{5}{2}i \\ \frac{1}{20} & \frac{5}{2} \end{pmatrix}$$

The image location can be found by demanding

$$B = 30 + \frac{5}{2}i = 0 \rightarrow i = -12 \text{ cm}$$

The magnification is

$$m = A = 1 + \frac{-12}{20} = 0.4$$

in agreement with the results from the mirror formula.

EXAMPLE 12.10

Lens-mirror combination (Figure 12–47). If a lens with a focal length f_1 is at a distance d in front of a mirror with a focal length f_2, the system matrix is given by

$$\begin{pmatrix} 1 & i \\ 0 & 1 \end{pmatrix} \begin{pmatrix} 1 & 0 \\ -\frac{1}{f_1} & 1 \end{pmatrix} \begin{pmatrix} 1 & d \\ 0 & 1 \end{pmatrix} \begin{pmatrix} 1 & 0 \\ -\frac{1}{f_2} & 1 \end{pmatrix} \begin{pmatrix} 1 & d \\ 0 & 1 \end{pmatrix} \begin{pmatrix} 1 & 0 \\ -\frac{1}{f_1} & 1 \end{pmatrix} \begin{pmatrix} 1 & o \\ 0 & 1 \end{pmatrix}$$

where o is the object distance in front of the lens and i is the image distance, also in front of the lens since the mirror reverses the direction of the ray. Let us assume that a lens having a focal length of 10 cm is placed at 15 cm in front of a concave mirror with a focal length of 20 cm. An object is placed at 20 cm from the lens. Determine the characteristics of the system.

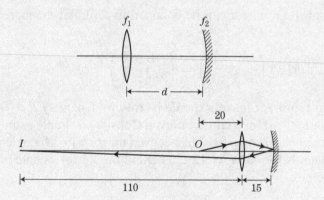

FIGURE 12–47

Lens-mirror system. The lower figure shows the ray tracing.

Solution

The matrix reduces to

$$\begin{pmatrix} 1 & i \\ 0 & 1 \end{pmatrix} \begin{pmatrix} 1 & 0 \\ -\frac{1}{10} & 1 \end{pmatrix} \begin{pmatrix} 1 & 15 \\ 0 & 1 \end{pmatrix} \begin{pmatrix} 1 & 0 \\ -\frac{1}{20} & 1 \end{pmatrix} \begin{pmatrix} 1 & 15 \\ 0 & 1 \end{pmatrix} \begin{pmatrix} 1 & 0 \\ -\frac{1}{10} & 1 \end{pmatrix} \begin{pmatrix} 1 & 20 \\ 0 & 1 \end{pmatrix}$$

$$= \begin{pmatrix} -\frac{13}{8} + \frac{7}{80}i & -\frac{55}{4} + \frac{i}{8} \\ \frac{7}{80} & \frac{1}{8} \end{pmatrix}$$

From $B = 0$, the image location can be found to be $i = 110$ cm to the left of the lens. The final magnification is

$$m = A = -\frac{13}{8} + \frac{7i}{80} = -\frac{13}{8} + \frac{7 \times 110}{80} = 8.0$$

The image is real and erect. The reader should check the results using the lens and mirror formulae successively.

Problems

1. Light beam falls on the water–glass boundary at an angle of 30° (Figure 12–48). Find the refraction angle θ.

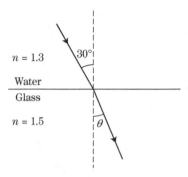

FIGURE 12–48

Problem 1.

2. Explain how heat haze is created on a hot day.

3. Show that the apparent depth of a water pool if one looks perpendicular to the surface is d/n where d is the true depth and n is the index of refraction of water.

4. A glass cube ($n = 1.5$) has a spot at the center. Assuming that the cube has an edge of 2 cm, find what parts of the cube face must be covered so that the spot cannot be seen at all.

5. *Fermat's principle.* Show that the time required for light to travel from A to B is a minimum when the angles θ_1 and θ_2 satisfy Snell's law (Figure 12–49).

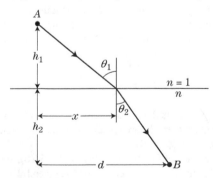

FIGURE 12–49

Problem 5.

(*Hint:* Find the time τ as a function of x. Calculate $d\tau/dx$. $d\tau/dx = 0$ yields Snell's law.)

6. Show that a mirror having a vertical size that is half of a person's height is sufficient for the person to view his whole body.

7. A spherical mirror has a focal length of $+20$ cm. An object is placed at 10 cm in front of the mirror. Find the image distance, magnification, and whether the image is (a) real or virtual and (b) erect or inverted.

8. Repeat Problem 7 for a mirror having $f = -20$ cm.

9. When an object is placed at 60 cm in front of a mirror, a magnification of -0.5 results. Find the focal length of the mirror and the image distance.

10. Does the image of a cubic object appear to be cubic when formed by a mirror?

11. Locate the position of the final image and find the magnification (Figure 12–50).

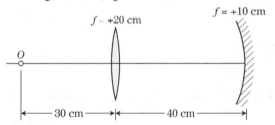

FIGURE 12–50

Problem 11.

12. Repeat Problem 11 by replacing the lens with a diverging lens having $f = -20$ cm.

13. Complete the table for four lenses, a, b, c, and d. The lengths are in centimeters.

Lens	a	b	c	d
f	+10			+40
o	+20	+10	+120	
i				
m		0.5	2	−0.6
Real?			No	
Erect?		Yes		

14. Find the location of the image (Figure 12–51).

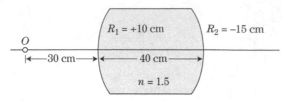

$R_1 = +10$ cm $R_2 = -15$ cm

O

$\leftarrow$—30 cm—$\rightarrow$$\leftarrow$———40 cm———$\rightarrow$

$n = 1.5$

FIGURE 12–51

Problem 14.

15. A solid glass sphere of radius R and index of refraction $n = 1.4$ is silvered over one hemisphere. An object is placed at a distance from the sphere as shown in Figure 12–52. Find the position of the final image. Neglect spherical aberration.

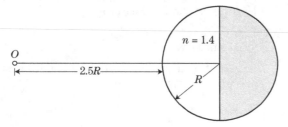

$n = 1.4$

O

$\leftarrow$————2.5R————$\rightarrow$

R

FIGURE 12–52

Problem 15.

16. Repeat Example 12.6 for a compound lens of focal length 20 cm. Take $R_2' = \infty$.

17. A single lens reflex camera can select lenses for different purposes.
  (a) If a 50-mm lens is attached, what is the minimum object distance that the camera can focus? Assume that the maximum lens–film distance is 6.0 cm.
  (b) The 50-mm lens is replaced with a telephoto lens that has a focal length of 800 mm. What is the magnification of an object far away relative to that of a 50-mm lens?
  (c) Repeat (a) assuming a lens with a focal length of 20 mm is placed in front of the 50-mm lens.

18. **(a)** Does the eyepiece of a telescope have to be a convergent lens?
  (b) Does the eyepiece of a microscope have to be a convergent lens?

19. If a mirror surface is parabolic, rays parallel to the axis will all converge at the same point irrespective of the height of the object. Prove this (see Figure 12–53).

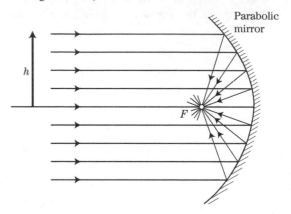

Parabolic mirror

h

F

FIGURE 12–53

Problem 19.

20. Derive Eq. (12.13).

21. Reflection and refraction of electromagnetic waves (light included) at a discontinuity in the medium can be analyzed with the aid of Maxwell's equations. One requirement resulting from using Maxwell's equations is that the tangential component of the electric field be continuous across the boundary. (We have already used this for the case of normal incidence in Chapter 10 where the electric fields were always tangent to the boundary surface.) For the incident wave, there are two possible field configurations depending on the orientation of the electric field, the field normal to the "incident plane" (Figure 12–54(a) and tangent to the incident plane [Figure 12–54(b)].
  (a) For the configuration in Figure 12–54(a), show that the reflected electric field is given by

$$E_r = -\frac{\sin(\theta_i - \theta_r)}{\sin(\theta_i + \theta_r)} E_i$$

where E_i is the incident electric field. Does this reduce to the known result of normal incidence, $\theta_i \to 0, \theta_r \to 0$?
  (b) For the configuration in Figure 12–54(b), show that the reflected electric field is given by

$$E_r = -\frac{\tan(\theta_i - \theta_r)}{\tan(\theta_i + \theta_r)} E_i$$

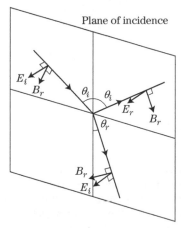
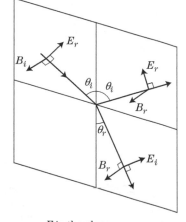

E normal to the plane
B in the plane

E in the plane
B normal to the plane

(a)   (b)

FIGURE 12–54

Problem 21.

Note: For case (b), the reflected field becomes zero when

$$\theta_i + \theta_r = \frac{\pi}{2}, \left(\tan \frac{\pi}{2} = \infty\right)$$

or when $\tan \theta_i = n$. This particular incident angle is called the *Brewster angle* and has important applications for the polarization of light. For typical glass having an index refraction of $n = 1.5$, the Brewster angle becomes 56.3°.

Hint: Apply the boundary condition for the electric field and conservation of energy.

22. In the eighteenth century, Bradley observed an apparent change in the angular location of distant stars by 22" when the earth is moving normal to the line of sight. Assuming that the earth's orbital velocity v about the sun is 3×10^4 m/s, estimate the velocity of light c using the aberration formula

$$\Delta \theta = \frac{v}{c}$$

Hint: If you run in rain, the path of rain drops is not normal but tilted.

23. A rainbow is caused by the prism action of spherical water drops. See Figure 12–55. The deviation angle for red light is about 42° and that for violet is about 40°. The index of refraction of water for red light is 1.33 and for violet light it is 1.35. Find the deviation angles δ between the incident white ray and red and violet refracted rays.

Hint: First show that the incident angle θ and deviation angle δ are related through

$$\sin \theta = n \left[\sin(\theta/2) + \frac{\delta}{4}\right]$$

Determine $\delta_{\max}$ by trial and error.

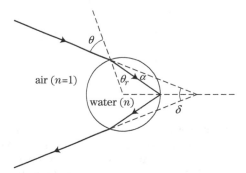

FIGURE 12–55

Problem 23.

24. Show that the deviation angle β due to double reflections by mirrors intersecting at the angle α is

given by $\beta = \pi - 2\alpha$ irrespective of the angle of incidence. (Retro-reflectors use $\alpha = 90°$, $\beta = 0$.)

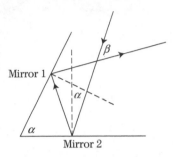

FIGURE 12–56

Problem 24.

25. A light beam falls on a glass slab of index of refraction n and thickness t at an incident angle θ. Show that the deviation distance δ as shown in Figure 12–57 is given by

$$\delta = t(\sin\theta - \cos\theta \tan\theta_r)$$

where θ_r is the refraction angle related to θ through Snell's law,

$$\sin\theta = n\sin\theta_r$$

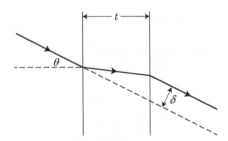

FIGURE 12–57

Problem 25.

26. Find the deviation angle δ when a light beam is incident at an angle 40° on a 60° prism as shown in Figure 12–58. The index of refraction of glass is $n_g = 1.5$. At what incident angle does the minimum deviation δ_{min} occur? What is δ_{min}?

27. A lens made of glass ($n = 1.5$) has a focal length of 15 cm in air. If it is placed in water that has an index of refraction of $n = 1.34$, what is the focal length?

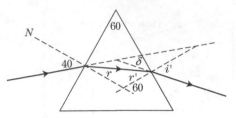

FIGURE 12–58

Problem 26.

28. The three-lens system in Figure 12–59 simulates a zoom lens. The position of the second lens d is adjustable while the distance between the first and third lens is fixed at 60 cm. For an object distance of 100 cm from the first lens, find the location of the image and the magnification when the second lens is at (a) 10 cm and (b) 30 cm from the first lens.

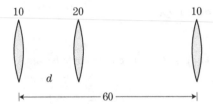

FIGURE 12–59

Problem 28.

29. A person with hyperopia or farsightedness wears contact lenses having $f = 31.25$ cm. (The inverse of the focal length in meters is called the Dioptric power. The contact lens has power of $+3.2$ D.) If the contact lenses are to be replaced with spectacle lenses worn 17 mm in front of the eyes, what should the focal length of the glasses be? Assume 22.5 mm for the size of the eye, which would be the normal focal length of the eye lens without effort.

30. (a) Find the effective focal length f_{eff} of a two-lens system with $f_1 = 10$ cm, $f_2 = -15$ cm with a separation of $d = 5$ cm.
   (b) Locate the principal planes of the two lenses.
   (c) The lens formula

$$\frac{1}{o} + \frac{1}{i} = \frac{1}{f_{eff}}$$

   holds provided that the object distance o is measured from the principal plane H1 and image distance i from H2. Find the location of the image and the magnification of an object placed at 30 cm from the first lens.

31. Find the effective focal length and the positions of the principal planes of the thick lens shown in Figure 12–60. The glass index of refraction is $n = 1.5$. If an object is placed at 30 cm from the first vertex, where is its image formed? What is the magnification? (As you will find, H2 is to the left of H1 in this problem.)

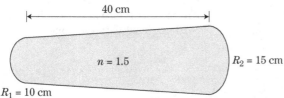

FIGURE 12–60

Problem 31.

32. The matrix of a biconcave thick lens is given by

$$\begin{pmatrix} 1.1333 & 1.3333 \\ 0.15667 & 1.0667 \end{pmatrix}$$

The radius of curvature of the first surface is -5 cm and the glass index is 1.5. Find the thickness of the glass and the other radius of curvature.

33. In the mirror-lens system shown in Figure 12–61, when viewed from the right, two images with magnifications ± 1.5 are observed at the same location. Determine the object location and the focal length of the mirror.

Hint: There are two solutions.

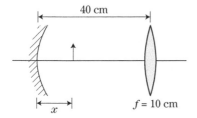

FIGURE 12–61

Problem 33.

34. A telescope consists of an objective lens with $f_o = 50$ cm and an eyepiece with $f_e = 2.5$ cm. If it is used to view an object 10 m away with the image to be formed at 25 cm, what is the separation distance of the lenses? What is the magnification?

35. What is the magnification of the telescope shown in Figure 12–62 for an object at infinity? The lens (erector) in between is to make the total magnification positive (e.g., rifle view finder). If the telescope is used to view an object 50 m away with its image formed at 25 cm, what adjustment has to be made? The numbers above the lenses indicate the focal lengths.

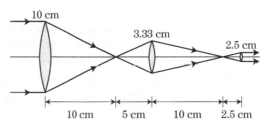

FIGURE 12–62

Problem 35.

36. An achromatic doublet has a focal length of $+8$ cm. The Crown component is an equiconvex lens with a common radius R_1 and the Flint component is a divergent lens with radii of curvature $-R_1$ and R_2. Determine R_2. Assume the following data: Crown glass $n_{Blue} = 1.512$, $n_{Red} = 1.51$; Flint glass $n_{Blue} = 1.63$, $n_{Red} = 1.62$. In the figure, R_2 shown in Figure 12–63 is positive but it could be negative.

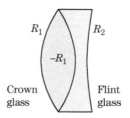

FIGURE 12–63

Problem 36.

37. A glass sphere of radius R has a concentric cubic void of side R as shown in Figure 12–64. Find the focal point for a parallel beam. Also find the effective focal length of the "lens."

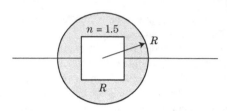

FIGURE 12–64

Problem 37.

38. A fish is at the center of a spherical bowl with radius R that is filled with water ($n = 4/3$). Find the location and magnification of its image.

39. What is the image of the square (2 cm by 2 cm) formed by the lens shown in Figure 12–65? What is the shape and area of the image?

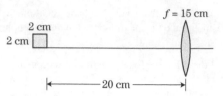

FIGURE 12–65

Problem 39.

40. In a four-lens system with 5 cm separation as shown in Figure 12–66, the first and third lenses have $f = +20$ cm and the second and fourth lenses have $f = -20$ cm. Find the focal point for a parallel incident ray.

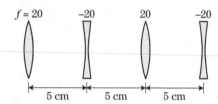

FIGURE 12–66

Problem 40.

Particle Nature of Light

 ## Introduction

Young's experiment clearly demonstrated that light is a wave phenomena since interference can be explained only in terms of the superposition of waves. In 1887, Hertz (the discoverer of electromagnetic waves) found that electrons can be released from a metal surface that was illuminated by light. Many people tried to explain this photoelectric effect in terms of classical theories involving the wave nature of light but no one was successful. Einstein introduced the concept of the photon to explain the photoelectric effect. Light behaves as if it were a collection of photon particles in the photoelectric effect. His theory could explain every aspect of the photoelectric effect. However, it should be emphasized that light also behaves as a wave. Both the wave and the particle nature can coexist in light.

Photoelectric Effect and Einstein's Photon Theory

The experimental arrangement to study the photoelectric effect is shown in Figure 13–1. If electrons are emitted from the metal surface, they can be collected by the anode and the needle on the ammeter deflects. The experimental results can be summarized as follows:

1. Electrons can be emitted only if the frequency of light is higher than a certain frequency that is called the *cutoff frequency*. The cutoff frequency depends on the metal that is used.

299

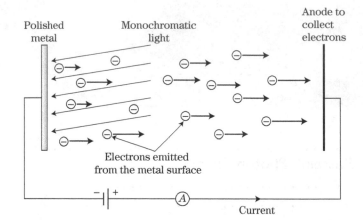

FIGURE 13–1

Arrangement for the photoelectric effect.

2. When released from the metal surface, electrons have a kinetic energy that increases with the frequency of the light. However, the energy does not depend on the intensity of the light.

The wave nature of light fails to explain the presence of a cutoff frequency. One would think that electrons are emitted whenever they acquire sufficient energy by absorbing the light energy. Then the actual frequency of the incident light should be immaterial. By the same token and considering the wave nature of light, we find that it is not consistent with the observation that the photoelectron energy does not depend on the intensity of the light.

Einstein proposed that monochromatic light with a frequency v is a collection of photons, each having an energy hv (J) where $h = 6.63 \times 10^{-34}$ J·sec is Planck's constant. He postulated that electrons are emitted from the metallic surface, being liberated by the photons rather than gradually absorbing the energy from the incident light. Electrons, however, are bound to the metal and some energy has to be given to the electrons before they are released. This binding energy is called the *work function* W(J). Therefore, the difference $hv - W$ would appear as the kinetic energy of the electron

$$\frac{1}{2}m_e v^2 = hv - W \qquad (13.1)$$

This relationship is illustrated in Figure 13–2. If the photon energy hv is less than the work function of the metal W, the electron cannot climb up the hill and will not be released from the metal surface. This explains the cutoff frequency and the dependence of the kinetic energy on the frequency.

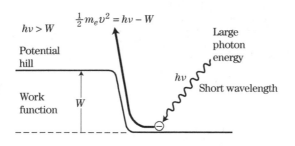

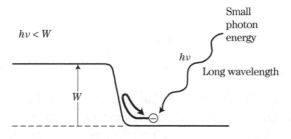

FIGURE 13–2

The electron absorbs the photon energy $h\nu$. If the energy is larger than the work function of the metal (potential energy), the electron will be released from the metal.

EXAMPLE 13.1

Sodium (Na) has a work function $W = 2.9 \times 10^{-19}$ J $(= 1.8$ eV). Find the cutoff frequency ν_c for photoelectric emission.

Solution

The cutoff frequency is

$$\nu_c = \frac{W}{h} = \frac{1.8 \times 1.6 \times 10^{-19}}{6.63 \times 10^{-34}} = 4.4 \times 10^{14} \text{ Hz}$$

The corresponding wavelength is

$$\lambda_c = \frac{c}{\nu_c} = \frac{3 \times 10^8}{4.4 \times 10^{14}} = 6.9 \times 10^{-7} \text{ m} = 6900 \text{ Å}$$

Therefore visible light (4000–7000 Å) can cause photoelectric emission from sodium. This particular wavelength corresponds to the color red.

EXAMPLE 13.2

A metal having a work function of 2.3 eV is illuminated with ulatraviolet radiation with a wavelength $\lambda = 3000$ Å. Calculate the maximum kinetic energy of the photoelectrons that are emitted from the surface.

Solution

The kinetic energy is computed from

$$\frac{1}{2}m_e v^2 = h\nu - W$$

where

$$h\nu = \frac{hc}{\lambda} = \frac{(6.63 \times 10^{-34}) \times (3 \times 10^8)}{3 \times 10^{-7}} = 6.63 \times 10^{-19}\,\text{J} = 4.1\,\text{eV}$$

Then

$$\frac{1}{2}m_e v^2 = 4.1 - 2.3 = 1.8\ \text{eV} = 3.0 \times 10^{-13}\,\text{J}$$

In Chapter 11, we learned that accelerated or decelerated charges can create electromagnetic radiation. For example, if an energetic electron hits a hard metal surface such as tungsten, X-rays can be created. In this case the electron experiences a large deceleration and its energy can be converted into that of the X-ray. This process may be called an inverse photoelectric effect. The more common terminology is *Bremsstrahlung* which is a German word for *Bremsen* (braking, deceleration) and *Strahlung* (radiation).

Hydrogen Atom

Neon signs use a gas discharge to create various colors. Incandescent lamps can create light for illumination. In this section, we give a short introduction to the quantum theory of radiation in order to explain the mechanism of radiation from atoms. If the radiation spectrum from a hydrogen gas is carefully analyzed, one finds that the spectrum is composed of many discrete lines rather than a continuous spectrum. As we have seen in the previous section, if the electron in the hydrogen atom somehow loses a certain amount of energy, the energy is released as electromagnetic radiation.

The simplest picture of the hydrogen atom is that one electron is revolving around a proton. Let the distance between the electron and proton be r. Then the potential energy of the electron-proton system is

$$U = -\frac{e^2}{4\pi\epsilon_0 r}\ \text{(J)} \tag{13.2}$$

The Coulomb force that keeps the electron from flying away is balanced with the centrifugal force,

$$\frac{m_e v^2}{r} = \frac{e^2}{4\pi\epsilon_0 r^2} \tag{13.3}$$

Then the kinetic energy of the electron is

$$\frac{1}{2}m_e v^2 = \frac{e^2}{8\pi\epsilon_0 r} \quad \text{(J)} \tag{13.4}$$

The total energy E of the atom is found from $U + \frac{1}{2}m_e v^2$,

$$E = -\frac{e^2}{8\pi\epsilon_0 r} \quad \text{(J)} \tag{13.5}$$

The question is: What is the radius r? According to classical dynamics, the radius r can be arbitrary and each hydrogen atom could have a different radius and a different energy. That this is not the case is clearly evident from the experimentally measured discrete spectrum of radiation from hydrogen gas since if the radius is arbitrary, then also is the energy and we would expect the radiation spectrum to be continuous. We conclude that the radius must be discrete in order to explain the discrete spectrum.

In 1913, Niels Bohr (a Danish physicist) proposed that the angular momentum of electrons in atoms should be discrete. For the case of the hydrogen atom, the angular momentum of the electron is determined from Eq. (13.3),

$$m_e vr = \sqrt{\frac{m_e e^2 r}{4\pi\epsilon_0}} \tag{13.6}$$

Bohr postulated that the angular momentum of the electron was quantized

$$m_e vr = n\frac{h}{2\pi} \quad (n = 1, 2, 3, \ldots) \tag{13.7}$$

where h is the Planck's constant. Then the radius r is given by

$$r = n^2 \frac{\epsilon_0 h^2}{\pi m_e e^2} = n^2 \times 5.3 \times 10^{-11} \text{ (m)} \tag{13.8}$$

and the energy becomes

$$E_n = -\frac{m_e e^4}{8\epsilon_0^2 h^2}\frac{1}{n^2} = -13.6\frac{1}{n^2} \text{ (eV)} \tag{13.9}$$

which is shown in Figure 13–3.

The state with $n = 1$ is the lowest energy level that the ordinary hydrogen atom has. The electron in this bottom state can be brought to higher energy levels with electrical or optical disturbances. This process is called *excitation*. When this excited electron falls down to a lower state, it releases an amount of energy that appears as radiation. Since the energy levels are discrete (Figure 13–4), the frequencies of the radiation are also discrete,

$$h\nu = 13.6 \left(\frac{1}{n^2} - \frac{1}{m^2}\right) \quad \text{(eV)}$$

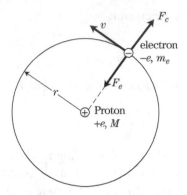

FIGURE 13–3

A model of the hydrogen atom: $F_e = e^2/4\pi\epsilon_0 r^2$ (Coulomb force) and $F_c = m_e v^2/r$ (centrifugal force).

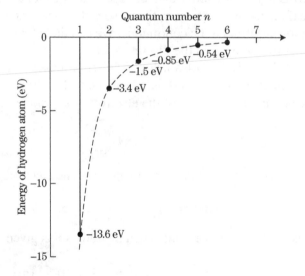

FIGURE 13–4

Discrete allowed energy states of a hydrogen atom.

EXAMPLE 13.3

The electron in a hydrogen atom is excited to the second energy level. When it falls down to the ground level, what value of radiation in terms of frequency should appear?

Solution

The frequency is computed to be

$$hv = \Delta E = 13.6\left(1 - \frac{1}{4}\right) = 10.2\ \text{eV} = 1.63 \times 10^{-18}\ \text{J}$$

$$v = \frac{\Delta E}{h} = \frac{1.63 \times 10^{-18}\ \text{J}}{6.63 \times 10^{-34}\ \text{J} \cdot \text{s}} = 2.46 \times 10^{15}\ \text{Hz}$$

The corresponding wavelength is

$$\lambda = \frac{c}{\nu} = \frac{3 \times 10^8}{2.46 \times 10^{15}} = 1.2 \times 10^{-7} \text{ m} = 1200 \text{ Å}$$

which is in the ultraviolet regime.

EXAMPLE 13.4

How much energy is required to completely free the electron in a hydrogen atom or to ionize the atom?

Solution

A free electron corresponds to $r = \infty$, or $n = \infty$. Then,

$$E = 13.6 \left(1 - \frac{1}{\infty} \right) = 13.6 \text{ eV}$$

This is the ionization potential of the hydrogen atom.

13.4   de Broglie Wave

In Chapters 3 and 10 we saw that waves, both mechanical and electromagnetic, carry momentum as well as energy. The relationship between them was

$$\text{momentum} = \frac{\text{wave energy}}{\text{wave velocity}} \tag{13.10}$$

Since a photon carries an energy $h\nu$ at a velocity c, we expect that the momentum p associated with a photon is

$$p = \frac{h\nu}{c} = \frac{h}{\lambda} \tag{13.11}$$

This is indeed correct and a more rigorous proof can be found using quantum mechanics.

Here we reverse the argument. We ask: Can we assign a wavelength λ defined through

$$\lambda = \frac{h}{p} \tag{13.12}$$

to any object having a momentum p? In 1924, the French scientist de Broglie argued that we can; that is, any physical object should have both a particle *and* a wave nature. Previously we saw that electron microscopes can have a much higher resolving power than conventional optical microscopes because the wavelength associated with energetic electrons is much shorter than that of visible light. For example, the momentum of a 100-keV electron is 1.7×10^{-22} kg m/sec and the de Broglie wavelength $\lambda_{de\,Broglie} = 3.9 \times 10^{-12}$ m. This is about 10^5 times shorter than the wavelength of visible light ($4 - 7 \times 10^{-7}$ m). If you recall that the resolving power is inversely proportional to the wavelength, it can be understood why electron microscopes can "see" better than optical microscopes.

13.5 — **Problems**

1. It is sometimes convenient to consider light (or electromagnetic waves) as a collection of photons. Estimate the number of photons emitted every second from a 100-W light bulb. Assume that the light is monochromatic and has a wavelength 5500 Å.

2. A laser beam ($\lambda = 6000$ Å) has a power density of 50 W/cm². What is the photon density associated with the beam?

3. Aluminum has a work function of 4.2 eV.
   (a) What is the cutoff frequency and wavelength for photoelectric emission?
   (b) Ultraviolet light of $\lambda = 1500$ Å falls on an aluminum surface. What is the maximum kinetic energy of emitted electrons?

4. A hydrogen atom is excited from an $n = 1$ to an $n = 4$ energy level.

   (a) What is the energy at each level?
   (b) What energy is required for the excitation?
   (c) If the electron falls down to a $n = 1$ level again, what is the wavelength of emitted radiation?

5. Find the electron orbit radii and energy levels of a singly ionized helium ion, which has one electron revolving around an α particle (two protons and two neutrons).

6. Calculate the de Broglie wavelength of a 200-keV electron.

7. Referring to the result of Example 10.2, answer the following questions.
   (a) What is the acceleration acting on the electron in a hydrogen atom at the ground state?
   (b) What is the approximate time constant for the electron to lose energy by radiation?

Fourier Analyses and Laplace Transformation

Introduction

In Chapter 2, we learned that the wave equation

$$\frac{\partial^2 f}{\partial t^2} = c_w^2 \frac{\partial^2 f}{\partial x^2}$$

can be satisfied with any function f as long as f depends on x and t in the form $x \pm c_w t$. Although we have studied many kinds of waves (mechanical and electromagnetic) in terms of sinusoidal waves such as $\xi_0 \sin(kx - \omega t)$, the waves do not have to have time harmonic variations. In fact, waves that we experience in daily life are usually not time harmonic. For example, radio waves emitted by a radio station are not pure sine waves. The pure sine wave, called the carrier, is modulated with another signal that contains the information. However, studying waves in terms of sinusoidal waves is extremely important because no matter how complicated a waveform is, it can be approximated with a collection of many sinusoidal waves, each term having a prescribed amplitude. The procedure for examining these more complicated waves is to perform a Fourier analysis of the wave. This is an extremely important mathematical tool in engineering and physics. We will also present the method of the closely related Laplace transform. This method can significantly simplify the solving of differential equations.

Sum of Sinusoidal Functions

Let us see how a sum of several sinusoidal functions creates a function that is not harmonic and is called an anharmonic function. We consider the following function

$$f(x) = \sin x + \tfrac{1}{3} \sin 3x + \tfrac{1}{5} \sin 5x + \cdots \qquad (14.1)$$

In Figure 14–1,

$$\sin x$$

$$\sin x + \tfrac{1}{3} \sin 3x$$

$$\sin x + \tfrac{1}{3} \sin 3x + \tfrac{1}{5} \sin 5x$$

$$\sin x + \tfrac{1}{3} \sin 3x + \tfrac{1}{5} \sin 5x + \tfrac{1}{7} \sin 7x$$

are shown. We can expect that as the number of terms in the sum increases, the summation approaches a square function. Or said another way, we expect that the square function can be expressed in terms of a sum of several sinusoidal harmonics, each term having a distinct amplitude. The Fourier analysis will yield the required amplitudes.

Consider another example, which is essentially the signal of an amplitude modulated (AM) radio. Let us examine the following function:

$$A \sin \omega t + \alpha \sin(\omega + \Delta\omega)t + \alpha \sin(\omega - \Delta\omega)t \qquad (14.2)$$

which is a sum of three sinusoidal waves. Using the identity

$$\sin \alpha + \sin \beta = 2 \sin \frac{\alpha + \beta}{2} \cos \frac{\alpha - \beta}{2}$$

we can rewrite Eq. (14.2) as

$$A \sin \omega t + 2\alpha \sin \omega t \cos \Delta\omega t = A \sin \omega t \left[1 + \frac{2\alpha}{A} \cos \Delta\omega t \right] \qquad (14.3)$$

which is shown in Figure 14–2 for the chosen numerical values of

$$2\alpha/A = 0.1 \quad \text{and} \quad \Delta\omega = 0.1\omega$$

This is the amplitude modulation of the sinusoidal carrier wave $A \sin \omega t$. The smaller frequency $\Delta\omega$ corresponds to the audio frequency which is at most $2\pi \times 20$ kHz.

An amplitude modulated radio signal from a broadcast station has a carrier frequency of 540 kHz that is indicated on your radio dial. Centered at this frequency, the station actually occupies the frequency band from $540 - 20$ kHz to $540 + 20$ kHz in order to transmit the audio signals. Each radio station is allotted this 40-kHz band that is centered at each carrier frequency and only the carrier frequency is indicated on the radio dial.

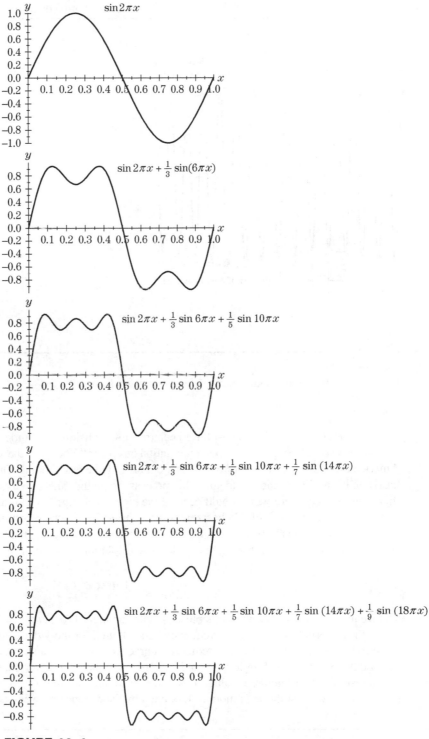

FIGURE 14–1

The sum of sinusoidal functions creates a periodic anharmonic function.

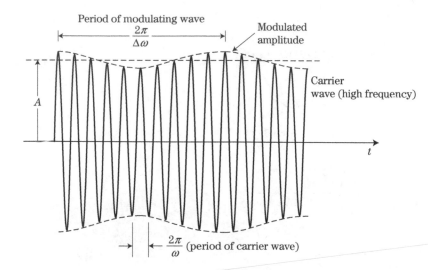

FIGURE 14–2

Amplitude modulation of a high-frequency carrier wave with a low frequency wave.

EXAMPLE 14.1

Estimate the frequency band required for a television station.

Solution

Video or picture signals in the past have been transmitted using amplitude modulation of the carrier wave although pulse code modulation has recently become the standard. In American standards, thirty picture frames are transmitted every second. This number is based on the ability of the eye to smoothly process the picture and also avoid any flickering of the picture which could occur if the pictures changed at a significantly slower rate. Each frame consists of 525 horizontal scans. Therefore, if the resolution in the horizontal direction is to be of the same order as that in the vertical. This is defined as the spacing between two neighboring lines. The time resolution must be of the order of

$$\frac{1}{30 \times (525)^2} \simeq 0.1 \ \mu\text{sec}$$

The frequency corresponding to this pulse is approximately $1/0.1 \ \mu\text{sec} = 10$ MHz.

In the American standard, the frequency band allotted for the video signals is 4 MHz, since only one-half of this value is required in practice because of a special transmission method that is known as Single Sideband. A narrower band would blur the picture because of the resulting poor picture resolution. A wider band does not improve the resolution since the vertical resolution is more or less fixed by the distance between adjacent scanning lines.

Fourier Series

Consider an arbitrary periodic function $f(t)$ with a period T (Figure 14–3):

$$f(t) = f(t + T) \tag{14.4}$$

Since both $\sin(2\pi nt/T)$ and $\cos(2\pi nt/T)$ where n is an integer are also periodic functions, we write

$$f(t) = a_0 + a_1 \cos\left(\frac{2\pi t}{T}\right) + a_2 \cos\left(\frac{4\pi t}{T}\right) + \cdots + b_1 \sin\left(\frac{2\pi t}{T}\right) + b_2 \sin\left(\frac{4\pi t}{T}\right) + \cdots$$

$$= a_0 + \sum_{n=1}^{\infty} a_n \cos\left(\frac{2\pi nt}{T}\right) + \sum_{n=1}^{\infty} b_n \sin\left(\frac{2\pi nt}{T}\right) \tag{14.5}$$

where the coefficients a_n and b_n are unknown coefficients that are to be determined for a given function $f(t)$.

It is convenient to introduce a new variable $x = 2\pi t/T$. Then

$$f(x) = a_0 + \sum_{n=1}^{\infty} a_n \cos(nx) + \sum_{n=1}^{\infty} b_n \sin(nx) \tag{14.6}$$

and the period of the function becomes 2π. To find the values for the coefficients a_n and b_n, let us recall the following definite integrals

$$\int_0^{2\pi} \cos mx \cos nx \, dx = \begin{cases} \pi, & m = m \\ 0, & m \neq n \end{cases}$$

$$\int_0^{2\pi} \sin mx \cos nx \, dx = 0$$

$$\int_0^{2\pi} \sin mx \sin nx \, dx = \begin{cases} \pi, & m = m \\ 0, & m \neq n \end{cases}$$

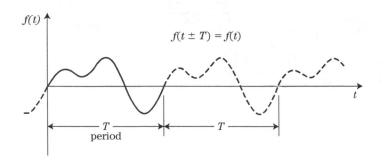

FIGURE 14–3

A periodic function of an arbitrary shape with a period T.

These can be proved easily by making use of the following trigonometric identities (see Appendix B).

$$\cos mx \cos nx = \tfrac{1}{2}[\cos(m+n)x + \cos(m-n)x]$$

$$\sin mx \sin nx = \tfrac{1}{2}[-\cos(m+n)x + \cos(m-n)x]$$

$$\sin mx \cos nx = \tfrac{1}{2}[\sin(m+n)x + \sin(m-n)x]$$

and integrating these quantities over the interval from 0 to 2π. Then multiplying Eq. (14.6) by $\cos nx$ and integrating the result from 0 to 2π, we find

$$a_n = \frac{1}{\pi} \int_0^{2\pi} f(x) \cos nx \, dx \quad (n = 1, 2, 3 \ldots) \tag{14.7}$$

Similarly, to find the coefficient b_n, we multiply Eq. (14.6) by $\sin nx$ and integrate

$$b_n = \frac{1}{\pi} \int_0^{2\pi} f(x) \sin nx \, dx \quad (n = 1, 2, 3 \ldots) \tag{14.8}$$

The coefficient a_0 can be found from

$$a_0 = \frac{1}{2\pi} \int_0^{2\pi} f(x) \, dx \tag{14.9}$$

This implies that a_0 corresponds to the average value of the function $f(x)$.

EXAMPLE 14.2

Find the Fourier series of the square wave shown in Figure 14–4.

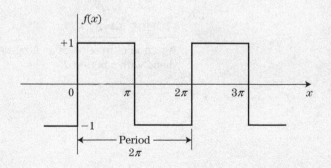

FIGURE 14–4

A periodic square wave function.

Solution

The average of $f(x)$ is zero which means that $a_0 = 0$. Also, the function $f(x)$ is an odd function in that $f(x) = -f(-x)$. Then the even components will have the coefficients $a_n = 0$ and the coefficients b_n are computed from

$$b_n = \frac{1}{\pi} \int_0^{2\pi} f(x) \sin nx \, dx$$

$$= \frac{1}{\pi} \left[\int_0^{\pi} \sin nx \, dx - \int_{\pi}^{2\pi} \sin nx \, dx \right]$$

$$= \frac{1}{\pi} \left[\frac{1}{n}(-\cos nx) \Big|_0^{\pi} + \frac{1}{n}(\cos nx) \Big|_{\pi}^{2\pi} \right]$$

$$= \frac{2}{n\pi} [1 - \cos n\pi]$$

Then the even terms $(b_2, b_4, b_6, \ldots)$ are zero and the odd terms are found to be

$$b_1 = \frac{4}{\pi}, \quad b_3 = \frac{4}{3\pi}, \quad b_5 = \frac{4}{5\pi}, \cdots$$

Therefore the square wave train can be Fourier expanded as

$$f(x) = \frac{4}{\pi} \left[\sin x + \tfrac{1}{3} \sin 3x + \tfrac{1}{5} \sin 5x + \cdots \right]$$

We previously guessed, as shown in Figure 14–1, that the function

$$\sin x + \tfrac{1}{3} \sin 3x + \tfrac{1}{5} \sin 5x + \cdots$$

would approach a square wave function as the number of terms is increased. This has now been shown to be a reasonable guess.

EXAMPLE 14.3

Find the Fourier series for the sawtooth function shown in Figure 14–5.

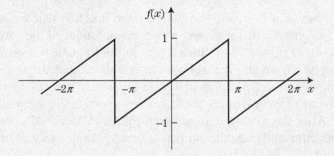

FIGURE 14–5

A periodic sawtooth function.

Solution

The average of $f(x)$ is zero. Then the coefficient $a_0 = 0$. Again the function is odd and the coefficients $a_n = 0$. The Fourier coefficient can be evaluated over an arbitrary period irrespective of location. Here it is convenient to choose the period from $-\pi$ to π.

$$b_n = \frac{1}{\pi} \int_{-\pi}^{\pi} \frac{x}{\pi} \sin nx \, dx$$

$$= \frac{1}{\pi^2} \left[-\frac{1}{n} x \cos nx \Big|_{-\pi}^{\pi} + \frac{1}{n} \int_{-\pi}^{\pi} \cos nx \, dx \right]$$

$$= \frac{1}{\pi^2} \left[-\frac{\pi}{n} \cos(nx) - \frac{\pi}{n} \cos(-n\pi) \right]$$

$$= \frac{2}{\pi} \frac{(-1)^{n-1}}{n} \quad (n = 1, 2, 3, \ldots)$$

Therefore

$$f(x) = \frac{2}{\pi} \left[\sin x - \tfrac{1}{2} \sin 2x + \tfrac{1}{3} \sin 3x - \cdots \right]$$

Note we have made use of integration by parts in order to evaluate the integrals,

$$\int x \sin x \, dx = -x \cos x + \int \cos x \, dx = -x \cos x + \sin x$$

Fourier Spectrum

We have seen that any periodic function can be expanded in terms of sinusoidal functions. The lowest frequency is determined by the period T as $\omega_0 = 2\pi/T$ and the higher harmonics have frequencies that are integer multiples of the fundamental frequency. Although we have only discussed the Fourier expansion of time-varying functions, the same procedure can be applied to any function that is periodic.

The amplitudes of the various harmonics a_n and b_n can be plotted as a function of frequency. For example, the square wave function that we studied in Example 14.2 can be characterized with the amplitudes of the various harmonics as shown in Figure 14–6. Such a plot in the frequency domain is called a *frequency spectrum* and it enables us to visualize the distribution of the various harmonic terms. Spectrum analyzers frequently used in communication research can directly display the frequency spectrum of a signal on an oscilloscope. Also, the grating optical spectrometer (Chapter 12) can be regarded as a spectrum analyzer that can perform the Fourier analysis in the wavelength domain.

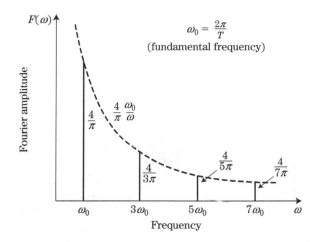

FIGURE 14–6

Fourier spectrum of a square wave.

EXAMPLE 14.4

Find the Fourier spectrum of the amplitude-modulated wave given by

$$f(t) = A(1 + a\cos\omega_a t)\sin\omega_c t, \quad a < 1$$

where ω_a is the modulating audio frequency and ω_c is the carrier frequency.

Solution

The function $f(t)$ can be rewritten as

$$A\sin\omega_c t + \tfrac{1}{2}aA[\sin(\omega_c - \omega_a)t + \sin(\omega_c + \omega_a)t]$$

Therefore we obtain the spectrum shown in Figure 14–7.

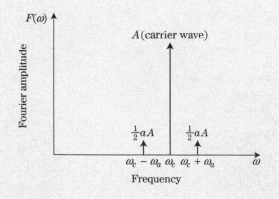

FIGURE 14–7

Fourier spectrum of an amplitude modulated wave.

EXAMPLE 14.5

Find the Fourier spectrum of a single pulse shown in Figure 14–8(a).

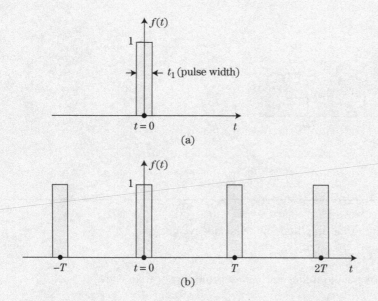

FIGURE 14–8

(a) A single rectangular pulse. (b) A series of periodic rectangular pulses.

Solution

The single pulse is not a periodic function and you may wonder how to Fourier-analyze such a function. There is a trick. We assume a pulse train with a period T as shown in Figure 14–8(b), Fourier-analyze it and then eventually let $T \rightarrow \infty$. Consider a pulse train shown above. Since the function is an even function, we are able to immediately determine that the coefficients $b_n = 0$. The other coefficients are found to be

$$a_0 = \text{ average of } f(t) = \frac{t_1}{T}$$

$$a_n = \frac{2}{T} \int_0^T f(t) \cos n\omega_0 t \, dt$$

where $\omega_0 = 2\pi/T$ is the fundamental frequency. Since

$$f(t) = \begin{cases} 1 & \left(0 < t < \dfrac{t_1}{2}\right) \\ 0 & \left(\dfrac{t_1}{2} < t < T - \dfrac{t_1}{2}\right) \\ 1 & \left(T - \dfrac{t_1}{2} < t < T\right) \end{cases}$$

we find

$$a_n = \frac{2}{T} \left[\int_0^{t_1/2} \cos n\omega_0 t\, dt + \int_{T-(t_1/2)}^{T} \cos n\omega_0 t\, dt \right]$$

$$= \frac{2}{T}\frac{1}{n\omega_0} \left[\sin\left(n\omega_0 \frac{t_1}{2}\right) - \sin n\omega_0\left(T - \frac{t_1}{2}\right) \right]$$

Recalling that $\omega_0 = 2\pi/T$ and using the trigonometric identity

$$\sin(\alpha - \beta) = \sin\alpha\cos\beta - \cos\alpha\sin\beta,$$

we find

$$a_n = \frac{2}{n\pi} \sin\left(n\pi\frac{t_1}{T}\right)$$

This spectrum is shown in Figure 14–9 for the case $T = 10t_1$. As T is increased, the spacing between adjacent lines decreases and in the limit of $T \to \infty$, we have a *continuous spectrum* rather than the *discrete spectrum* (Figure 14–10).

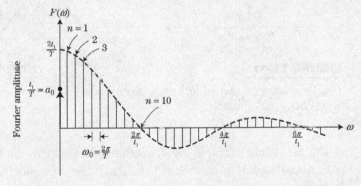

FIGURE 14–9

Fourier spectrum of a pulse with $T/t_1 = 10$.

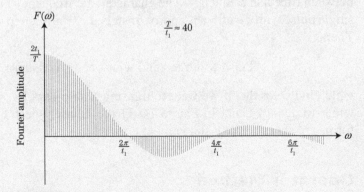

FIGURE 14–10

As T increases, the spectrum becomes continuous rather than discrete. Here the case $T/t_1 = 40$ is shown.

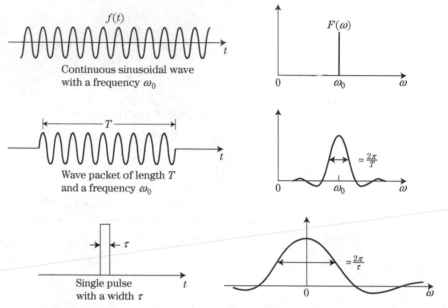

FIGURE 14–11

Fourier spectra for several functions. Note that a smaller time spread results in a wider frequency spread.

The preceding example indicates that even nonperiodic functions can be Fourier-analyzed. If a function is periodic, its Fourier spectrum is discrete. If a function is nonperiodic, its Fourier spectrum is, in general, continuous. The preceding example also reveals an interesting fact concerning the relationship between time and frequency. The characteristic frequency bandwidth for the single pulse with a width τ is approximately $2\pi/\tau$ as seen from Figure 14–11. Then

$$\text{Pulse width} \times \text{band width} \simeq 2\pi \, (\text{constant}) \tag{14.10}$$

which indicates that if we want to transmit a narrow pulse, we have to use a wider-frequency band. In Figure 14–11 the Fourier spectra for some waveforms are schematically shown.

14.5 Operator Method

Sinusoidal waves are usually described with

$$A\cos(kx - \omega t) \quad \text{or} \quad A\sin(kx - \omega t)$$

as we have seen in previous chapters. Since

$$\cos \alpha = \text{Re } e^{i\alpha} \quad \text{and} \quad \sin \alpha = \text{Im } e^{i\alpha} \tag{14.11}$$

it is more convenient to use the exponential function

$$A e^{i(kx - \omega t)} \tag{14.12}$$

rather than using the trigonometric functions to describe a wave. When we need a cosine wave, we take its real part of the exponential. For a sine wave, we take the imaginary part of the exponential.

Writing a wave in the form Eq. (14.12) greatly simplifies wave analysis. Since

$$\frac{\partial}{\partial x} e^{i(kx - \omega t)} = ik e^{i(kx - \omega t)}$$

$$\frac{\partial}{\partial t} e^{i(kx - \omega t)} = -i\omega e^{i(kx - \omega t)}$$

we can simply replace $\partial/\partial x$ with ik and $\partial/\partial t$ with $-i\omega$. For example, the wave equation

$$\frac{\partial^2 \xi}{\partial t^2} = c_w^2 \frac{\partial^2 \xi}{\partial x^2}$$

can readily be converted into

$$(-i\omega)^2 \xi = (ik)^2 c_w^2 \xi$$

or

$$\omega^2 = c_w^2 k^2$$

which immediately yields the dispersion relation $\omega = \pm c_w k$.

EXAMPLE 14.6

Find the natural oscillation frequency and the damping coefficient of an *LCR* circuit.

Solution

The differential equation that describes the charge on the capacitor was obtained in Chapter 1 to be

$$\frac{d^2 q}{dt^2} + \frac{R}{L} \frac{dq}{dt} + \frac{1}{LC} q = 0$$

Let us assume a solution of the form $q_0 e^{-i\omega t}$ where ω may be a complex number. Using $d/dt = -i\omega$, we obtain

$$-\omega^2 q - i\frac{R}{L}\omega q + \frac{1}{LC}q = 0$$

or

$$\omega^2 + i\frac{R}{L}\omega - \frac{1}{LC} = 0$$

Solving for the frequency ω, we find

$$\omega = \frac{1}{2}\left[-i\frac{R}{L} \pm \sqrt{\frac{4}{LC} - \left(\frac{R}{L}\right)^2}\right] \tag{14.13}$$

The complex frequency indicates that the oscillation damps as

$$e^{\gamma t}$$

where

$$\gamma = \text{Im }\omega = -\frac{R}{2L} \quad (< 0) \tag{14.14}$$

The oscillation frequency is

$$\text{Re }\omega = \sqrt{\frac{1}{LC} - \left(\frac{R}{2L}\right)^2}, \quad \left[\frac{1}{LC} > \left(\frac{R}{2L}\right)^2\right] \tag{14.15}$$

Electromagnetic waves can propagate without damping if the medium is loss-free as in an ideal vacuum. In practice, all dielectric materials have a finite conductivity although it is usually much smaller than in metals. In Chapter 9, we found that a medium with losses has a finite resistance and a transmission line consisting of a conductance in parallel with the capacitor as shown in Figure 14–12 could be used. Let the conductance per unit element be $G/\Delta x$. The characteristic impedance is then

$$Z_c = \sqrt{\frac{-i\omega L/\Delta x}{(-i\omega C + G)/\Delta x}} \tag{14.16}$$

In the limit of $G \to 0$, this reduces to the lossless case. In general, the characteristic impedance should be given by

$$Z_c = \sqrt{\frac{\text{reactance per unit length}}{\text{susceptance per unit length}}} = \sqrt{\frac{(-i\omega L + R)/\Delta x}{(-i\omega C + G)/\Delta x}} \tag{14.17}$$

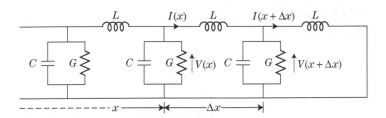

FIGURE 14–12

A lossy medium can be modeled with a conductance in parallel with a capacitance. In full generality, a resistance should also be placed in series with the inductance.

For waves in unbounded space such as in air, the characteristic impedance becomes

$$Z_c = \sqrt{\frac{-i\omega\mu_0}{-i\omega\varepsilon + \sigma}} \tag{14.18}$$

where σ is conductivity $(1/\Omega\,\text{m})$. In the presence of loss, the characteristic impedance becomes complex.

The propagation velocity in the presence of the finite conductivity can be found as follows. Kirchhoff's voltage and current laws (Figure 14–12) yield

$$V(x, t) = L\frac{\partial I(x, t)}{\partial t} + V(x + \Delta x, t) \tag{14.19}$$

$$I(x, t) = C\frac{\partial V(x + \Delta x, t)}{\partial t} + GV(x + \Delta x, t) + I(x + \Delta x, t) \tag{14.20}$$

from which we obtain for the current I

$$\frac{\partial^2 I}{\partial x^2} = \frac{LC}{(\Delta x)^2}\frac{\partial^2 I}{\partial t^2} + \frac{LG}{(\Delta x)^2}\frac{\partial I}{\partial t} \tag{14.21}$$

where the independent variables are not explicitly stated. Substituting $L/\Delta x = \mu_0$, $C/\Delta x = \varepsilon$, and $G/\Delta x = \sigma$, we obtain

$$\frac{\partial^2 I}{\partial x^2} = \varepsilon\mu_0\frac{\partial^2 I}{\partial t^2} + \mu_0\sigma\frac{\partial I}{\partial t} \tag{14.22}$$

This is not the conventional wave equation because of the presence of the term having only the first partial derivative with respect to time.

Let us apply the operator method to find the propagation velocity. Replacing $\partial/\partial x$ with ik and $\partial/\partial t$ with $-i\omega$, we find

$$k^2 = \varepsilon\mu_0\omega^2 + i\omega\mu_0\sigma \tag{14.23}$$

Then the propagation velocity becomes

$$\frac{\omega}{k} = \frac{1}{\sqrt{\varepsilon\mu_0}}\frac{1}{\sqrt{1 + \frac{i\sigma}{\omega\varepsilon}}} \tag{14.24}$$

which is complex! The complex velocity should receive the following inter-
pretation. Since the propagation constant

$$k = \sqrt{\varepsilon \mu_0 \omega} \left(1 + \frac{i\sigma}{\omega \varepsilon}\right)^{1/2} \tag{14.25}$$

is complex for a real frequency ω, the function e^{ikx} exponentially decreases
as the wave propagates. In taking the square root, there is no ambiguity in
choosing the proper sign since the energy of the wave will be transformed
into thermal energy where it is lost. The wave will decrease to a value of
$1/e \simeq 37\%$ of its initial value in a distance that is called the damping length
or the *e-folding length* that is given by the imaginary part of k

$$\delta = \frac{1}{\operatorname{Im} k} \tag{14.26}$$

This can also be written as

$$1 + i \frac{\sigma}{\omega \varepsilon} = \sqrt{1 + \left(\frac{\sigma}{\omega \varepsilon}\right)^2} \, e^{i\theta}$$

where

$$\theta = \tan^{-1} \frac{\sigma}{\omega \varepsilon}$$

The imaginary part of the wave number $\operatorname{Im} k$ is found to be

$$\operatorname{Im} k = \omega \sqrt{\varepsilon \mu_0} \left[1 + \left(\frac{\sigma}{\omega \varepsilon}\right)^2\right]^{1/4} \sin \frac{\theta}{2} \tag{14.27}$$

and the damping length becomes

$$\delta = \frac{1}{\omega \sqrt{\varepsilon \mu_0}} \left[1 + \left(\frac{\sigma}{\omega \varepsilon}\right)^2\right]^{-1/4} \frac{1}{\sin(\theta/2)} \tag{14.28}$$

This quantity is also called the *skin depth* and it is a measure of how deep
an electromagnetic wave can penetrate into a lossy material assuming normal
incidence of the wave.

The propagation phase velocity can be found from

$$\frac{\omega}{\operatorname{Re} k} = \frac{1}{\sqrt{\varepsilon \mu_0}} \left[1 + \left(\frac{\sigma}{\omega \varepsilon}\right)^2\right]^{-1/4} \frac{1}{\cos(\theta/2)} \tag{14.29}$$

EXAMPLE 14.7

Calculate the skin depth in soil for 1-MHz (a typical AM radio frequency) electromagnetic waves. Assume that the conductivity of soil is 10^{-2} S/m.

Solution

From Eq. (14.28)

$$\delta = \frac{3 \times 10^8}{2\pi \times 10^6}\left[1 + \left(\frac{10^{-2}}{2\pi \times 10^6 \times 8.85 \times 10^{-12}}\right)^2\right]^{-1/4}\frac{1}{\sin(\theta/2)} = \frac{3.6}{\sin(\theta/2)}$$

where

$$\theta = \tan^{-1}\left(\frac{10^{-2}}{2\pi \times 10^6 \times 8.85 \times 10^{-12}}\right) = \tan^{-1}(180) \simeq \frac{\pi}{2}\text{ rad}$$

Then $\delta = 5.0$ m, which is much less than the wavelength of the AM radio which has $\lambda = 300$ m. Therefore, earth may be regarded to be a good conductor.

In the presence of loss (or finite conductivity), electromagnetic waves necessarily become dispersive since, as Eq. (14.29) indicates, the phase velocity depends on the frequency ω. A dissipative medium is also dispersive. However, a dispersive media is not always dissipative.

14.6   Laplace Transform

In solving differential equations, we usually guess a possible solution and substitute it back into the differential equation to see if the solution is correct or not. In most cases, we depend on our own experience and intuition.

Let us consider a simple case

$$\frac{d^2 f}{dx^2} - 4f = 0 \tag{14.30}$$

The function $Ae^{-2x} + Be^{2x}$ is a solution of the differential equation. To obtain a more general exponential solution, we assume a solution of the form e^{Dx} where D is as yet unknown. Substituting the assumed solution into Eq. (14.30), we find

$$(D^2 - 4)f = 0 \tag{14.31}$$

and this can be satisfied only if D has the following values

$$D = \pm 2 \tag{14.32}$$

This is already one significant step to converting the differential equation into an algebraic equation.

The Laplace transform exactly does this. It can convert a differential equation into a simple algebraic equation. What is more, the Laplace transform can automatically take into account any prescribed initial or boundary conditions in a natural manner.

The Laplace transform $F(s)$ of a function $f(t)$ is defined by

$$F(s) = \int_0^\infty e^{-st} f(t)dt \tag{14.33}$$

This has an interesting property. The Laplace transform of df/dt is

$$\int_0^\infty e^{-st} \frac{df}{dt} dt = e^{-st} f \Big|_0^\infty + s \int_0^\infty e^{-st} f dt = -f(0) + sF(s) \tag{14.34}$$

where $f(0)$ is the value of the function $f(t)$ evaluated at $t = 0$. Similarly,

$$\int_0^\infty e^{-st} \frac{d^2 f}{dt^2} dt = s^2 F(s) - (sf(0) + f'(0)) \tag{14.35}$$

where $f'(0)$ is the value of the derivative df/dt evaluated at $t = 0$. Therefore, differentiation of a function in time corresponds to multiplying the function by s. Using these properties, we can convert the following differential equation

$$\frac{d^2 f}{dt^2} + A \frac{df}{dt} + Bf = 0$$

into the following algebraic equation

$$s^2 F(s) + As F(s) + B F(s) = sf(0) + f'(0) + Asf(0) \tag{14.36}$$

This can be easily solved for $F(s)$. To find the actual time-dependent solution $f(t)$, we just have to find the appropriate inverse Laplace transform to convert $F(s)$ into $f(t)$. Laplace and inverse Laplace transforms are tabulated and a few of them are listed in Table 14-1. More extensive tables can be found in the literature.

Table 14–1 Short table of Laplace transforms

	$f(t)$	$F(s)$
(1)	1 (step function)	$1/s$
(2)	t^n (n integer)	$\dfrac{n!}{s^{n+1}}$
(3)	e^{at}	$\dfrac{1}{s-a}$
(4)	$e^{at} f(t)$	$F(s-a)$
(5)	$\sin \omega t$	$\dfrac{\omega}{s^2 + \omega^2}$
(6)	$\cos \omega t$	$\dfrac{s}{s^2 + \omega^2}$
(7)	$\delta(t)$ (delta function)	1
(8)	$\ln t$	$-\dfrac{1}{s}(\ln s + 0.5772 \cdots)$

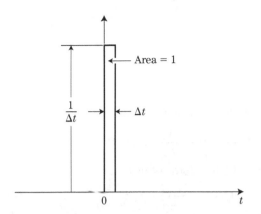

FIGURE 14–13

Definition of a $\delta(t)$ function.

The delta function in row (7) needs some explanation. It is extremely peculiar but very useful in practice. There are several ways to define the delta function. Here we use the following definition (Figure 14–13):

$$\delta(t) = \lim_{\Delta t \to 0} \frac{1}{\Delta t} \qquad (14.37)$$

The area under the delta function is unity as can be easily seen in Figure 14–13. Also, for any function $f(t)$,

$$\int_{-\infty}^{\infty} \delta(t) f(t)\, dt = f(0) \qquad (14.38)$$

The delta function is also known as the *impulse function*. A short disturbance given to mechanical or electrical systems can be well approximated with the delta function. The delta function was originally postulated at the time of the development of quantum mechanics and it is frequently called the Dirac delta function. Some twenty years after its usage in this discipline, it was mathematically proved to be a valid function.

EXAMPLE 14.8

A capacitor charged to q_0 is suddenly connected to a resistor R (Figure 14–14). Find the charge $q(t)$.

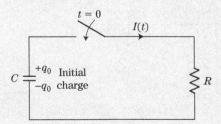

FIGURE 14–14

An R C circuit with the capacitor initially charged.

Solution

The differential equation for the charge q is

$$\frac{dq}{dt} + \frac{1}{RC}q = 0$$

This equation can be written using the Laplace transformation as

$$sQ(s) - q_0 + \frac{1}{RC}Q(s) = 0$$

where q_0 is the initial charge. Solving for $Q(s)$, we find

$$Q(s) = \frac{q_0}{s + 1/RC}$$

Using row (3) from Table 14–1, the inverse Laplace transform $q(t)$ can be found to be

$$q(t) = q_0 e^{-t/RC}$$

which is the familiar exponential damping.

EXAMPLE 14.9

Solve the differential equation for the *LCR* circuit in Example 14.6. Assume that the capacitor has an initial charge q_0 and

$$\frac{1}{LC} > \left(\frac{R}{2L}\right)^2$$

is satisfied.

Solution

The differential equation

$$\frac{d^2q}{dt^2} + \frac{R}{L}\frac{dq}{dt} + \frac{1}{LC}q = 0$$

can be Laplace transformed to yield

$$s^2 Q - (sq_0 + q'(0)) + \frac{R}{L}[sQ - q_0] + \frac{1}{LC}Q = 0$$

where $q'(0)$ is the initial current which is zero since the inductor will not permit an instantaneous change of current. Then

$$Q(s) = \frac{(s + R/L)q_0}{s^2 + (R/L)s + 1/LC} = \frac{[s + R/2L]q_0 + (R/2L)q_0}{(s + R/2L)^2 + 1/LC - (R/2L)^2}$$

Using (4), (5), and (6) from Table 14–1, we find

$$q(t) = q_0 e^{-(R/2L)t}\left[\cos \omega t + \frac{R}{2\omega L}\sin \omega t\right]$$

where

$$\omega = \sqrt{\frac{1}{LC} - \left(\frac{R}{2L}\right)^2}$$

This solution was also obtained as a problem in Chapter 1.

EXAMPLE 14.10

An ac voltage source $V_0 \sin \omega t$ is suddenly connected to an RL circuit shown in Figure 14–15. Find the current $I(t)$.

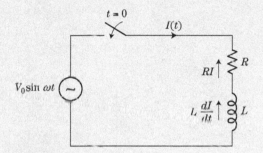

FIGURE 14–15

Solution

Kirchhoff's voltage law yields

$$L\frac{dI(t)}{dt} + RI(t) = V_0 \sin \omega t$$

which can be Laplace transformed to yield

$$sLi(s) - Li(0) + Ri(s) = V_0 \frac{\omega}{s^2 + \omega^2}$$

Note that the initial current is zero because of the inductance which resists sudden change in the current. Then

$$i(s) = \frac{\omega V_0}{(sL + R)(s^2 + \omega^2)}$$

$$= \frac{\omega V_0}{L} \frac{1}{\omega^2 + (R/L)^2} \left[\frac{1}{s + R/L} - \frac{s}{s^2 + \omega^2} + \frac{R/L}{s^2 + \omega^2}\right]$$

The current $I(t)$ can be found by inverse Laplace transforming $i(s)$ as

$$I(t) = \frac{\omega V_0}{L} \frac{1}{\omega^2 + (R/L)^2} \left[e^{-(R/L)t} + \frac{R}{\omega L} \sin \omega t - \cos \omega t\right]$$

After many time constants, the exponential term decreases to 0 and the current oscillates in time and is consistent with the results obtained from ac circuit theory.

Problems

1. High-tone instruments (violin, flute, etc.) can play fast, but low-tone instruments (bass, tuba, etc.) cannot. Explain why in terms of Fourier spectra.

2. Estimate the Fourier spectrum of the wave packet shown in Figure 14–16.

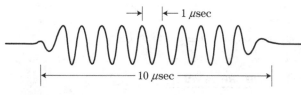

FIGURE 14–16

Problem 2.

3. Calculate the Fourier series of the triangle wave shown in Figure 14–17.

4. Since $\cos(n\omega_0 t)$ is an even function of n, the Fourier series in Example 14.5 may be written as

$$f(t) = \sum_{n=-\infty}^{\infty} a_n \cos(n\omega_0 t)$$

where

$$a_n = \frac{1}{\pi n} \sin\left(\frac{n\omega_0}{2} t_1\right), \quad \omega_0 = \frac{2\pi}{T}$$

In the limit of $T \to \infty$, the summation over n can be replaced by integration ($n\omega_0 \to \omega$)

$$\lim_{\omega_0 \to 0} \sum_{n=\infty}^{\infty} a_n \cos(n\omega_0 t) \to \int_{-\infty}^{\infty} \frac{F(\omega)}{2\pi} \cos(\omega t)\, d\omega$$

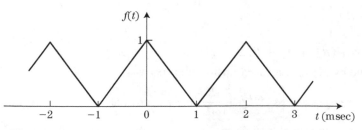

FIGURE 14–17

Problem 3.

(a) Show that $F(\omega)$ for the single pulse in Example 14.5 becomes

$$F(\omega) = \frac{2}{\omega} \sin\left(\frac{\omega}{2} t_1\right)$$

$F(\omega)$ is called the Fourier transform of $f(t)$ and in general can be calculated from

$$F(\omega) = \int_{-\infty}^{\infty} f(t) e^{i\omega t}\, dt$$

The inverse Fourier transform yields $f(t)$,

$$f(t) = \frac{1}{2\pi} \int_{-\infty}^{\infty} F(\omega) e^{-i\omega t}\, d\omega$$

(b) Find the Fourier transform of $f(t) = e^{-t^2/a^2}$ (see Figure 14–18).

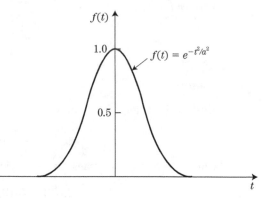

FIGURE 14–18

Problem 4.

5. A certain wave in a plasma (ionized gas) is described by the following differential equation

$$\frac{\partial^2}{\partial t^2} \left(\frac{\partial^2}{\partial x^2} - k_D^2 \right) \xi = -\omega_{pi}^2 \frac{\partial^2 \xi}{\partial x^2}$$

where k_D and ω_{pi} are constants. Find the dispersion relation and plot ω/ω_{pi} as a function of k/k_D.

6. A 50-Ω coaxial cable has polyethylene dielectric that has $\sigma/\omega\varepsilon \simeq 2 \times 10^{-4}$ at 1 MHz. Assuming $\varepsilon = 2.3\varepsilon_0$, calculate the e-folding damping length at 1 MHz.

7. The mass in a mass-spring oscillation system is suddenly hit by a hammer and instantly acquires a momentum p. Find the solution for the displacement from the equilibrium position.

   Hint: The force given by the hammer is $p\delta(t)$, where $\delta(t)$ is the delta function, whose Laplace transform is 1.

CHAPTER 15

Nonlinear Waves, Solitons, Shocks, and Chaos

15.1 Introduction

Up to now, we have studied linear oscillations and linear waves. This is a drastic simplification on our part in order to understand nature. Nonlinearity is a more common state in nature and, as will be demonstrated in this chapter, the inclusion of nonlinearity will have dramatic effects on simple oscillations and on the propagation of the linear waves that were examined in previous chapters. For example, the inclusion of nonlinear effects will lead to certain nonlinear wave equations whose wave solutions will be significantly different from those that were obtained using the linear model. Solitons and shocks are two of the nonlinear waves that will be examined. Finally, the subject of chaos is introduced.

15.2 Nonlinear Oscillations

In Chapter 1, we discussed linear oscillatory phenomena and examined, in particular, the simple pendulum (Figure 15–1). It was shown there that the equation describing the oscillation could be written as

$$\frac{d^2\theta}{dt^2} = -\frac{g}{l}\sin\theta \tag{15.1}$$

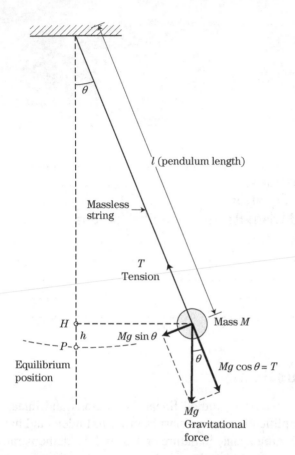

FIGURE 15–1

Pendulum having a mass M
and a length l.

This equation is actually a nonlinear equation in that if the angle of oscillation θ is large, the simplifying approximation of assuming that $\theta \ll 1$ would be violated and our previous analysis is no longer valid. Numerical methods must be employed to solve this particular equation although, as we will see, certain mathematical techniques can be applied to obtain approximate analytical solutions that are valid in certain regimes. Recall that if the approximation that $\sin\theta \simeq \theta$ were made, it would be possible to approximate Eq. (15.1) as

$$\frac{d^2\theta}{dt^2} = -\frac{g}{l}\theta \tag{15.2}$$

and obtain the simple oscillatory solution that was described in Chapter 1. Numerical solutions with identical initial conditions for both the linear and the nonlinear equations are shown in Figure 15–2.

Figure 15–2 illustrates the solutions that would be obtained from a numerical solution of Eqs. (15.1) and (15.2). Identical initial conditions have been used in both figures so it is possible to detect differences between the

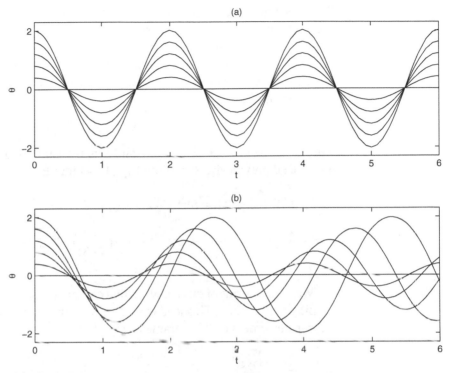

FIGURE 15–2

Numerical solutions for the angular oscillation of a pendulum with increasing initial conditions. (a) Solution of the linear equation Eq. (15.2). (b) Solution of the nonlinear equation Eq. (15.1).

two equations. As expected, the oscillation from the linear equation only has the amplitude of the oscillation increase as the initial amplitude increases. However, the oscillation calculated from the nonlinear equation indicates that the frequency of the oscillation depends upon the initial amplitude. It should be noted that the two oscillations are almost equivalent for the smallest value of the initial condition. The reader could experimentally verify this point by using a child's swing at a park.

There are analytical techniques that can be used to approximately solve Eq. (15.1) and one of them is introduced here. We relax the requirement that $\theta \ll 1$ which led to the linearized Eq. (15.2) to be $\theta < 1$. Referring to Chapter 3, we can replace the sin function with the first two terms of a Taylor series expansion of the sin function

$$\frac{d^2\theta}{dt^2} \simeq \frac{g}{l}\left(\theta - \varepsilon\frac{\theta^3}{3!}\right)$$

(15.3)

We have introduced a parameter ε in this equation that is called a "book-keeping parameter" to clearly separate the nonlinear term from the linear term. Eq. (15.3) is still formidable as it stands. We shall initially illustrate one powerful *perturbation* technique here although, there are others.

The philosophy of the technique is to expand the dependent variable into a number of terms

$$\theta = \theta^{(0)} + \varepsilon\theta^{(1)} + \varepsilon^2\theta^{(2)} + \cdots \qquad (15.4)$$

where $\theta^{(0)} \gg \theta^{(1)} \gg \theta^{(2)} \gg \cdots$ and ε is the same bookkeeping parameter introduced previously. Substitute Eq. (15.4) into Eq. (15.3) and obtain

$$\frac{d^2}{dt^2}\{\theta^{(0)} + \varepsilon\theta^{(1)} + \varepsilon^2\theta^{(2)} + \cdots\}$$

$$= -\frac{g}{l}\left\{[\theta^{(0)} + \varepsilon\theta^{(1)} + \varepsilon^2\theta^{(2)} + \cdots] - \frac{\varepsilon}{3!}[\theta^{(0)} + \varepsilon\theta^{(1)} + \varepsilon^2\theta^{(2)} + \cdots]^3\right\}$$

$$(15.5)$$

We collect terms of this equation that contain the same powers of this book-keeping parameter ε. This leads to the following equations that are defined by the various orders of this parameter $\varepsilon^0, \varepsilon^1, \varepsilon^2, \cdots$

$$\varepsilon^0: \quad \frac{d^2\theta^{(0)}}{dt^2} = \frac{-g}{l}\theta^{(0)} \qquad (15.6)$$

$$\varepsilon^1: \quad \frac{d^2\theta^{(1)}}{dt^2} = \frac{-g}{l}\left\{\theta^{(1)} - \frac{1}{3!}[(\theta^{(0)})^3]\right\} \qquad (15.7)$$

$$\varepsilon^2: \quad \frac{d^2\theta^{(2)}}{dt^2} = \frac{-g}{l}\left\{\theta^{(2)} - \frac{1}{3!}[3(\theta^{(0)})^2\theta^{(1)}]\right\} \qquad (15.8)$$

and so on. We could include more equations but would soon be lost in a sea of algebra.

The solution of Eq. (15.6), which is the same linear equation that was examined in Chapter 1, can be immediately written as

$$\theta^{(0)} = A\sin\omega t + B\cos\omega t \qquad (15.9)$$

where $\omega = \sqrt{g/l}$. This is the solution that one would obtain from a linear analysis of a nonlinear equation. If we now substitute Eq. (15.9) into Eq. (15.7) to solve for the first-order correction $\theta^{(1)}$, certain difficulties will be soon encountered as shown below. To simplify the calculation, we choose the constant of integration $B = 0$ in $\theta^{(0)}$. Hence Eq. (15.7) is written as

$$\frac{d^2\theta^{(1)}}{dt^2} + \frac{g}{l}\theta^{(1)} = \frac{-1}{3!}\frac{g}{l}[A\sin\omega t]^3 = \frac{1}{3!}\frac{g}{l}A^3\left[\frac{-3}{4}\sin\omega t + \frac{1}{4}\sin 3\omega t\right]$$

$$(15.10)$$

where a trigonometric identity has been employed to expand the nonlinear term. Recall that $\omega = \sqrt{g/l}$ and note also that there is now a forcing term on the right-hand side of the equation that oscillates at the *same* frequency as the natural frequency of oscillation of the terms on the left-hand side of the equation. The forcing term that oscillates at $3\omega = 3\sqrt{g/l}$ creates no problem as far as the natural frequency of oscillation of $\theta^{(1)}$ at the frequency $\omega = \sqrt{g/l}$. The solution for $\theta^{(1)}$ is straightforward and we find

$$\theta^{(1)} = \hat{A}\sin\omega t + \hat{B}\cos\omega t + \frac{1}{3!}\frac{gA^3}{8l}\sin 3\omega t - \frac{1}{3!}\frac{gA^3}{8l}\omega t\cos\omega t \quad (15.11)$$

Note that as time increases, the last term in Eq. (15.11) increases in time, which implies that the first-order correction term $\theta^{(1)}$ eventually becomes comparable with or even larger than the lowest-order term $\theta^{(0)}$, a violation of the assumption given following Eq. (15.4). This term is called a *secular* term. A similar result would ensue for the terms $\theta^{(2)}, \theta^{(3)}$, and so on. Let us here suggest a technique to get out of this dilemma, that is, the development of a secular-free perturbation expansion. The technique is also called a *multiple time scale* perturbation expansion and it has found extensive use in the study of nonlinear oscillations.

In this technique, we define two (*multiple*) different timescales, a fast timescale t and a slow timescale $\tau = \varepsilon t$, where the same bookkeeping parameter has been employed. Since this parameter has been assumed to be small, the second timescale will be much slower than the original timescale. We are able to treat these two timescales as *independent variables*. Therefore, we must use partial differential notation and write the temporal derivative using the chain rule

$$\frac{d}{dt} = \frac{\partial}{\partial t} + \frac{\partial}{\partial \tau}\frac{\partial \tau}{\partial t} = \frac{\partial}{\partial t} + \varepsilon\frac{\partial}{\partial \tau} \quad (15.12)$$

Hence we obtain from Eq. (15.3)

$$\left(\frac{\partial^2}{\partial t^2} + 2\varepsilon\frac{\partial}{\partial t}\frac{\partial}{\partial \tau} + \varepsilon^2\frac{\partial^2}{\partial \tau^2}\right)\theta = \frac{-g}{l}\left(\theta - \varepsilon\frac{\theta^3}{3!}\right) \quad (15.13)$$

Substitute Eq. (15.4) into Eq. (15.13) and equate the terms containing the same powers of the bookkeeping parameter ε

$$\varepsilon^{(0)}: \frac{\partial^2\theta^{(0)}}{\partial t^2} = -\frac{g}{l}\theta^{(0)} \quad (15.14)$$

$$\varepsilon^{(1)}: \frac{\partial^2\theta^{(1)}}{\partial t^2} + 2\frac{\partial}{\partial t}\frac{\partial\theta^{(0)}}{\partial \tau} = -\frac{g}{l}\left\{\theta^{(1)} - \frac{1}{3!}(\theta^{(0)})^3\right\} \quad (15.15)$$

and so on.

The solution of Eq. (15.14) is as before

$$\theta^{(0)} = \frac{1}{2}(A\varepsilon^{i\omega t} + A^*\varepsilon^{-i\omega t}), \tag{15.16}$$

where $\omega = \sqrt{g/l}$ and we have used the exponential notation for convenience. We now assume that the constant of integration A may change on the slow time scale $\tau = \varepsilon t$. That is, $A = A(\tau)$. The notation "*" indicates the complex conjugate. Substitute Eq. (15.16) into Eq. (15.15) and obtain

$$\frac{\partial^2 \theta^{(1)}}{\partial t^2} + \omega^2 \theta^{(1)} = \frac{\omega^2}{8 \times 3!}[A^3 e^{i3\omega t} + 3A^2 A^* e^{i\omega t} + 3AA^{*2} e^{-i\omega t} + A^{*3} e^{-i3\omega t}]$$
$$-i\left[\omega\frac{\partial A}{\partial \tau}e^{i\omega t} - \omega\frac{\partial A^*}{\partial \tau}e^{-i\omega t}\right] \tag{15.17}$$

We shall choose the slow time dependence of the constant A such that the terms multiplying $e^{+i\omega t}$ and $e^{-i\omega t}$ are separately set equal to zero. We will therefore eliminate the secularity-causing terms in the equation by choosing the slow time dependence of the amplitude to satisfy the following equation

$$\frac{\partial A}{\partial \tau} + i\frac{\omega}{16}|A|^2 A = 0 \tag{15.18}$$

along with the complex conjugate of the amplitude satisfying the following

$$\frac{\partial A^*}{\partial \tau} - i\frac{\omega}{16}|A|^2 A^* = 0 \tag{15.19}$$

The solution of Eq. (15.18) is

$$A \Rightarrow A(\tau) = A_0 e^{-i(\omega/16)|A|^2 \tau} \tag{15.20}$$

where A_0 is the initial amplitude. A similar result is found for the complex conjugate of the amplitude. The first-order perturbation term becomes

$$\theta^{(1)} = \frac{1}{2}\left\{A_0 e^{i\omega(1-|A|^2/16)t} + A_0^* e^{-i\omega(1-|A|^2/16)t}\right\} \tag{15.21}$$

We note that the frequency of oscillation decreases and the deviation of the resonant frequency depends on the square of the amplitude of the signal. The oscillation is essentially "detuned" to lower frequencies with the detuning being proportional to the signal's power. This detuning is observed in the numerical solution shown in Figure 15–2.

One could conjecture at this stage that this "detuning process" could have profound implications if there were a coupling between two oscillatory systems, one with a natural frequency of oscillation ω_1 being much higher than the second whose frequency of oscillation is ω_2, that is, $\omega_1 \gg \omega_2$. If the first system is excited at a frequency $\omega_{excite} \simeq \omega_1 \gg \omega_2$, the system may "detune" enough such that the difference is approximately ω_2 and the high-frequency signal will "parametrically" excite the signal at ω_2. The coupling between electrons and the more massive ions in a gaseous plasma, each of the species having their own characteristic resonant frequency, is an example that exhibits such a behavior both in theory and in laboratory experiments.

15.3 Nonlinear Wave Equation

Our first encounter with nonlinearity using a perturbation expansion and particularized for the examination of oscillations will also have important effects on wave propagation. The resulting nonlinear waves that will be examined here are called *solitons*. These nonlinear waves will be introduced using the same transmission line model that was examined earlier when the subject of dispersion was discussed with one major modification. In the transmission line model, the linear shunt capacitor will be replaced with a nonlinear shunt capacitor, which is a reverse biased PN junction diode in which the capacitance depends upon the amplitude of the localized voltage perturbation that appears across it. This element is therefore a nonlinear voltage-dependent capacitor. The resulting equation can also be used to describe the propagation of nonlinear water waves in a shallow water channel where the normalized dependent variable is the ratio of the amplitude of the perturbation in the water divided by the depth of the channel.

The first problem is to derive a nonlinear wave equation for electromagnetic wave propagation on the nonlinear-dispersive transmission line, a section of which is shown in Figure 15–3. In this transmission line model, C_N is a nonlinear capacitor that is the reverse-biased *PN* junction diode, called a *varactor* or *varicap*. The value of the capacitance at a particular section will change as the nonlinear wave propagates. In addition, the capacitance between the turns of the wire will be represented with the capacitor C_S that is in parallel with the inductance. The inclusion of this capacitor introduces dispersion to the transmission line, as we have already described in Chapter 9.

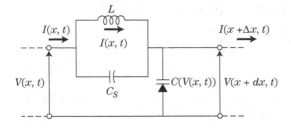

FIGURE 15–3

A typical section of a distributed nonlinear-dispersive transmission line. The nonlinear capacitor $C(V(x, t))$ is a reverse biased PN diode or varicap. The units of the elements are: $L \rightarrow$ H/m, $C(V(x, t)) \simeq [C_0 + (dC(V)/dV)V(x, t)] \simeq [C_0 - C_N V(x, t)] \rightarrow$ F/m, and $C_S \rightarrow$ F-m

Following the procedure outlined in Chapter 9, one can write the set of partial differential equations for the currents and voltages as

$$\frac{\partial I}{\partial x} + \frac{\partial Q(V)}{\partial t} = 0 \tag{15.22}$$

$$\frac{\partial V}{\partial x} + L\frac{\partial I'}{\partial t} = 0 \tag{15.23}$$

$$\frac{\partial^2 V}{\partial x \partial t} + \frac{1}{C_S}(I - I') = 0 \tag{15.24}$$

where the current through the nonlinear capacitor is given by $\partial Q(V)/\partial t$ and $Q(V)$ is the voltage-dependent charge on the nonlinear capacitor at that particular point. From Eqs. (15.22)–(15.24), we can eliminate the two currents I and I' and write

$$C_S\frac{\partial^4 V}{\partial x^2 \partial t^2} + \frac{1}{L}\frac{\partial^2 V}{\partial x^2} - \frac{\partial^2 Q(V)}{\partial t^2} = 0 \tag{15.25}$$

The voltage dependence of $Q(V)$ (Coulomb/length) must be specified before we can proceed. The simplest choice is to expand $Q(V)$ in a Taylor series and retain only the first two terms

$$Q(V) \simeq C_0 V - C_N V^2 \tag{15.26}$$

where C_0 has the units of farads/meter and the coefficient C_N has the units of farads/(volt-meter). Therefore Eq. (15.25) can be written as

$$\frac{1}{L}\frac{\partial^2 V}{\partial x^2} - C_0\frac{\partial^2 V}{\partial t^2} + C_S\frac{\partial^4 V}{\partial x^2 \partial t^2} + C_N\frac{\partial^2 V^2}{\partial t^2} = 0 \tag{15.27}$$

We can recognize some features of Eq. (15.27). The first two terms are identical to the wave equation for the linear transmission line in Chapter 9. As such, the velocity of propagation of a nondispersive linear wave was given by $c_w = 1/\sqrt{LC_0}$ if the other two terms were absent. The third term accounts for the dispersion introduced by the capacitor C_S that is in parallel with the inductor. The fourth term is the nonlinear term, which will be important in this section. If the nonlinear term can be neglected and $V \simeq V_0 e^{i(kx-\omega t)}$, one obtains the dispersion relation

$$\frac{\omega}{k} = \frac{1}{\sqrt{LC_S k^2 + LC_0}} \simeq \frac{1}{\sqrt{LC_0}}\left(1 - \frac{C_S}{2C_0}k^2\right) \tag{15.28}$$

In the long-wave length limit ($k \equiv 2\pi/\lambda \to 0$), the velocity approaches the linear velocity $c_w \to 1/\sqrt{LC_0}$. In the shortwave length limit ($k \equiv 2\pi/\lambda \to \infty$), there is an upper cutoff frequency $1/\sqrt{LC_S}$. This is the resonant frequency of the parallel resonant "tank" circuit in the series arm. For frequencies above this value, the wave will not propagate. A similar dispersion curve can be obtained in plasma physics as shown in Figure 15–4.

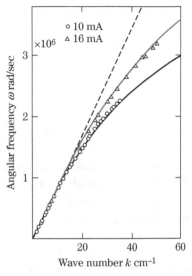

FIGURE 15–4

Experimental results of the dispersive nature of ion acoustic waves in a plasma that have a similar dispersion relation as the transmission line described in Figure 15–3. [H. Tanaca, A. Hirose, and M. Koganei, *Physical Review* Vol. 161, p. 94 (1967), used with permission]

To derive a nonlinear wave equation, we shall introduce and make use of a technique entitled the *reductive perturbation technique*. This technique is different than the multiple timescale perturbation technique outlined previously, but it is useful in understanding nonlinear wave phenomena. We define two new independent variables

$$\xi = \varepsilon^{1/2} \left[x - \frac{t}{\sqrt{LC_0}} \right],$$

$$\tau = \varepsilon^{3/2} \frac{t}{\sqrt{LC_0}}$$

(15.29)

where ε is the same bookkeeping parameter that was used previously. The first variable ξ states that we shall transform the wave propagation variables to the wave frame and then examine the higher-order temporal deviations occurring in this frame of reference using the second variable τ. The ordering of the variables $\varepsilon^{1/2}$ and $\varepsilon^{3/2}$ reflects the dispersion found in the weakly dispersive limit (small k) in Eq. (15.28). These two variables are considered to be independent variables just as we considered two different timescales in the previous section. Using the chain rule for differentiation, we write:

$$\frac{\partial}{\partial t} = \frac{\partial \xi}{\partial t} \frac{\partial}{\partial \xi} + \frac{\partial \tau}{\partial t} \frac{\partial}{\partial \tau} = \frac{-\varepsilon^{1/2}}{\sqrt{LC_0}} \frac{\partial}{\partial \xi} + \varepsilon^{3/2} \frac{\partial}{\partial \tau}$$

$$\frac{\partial}{\partial x} = \frac{\partial \xi}{\partial x} \frac{\partial}{\partial \xi} + \frac{\partial \tau}{\partial x} \frac{\partial}{\partial \tau} = \varepsilon^{1/2} \frac{\partial}{\partial \xi} + 0$$

(15.30)

The voltage is expressed in a normal perturbation series

$$V = \varepsilon u^{(1)} + \varepsilon^2 u^{(2)} + \cdots$$

(15.31)

Collecting all terms that have similar powers in the ordering parameter ε in Eq. (15.27) which has been transformed to the new variables ξ and τ, we find that the dependent variable $u^{(1)}$ at the order (ε^3) satisfies the equation

$$\sqrt{LC_0}\frac{\partial u^{(1)}}{\partial \tau} + \frac{C_N}{C_0}u^{(1)}\frac{\partial u^{(1)}}{\partial \xi} + \frac{C_S}{2C_0}\frac{\partial^3 u^{(1)}}{\partial \xi^3} = 0 \qquad (15.32)$$

One can identify three different physical phenomena that are occurring in the three terms in Eq. (15.32). The first linear term reflects the fact that the perturbation $u^{(1)}$ is propagating with a constant velocity. The second term is the product of the amplitude of the perturbation times the derivative of the perturbation amplitude and is thus a nonlinear term. The third term reflects that the media, which in this case is the transmission line, is dispersive.

This equation is the first of an extensive series of equations that are called nonlinear evolution equations and it was originally obtained by Korteweg and de Vries to describe propagating perturbations in the elevation of the water in a shallow canal (see Figure 15–5). The elevation of the perturbation was normalized by the actual depth of the water in the canal. If this normalized amplitude is small, the nonlinear terms may be neglected. This approximation cannot be made in the present case. The perturbations that were excited from the front of a barge that suddenly stopped, separated from the barge and were observed to continue to propagate in the canal for long distances. This observation was reported in 1834 by John Scott Russell who wrote:

I was observing the motion of a boat which was rapidly drawn along a narrow channel by a pair of horses, when the boat suddenly stopped—not so the mass of water in the channel which it had put in motion; it accumulated round the prow of the vessel in a state of violent agitation, then suddenly

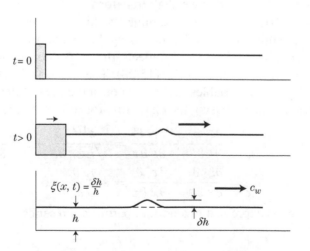

FIGURE 15–5

Perturbations in the elevation of water in a channel can be excited with the horizontal motion of the plunger.

leaving it behind, rolled forward with great velocity, assuming the form of a large solitary elevation, a rounded, smooth and well-defined heap of water, which continued its course along the channel apparently without change of form or diminution of speed. I followed it on horseback, and overtook it still rolling on at a rate of some eight or nine miles an hour [14 km/h], preserving its original figure some thirty feet [9 m] long and a foot to a foot and a half [300-450 mm] in height. Its height gradually diminished, and after a chase of one or two miles [2–3 km] I lost it in the windings of the channel. Such, in the month of August 1834, was my first chance interview with that singular and beautiful phenomenon which I have called the Wave of Translation.

He also wrote "The great primary waves of translation cross each other without change of any kind in the same manner as the small oscillations produced on the surface of a pool by a falling stone." These observations were to the British Association for the Advancement of Science.

Eq. (15.32) is now known as the Korteweg de Vries or KdV equation. Although it was derived at the end of the nineteenth century to describe the "great wave of translation" that was observed by Russell after his infamous horseback ride following the propagating wave in a canal, it is still extensively studied as it admits solutions that are of current interest, solitons. The acronym *soliton* was given to the solitary wave solution of the KdV equation that possessed certain additional mathematical properties

To find the solution of the KdV equation, it is convenient to look for a "wave of permanent profile," that is, a wave solution that does not change its shape as the wave propagates. To do this, we combine the first two terms in Eq. (15.32) by combining the two independent variables into one new independent variable $\zeta = \xi - (1/\sqrt{LC_0})\tau$. This in essence implies that we are looking for a wave of constant profile. In addition, the dependent variable $u^{(1)}$ with its superscript will be replaced with U to simplify the notation,

$$-\frac{dU}{d\zeta} + \frac{C_N}{C_0}U\frac{dU}{d\zeta} + \frac{C_S}{2C_0}\frac{d^3U}{d\zeta^3} = 0 \tag{15.33}$$

This ordinary differential equation can be integrated term by term once to yield

$$-U + \frac{C_N}{2C_0}(U)^2 + \frac{C_S}{2C_0}\frac{d^2U}{d\zeta^2} = k_1 \tag{15.34}$$

where k_1 is a constant integration. We shall look for a pulse solution that satisfies the conditions that $U \to 0, dU/d\zeta \to 0$ and $d^2U/d\zeta^2 \to 0$ as $|\zeta| \to \infty$. Hence the constant of integration $k_1 = 0$. The integral of Eq. (15.34) is found by multiplying Eq. (15.34) by the factor $(dU/d\zeta)d\zeta$ and integrating the resulting equations. We find

$$-\frac{U^2}{2} + \frac{C_N}{6C_0}U^3 + \frac{C_S}{4C_0}\left(\frac{dU}{d\zeta}\right)^2 = k_2 \tag{15.35}$$

where once again the constant integration will have the value $k_2 = 0$. Eq. (15.35) can be written as

$$d\zeta = \frac{dU}{\sqrt{\frac{2C_N}{3C_S}U^3 - \frac{2C_0}{C_S}U^2}} \tag{15.36}$$

The integral of Eq. (15.36) can be shown to be

$$U = U_0\text{sech}^2\left[\sqrt{\frac{C_N U_0}{6C_S}}\left(\zeta - \frac{C_N U_0}{3C_0}\tau\right)\right] \tag{15.37}$$

Transforming this equation back to laboratory coordinates using $\zeta = \xi - (1/\sqrt{LC_0})\tau$ and Eq. (15.29), Eq. (15.37) is finally written as

$$U(x,t) = U_0\text{sech}^2\left[\sqrt{\frac{C_N U_0}{6C_S}}\left(x - \left(1 + \frac{C_N U_0}{3C_0}\right)\frac{t}{\sqrt{LC_0}}\right)\right] \tag{15.38}$$

The velocity of propagation c of this "solitary wave" (so called because it is a single pulse-shaped object as shown in Figure 15–6) is given by

$$c = \left[1 + \frac{C_N}{3C_0}U_0\right]\frac{1}{\sqrt{LC_0}} \tag{15.39}$$

which indicates that its velocity increases with amplitude. The half-width of this pulse can be computed from $U/U_0 = 1/2 = \text{sech}^2(0.88)$. Hence

$$\Delta t|_{\text{half}} = 0.88 \times 2\sqrt{\frac{6C_S LC_0}{C_N}}\frac{1}{\sqrt{U_0}} \tag{15.40}$$

Therefore, the product of the amplitude of the pulse times the square of the half-width of the pulse is a constant. This particular property is extremely important in understanding the properties of a solitary wave and/or a soliton, which will be described later in this chapter.

There are other models that could have been chosen that would lead to different nonlinear wave equations. One of them is called the nonlinear

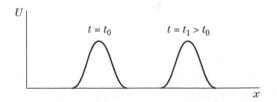

FIGURE 15–6

Propagation of a solitary wave pulse.

Schrodinger equation (NLS equation). A related mechanical transmission line consisting of a slender strand of rubber, say from a rubber band, to which pins are inserted on a regular interval would lead to another nonlinear equation that has been given the name of the sine Gordon equation. This equation would describe the angular rotation of the pins that propagates on the rubber band and is frequently encountered in the study of Josephson junctions. These two equations and other equations that are considered to be members of the nonlinear evolution equations have received considerable attention elsewhere and will not be further discussed here. We shall confine our discussion here to the KdV equation along with a short description of its extension to higher spatial dimensions.

Fermi, Pasta, and Ulam Recurrence Phenomena

The easiest way to jump onto the "soliton frame of reference" is to first understand the Fermi, Pasta, and Ulam (FPU) recurrence phenomena. In one of the earliest numerical calculations performed on a large computer (it was called the "Maniac" computer at Los Alamos), Enrico Fermi, John Pasta, and Stan Ulam examined the nonlinear equations that would model a crystal structure. Starting from the initial conditions of a sine wave excitation, they expected that due to the nonlinearity in the model, harmonics of the original signal would be generated and harmonics of these harmonics would be further generated, and that eventually the response would be completely randomized and one could assume that "thermalization" had occurred. To their surprise, they found that after several iterations in the numerical calculations, the original sine wave signal that was used to excite the model reappeared. This is now known as the FPU recurrence phenomena, but at the time, it was unexplained. The fact that a finite difference numerical scheme was used in the computer program introduced a numerical "dispersion" in that there was a minimum length that could be studied and it was dictated by the finite difference length in the numerical scheme. Hence we should be aware that the problem that was numerically analyzed is *both* nonlinear and dispersive.

It is possible to explain the FPU recurrence phenomena using a simple model that incorporates both dispersion and nonlinearity as shown below. We will focus our attention only on the KdV equation, which we have shown will describe the nonlinear dispersive transmission line. We write the approximate dispersion relation in dimensionless units as

$$\omega \simeq k(1 - bk^2) \text{ or } k \simeq \omega(1 + a\omega^2) \tag{15.41}$$

which is shown in Figure 15–7. This approximation assumes that the transmission line is weakly dispersive.

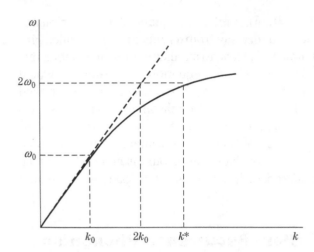

FIGURE 15–7

Linear dispersion curves in the long wavelength limit of Eq. (15.41). The dashed line is the curve $\omega = k$.

Let us assume a wave of the form

$$\phi = \phi_0 \sin(\omega_0 t - k_0 x) \tag{15.42}$$

is excited at $x = 0$ and propagates on the nonlinear dispersive transmission line. Because of the nonlinearity, harmonics of this signal will be generated at the frequencies and the corresponding propagation constants $2\omega_0, 2k_0$; $3\omega_0$, $3k_0$, ... From Figure 15–7, it is noted that a wave with the frequency $2\omega_0$ and propagation constant $2k_0$ does *not* satisfy the dispersion relation (15.41) and will therefore not be able to propagate on the transmission line. However, a signal at the frequency $2\omega_0$ with a propagation constant k^* satisfies the dispersion relation and can therefore propagate. This implies that two signals will exist at the same point in space, one that is the harmonic of the incident signal which cannot propagate and another that can propagate on the transmission line since it satisfies the dispersion relation. These signals are

$$\phi_a = \phi_{a0} \sin(2\omega_0 t - 2k_0 x)$$
$$\phi_b = \phi_{b0} \sin(2\omega_0 t - k^* x) \tag{15.43}$$

Assuming that the amplitudes of the two signals are the same ($\phi_{a0} = \phi_{b0}$), the sum of the two is

$$\phi_a + \phi_b = \phi_{a0}(\sin(2\omega_0 t - 2k_0 x) + \sin(2\omega_0 t - k^* x))$$

$$= 2\phi_{a0} \sin\left(\frac{4\omega_0 t - (2k_0 + k^*)x}{2}\right) \times \cos\left(\frac{(2k_0 - k^*)x}{2}\right) \tag{15.44}$$

where a trigonometric identity has been employed. This is an amplitude-modulated wave that is a minimum at $x = L$, where

$$\frac{(2k_0 - k^*)L}{2} = -n\pi \tag{15.45}$$

Using the approximate value for the propagation constant, we write

$$\frac{\left(2\omega_0 - 2\omega_0 \left(1 + a(2\omega_0)^2\right)\right) L}{2} = -n\pi \tag{15.46}$$

where n is an integer. This can be written as

$$L \simeq \frac{n\pi}{4a\omega_0^3} \tag{15.47}$$

A transmission line consisting of 50 identical sections has been constructed. A typical section is shown in Figure 15–8. The nonlinear capacitor was a varactor diode and the measured dispersion curves for two different values of a dc bias voltage were obtained. Results obtained from the laboratory experiment using the nonlinear dispersive transmission line, shown in Figure 15–9, verify the prediction of the recurrence length on the frequency as obtained in Eq. (15.47).

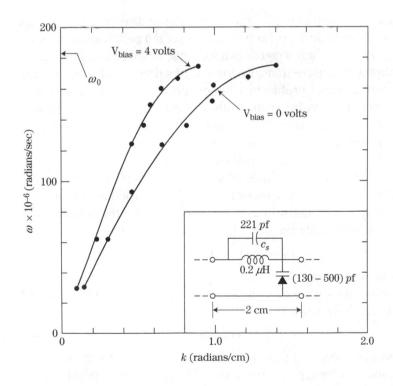

FIGURE 15–8

Measured dispersion curves for wave propagation on a nonlinear transmission line for two different values of the DC bias voltage.

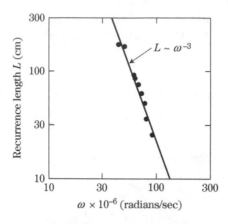

FIGURE 15–9

Measured recurrence length using the transmission line shown in Figure 15–8.

Additional modes could be included using this analysis, but the algebra would soon become horrendous and will not be attempted here. This is an initial example illustrating the delicate balance between nonlinearity and dispersion that is required to understand solitons. In the following section, a brief summary of some of the unique properties of a soliton are presented.

KdV Soliton Properties

Although we have obtained a nonlinear wave equation and might expect that its solitary wave solution could distort as it propagates or if it collides with another solitary wave, there is a solution that does not change its shape as it propagates along the nonlinear dispersive transmission line or suffers a collision. Rather than use the correct sech^2 profile that we obtained in Eq. (15.38), it will be simpler to model the pulse shape with a rectangle. This simplification will provide intuition concerning the soliton without resorting to the heavy mathematics that results from the more exact analysis. Rest assured that the following results do survive close mathematical scrutiny which is beyond the scope of this book. The amplitude of a rectangular pulse will be A_j and the width will be W_j where the subscript j specifies a particular soliton. The velocity of propagation of the solitary wave depends upon the amplitude of the solitary wave. We rewrite Eq. (15.39) as

$$c_j = c_0(1 + A_j) \tag{15.48}$$

The second important property is that the product of the amplitude of the solitary wave times the square of its half-width is a constant as was obtained in Eq. (15.40), which we write as

$$A_j W_j^2 = \text{constant} \tag{15.49}$$

The first property obtained from Eq. (15.48) is that the velocity of propagation of the solitary wave is proportional to the amplitude of the perturbation.

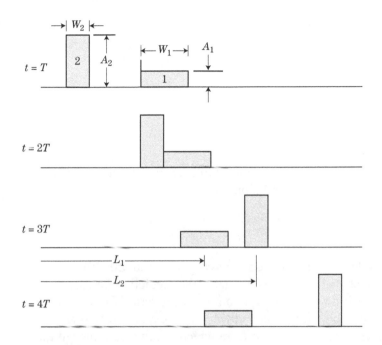

FIGURE 15–10

The overtaking collision of two solitons.

Therefore one would conclude that if two different solitary waves were launched on a transmission line that the perturbation with a larger amplitude would propagate with a greater velocity. This is shown qualitatively in Figure 15–10. If the larger amplitude perturbation were excited after the perturbation with a smaller amplitude, the larger one would catch up and collide with the smaller one. In fact, it would eventually pass through the smaller amplitude one. The fact that the two solitary waves survive this collision was initially used as a defining characteristic of these particular solitary waves and resulted in the coining of the word "*soliton*" to describe solitary waves that possessed this peculiar property. Later mathematical analysis has generalized these results. An animation of the overtaking collision of two solitons can be found at: http://physics.usask.ca/~hirose/ep225/animation/soliton/anim-soliton.htm

The trajectories of the two solitons are shown in Figure 15–11. There will be a shift in the phase following the collision since the collision is non-destructive. The details of what is happening during the actual collision are not considered here. We only examine the regions before and after the collision where the solitons are propagating as if they are completely unaware of each other. With reference to Figures 15–10 and 15–11, we note that the two solitons cannot overlap and must be separated by a finite distance D.

$$D = D_1 + D_2 \geq \frac{W_1}{2} + \frac{W_2}{2} \tag{15.50}$$

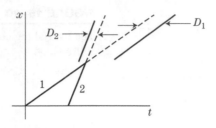

FIGURE 15–11

The trajectories of the two solitons depicted in Figure 15–10.

Incorporating the soliton property defined in Eq. (15.49), we find that the ratio of the two distances satisfies the relation that

$$\frac{D_1}{D_2} = \frac{W_1/2}{W_2/2} = \sqrt{\frac{A_2}{A_1}} \tag{15.51}$$

Although this derivation, based on the assumption that the soliton can be described with a rectangular pulse is heuristic, the prediction that the separation distance is proportional to the square root of the ratio of the amplitudes is in agreement with both numerical and laboratory experiments as shown in Figure 15–12. The spatial duration ΔL, in which the two solitons are interacting, can also be specified in terms of the KdV soliton property relating the

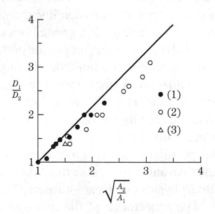

FIGURE 15–12

The laboratory and numerical experimental confirmation of Eq. (15.51). Experimental confirmation can be found at:

1. W. Aossey, et al., *Physical Review A*, Volume 45, p. 2606, (1992).
2. N. J. Zabusky and M. D. Kruskal, *Physical Review Letters*, Volume 15, p. 240, (1965).
3. G. L. Lamb, *Elements of Soliton Theory*, Wiley-Interscience, New York, p. 118 (1980).

soliton width and its amplitude,

$$\Delta L = L_2 - L_1 = (c_2 - c_1)\Delta T \geq W_2 + W_1 \qquad (15.52)$$

where ΔT is the temporal duration of the interaction. This spatial duration can be written in terms of the amplitudes of the two solitons and Eq. (15.49),

$$\Delta L \geq \text{constant} \left(\frac{1}{\sqrt{A_2}} + \frac{1}{\sqrt{A_1}} \right) \qquad (15.53)$$

The actual number of solitons that are launched on the transmission line with an arbitrary excitation potential can be calculated very rigorously using Inverse Scattering Theory. This formal procedure finds that the number of excited solitons N is related to the number of bound states that one would calculate from the Schrödinger equation if the potential well that was used was also the excitation signal for the KdV equation.

Rather than use this formal procedure, the intuitive approach described above will be used to predict the number of solitons. We assume a rectangular potential well that has the dimensions $L \times \sqrt{B}$ as shown in Figure 15–13. The number of solitons that will be excited is equal to the number of small square rectangles that can be inserted into this large rectangular box from which we find that

$$\text{Number of solitons} = N + 1 \qquad (15.54)$$

The first attempt to generalize the one-dimensional soliton propagation was to incorporate higher dimensions into the equation. The KdV equation was generalized to the KP equation, so named in honor of the two Russian scientists, Kadomtsev and Petviashvili, who were the first to derive it. They examined the effects found by including small loss terms that would occur in directions perpendicular to the dominant direction of propagation of the

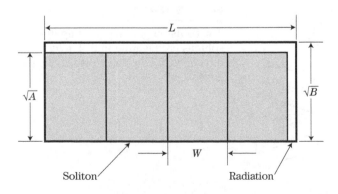

FIGURE 15–13

The number of solitons that are excited is determined from the number of small rectangles that can fit within the larger rectangle. Excess space indicates radiation of linear waves.

soliton. The equation that they obtained can be written as

$$\frac{\partial \left[\sqrt{LC_0}\dfrac{\partial u^{(1)}}{\partial \tau} + \dfrac{C_N}{C_0}u^{(1)}\dfrac{\partial u^{(1)}}{\partial \xi} + \dfrac{C_S}{2C_0}\dfrac{\partial^3 u^{(1)}}{\partial \xi^3} \right]}{\partial \xi} + \frac{1}{2}\frac{\partial^2 u^{(1)}}{\partial \psi^2} = 0 \quad (15.55)$$

where ψ is perpendicular to the dominant direction of propagation. In writing this equation in this format, it is clear that the term within the square brackets is the KdV equation.

An interesting observation that has been made concerning this equation is that it admits a *resonance* at a particular angle. This resonance can be obtained using the equations

$$\omega_1 + \omega_2 = \omega_3$$
$$\mathbf{k}_1 + \mathbf{k}_2 = \mathbf{k}_3 \qquad\qquad (15.56)$$

where now the propagation constants are actually vectors. These two equations represent conservation of energy and conservation of momentum and are used frequently in quantum mechanics. Planck's constant $\hbar = h/2\pi$ has been removed from these two equations. Assume that two equi-amplitude solitons are simultaneously excited on a two-dimensional transmission line and propagate in different directions. These two-dimensional solitons collide at an arbitrary angle as shown in Figure 15–14. The two components of the wave vector that are perpendicular to the dominant direction of propagation cancel and only the components in the direction of propagation need be considered.

At resonance, the amplitude of the newly created soliton A_{new} can be calculated in terms of the amplitude of the colliding solitons A_{old}:

$$A_{new}W_{new}^2 = \text{constant} = A_{old}W_{old}^2 \qquad\qquad (15.57)$$

FIGURE 15–14

The collision of two equi-amplitude solitons propagating in different directions. At the exact resonance, the distance $BC \to \infty$.

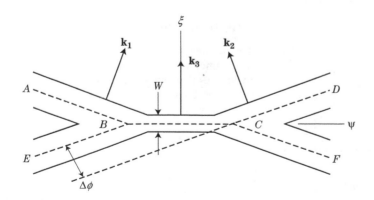

and the conservation of energy that states that

$$A_{new}^2 W_{new} = 2A_{old}^2 W_{old} \qquad (15.58)$$

From these two equations, we find that the amplitude of the new soliton satisfies the relation that

$$A_{new} = 4A_{old} \qquad (15.59)$$

We note that there will be a phase shift of $\Delta\phi$ in that the equiphase front appears to be shifted after the collision of the two solitons. The reader may have encountered this resonance phenomena in examining the amplitude of two waves approaching a beach from slightly different directions where the amplitude of the newly created wave has some localized increase. Numerical simulations and observations of the tsunami that was excited by local earthquakes in the Pacific Ocean has been considered to be a KdV soliton, although this is a controversial topic and is open to other interpretations. The oblique collision of two sections of a tsunami caused by the irregular shape of the excitation also appears to experience this resonance enhancement.

Finally, it is possible to describe the spatial evolution of a soliton that is propagating in a radial direction after it was excited from a small spherical source at the origin of a coordinate system. It is reasonable to assume that the energy that is supplied from the source shown in Figure 15–15 must appear in the spherically radiating soliton that was excited by the source.

We write this energy as

$$A_{spherical}^2 4\pi r^2 \Delta r = \text{ constant} \qquad (15.60)$$

The spherical KdV soliton also satisfies the soliton requirement that

$$A_{spherical} \Delta r^2 = \text{ constant} \qquad (15.61)$$

FIGURE 15–15

A soliton propagating in spherical coordinates.

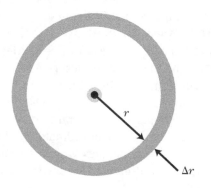

Eliminating the width Δr between these two equations, we find that the amplitude of a spherically radiating soliton will decrease in space as

$$A_{spherical} \simeq \frac{\text{constant}}{r^{4/3}} \qquad (15.62)$$

The amplitude of a spherically radiating linear wave decreases as $1/r$. Therefore, it is possible to separate a linear spherical wave from a nonlinear spherical soliton based on the radial amplitude variation.

It should be noted that all of the results presented in this section using our intuitive model have been verified theoretically using the inverse scattering method and experimentally with either a nonlinear dispersive transmission line, a water tank, or a plasma device. This is a case where the simple model using a "back of the envelope calculation" can bring forth many nuggets of truth.

Shocks

Closely allied to the study of solitons is the study of shocks. The reader has probably read articles concerning the potential noise problems of supersonic aircraft flying over inhabited land and may have wondered if there was a mathematical model to describe the phenomena. The early work of Mach demonstrated that if a projectile passed through a medium with a velocity greater than the velocity of sound, a cone of sound would be radiated at an angle from the projectile. This angle is related to the velocity of sound and the projectile and is

$$\frac{\text{sound velocity}}{\text{projectile velocity}} = \sin\theta \qquad (15.63)$$

Shocks can be easily created in nature. The projectile may be a fast pusher, such as an eruption of a volcano or an explosion of a nuclear bomb in which energy is rapidly released at one point in space and time and rapidly expands—in fact, it expands faster than the velocity of sound. The resulting transition between the undisturbed region ahead of the transition and the disturbed region trailing the transition is called a *shock* if a set of equations can describe the physics ahead of the shock and the same set describes the region behind the shock. They must be connected in some sense through the region of the shock.

Using the transmission line that was previously described to study soliton propagation, we illustrate the propagation of a ramp function using an oscilloscope located at increasing values of the coordinate x in Figure 15–16. The initial voltage ramp steepens as it propagates which is indicative of a shock formation. In addition, there are trailing oscillations in the signals behind the

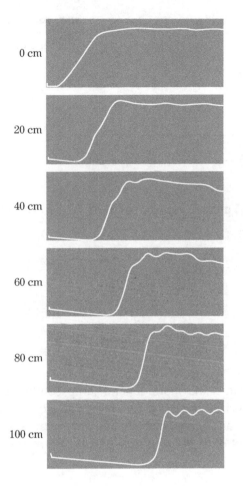

0 cm

20 cm

40 cm

60 cm

80 cm

100 cm

FIGURE 15–16

The observed experimental formation of a shock at increasing distances from the voltage source along a transmission line. The oscilloscope pictures display the detected amplitude on the vertical scale and time on the horizontal scale for each picture.

voltage transition. The shock may break up into a number of solitons in the experiment since there was no added dissipation mechanism in this transmission line. A true shock requires that there be a dissipation of the energy. For example, if the gas behind a shock front is heated, this thermodynamic energy cannot be recovered and as far as the shock wave is concerned this energy is lost.

It is probably best to suggest one important transition feature of shock using an equation for sound waves. A simple first-order equation

$$\frac{\partial \rho_v}{\partial t} + \frac{\partial (\rho_v v)}{\partial x} = 0 \tag{15.64}$$

can be used to illustrate the derivation of such jump conditions.

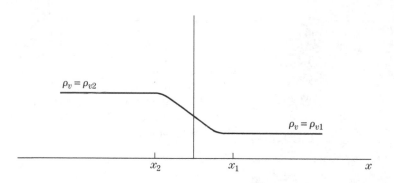

FIGURE 15–17

Structure of a shock.

In Figure 15–17 a shock transition is illustrated. Let us integrate Eq. (15.64) across this change

$$\int_{x_1}^{x_2} \frac{\partial \rho_v}{\partial t} dx + \int_{x_1}^{x_2} \frac{\partial (\rho_v v)}{\partial x} dx = 0$$

$$\frac{\partial \int_{x_1}^{x_2} \rho_v dx}{\partial t} + \rho_v v|_{x_1}^{x_2} = 0$$

(15.65)

If the transition does not change in time, we are left with the jump condition,

$$\rho_v v|_{x_2} - \rho_v v|_{x_1} = 0 \qquad (15.66)$$

More complicated and more general jump conditions can be obtained for sets of equations. In particular for gases, one would use the equations of continuity, motion, and energy and integrate them across the shock front. This fitting of the discontinuity is called satisfying the Rankine-Hugoniot relations.

Chaos

Chaos, as the name implies, may be the final state all linear and nonlinear waves reach as the wave amplitude increases and waves with different frequencies and propagation vectors interact, leading eventually to a turbulent or a chaotic state in nature. Analyzing a certain algebraic equation called the *logistic equation* provides a good introduction to the subject of chaos.

The logistic equation, which is a simple algebraic equation, relates a quantity $x(k+1)$ at one point in time or space to its value at a previous or adjacent point in time or space. We will continually refer to $x(k)$ as a quantity since it can refer to many different entities that may occur in nature. The value of $x(k)$ where k is an integer will be normalized so that $x(k) < 1$ for any value of k. The logistic equation is

$$x(k+1) = rx(k)[1 - x(k)] \qquad (15.67)$$

Only integer values of k are used in the equations so the logistic equation corresponds to a difference equation with a discrete separation in the iterations rather than a differential equation with a smooth transition. There may be a reduction or an increase in the quantity $x(k+1)$ from its previous value $x(k)$ that is governed by the parameter r. Different values of this parameter in the range $0 \leq r \leq 4$ will have very important consequences. The results of a numerical solution of Eq. (15.67) for various values of this parameter are shown in Figure 15–18. In all of the figures, the initial condition was arbitrarily chosen to have the value $x(1) = 0.8$. This is not a unique choice and other values would have led to the same conclusions cited here.

For small values of the parameter $r < 1$, the quantities $x(k)$ will approach zero as one might expect from an examination of Eq. (15.67). Interesting results will be found for larger values of this parameter. For example, we note

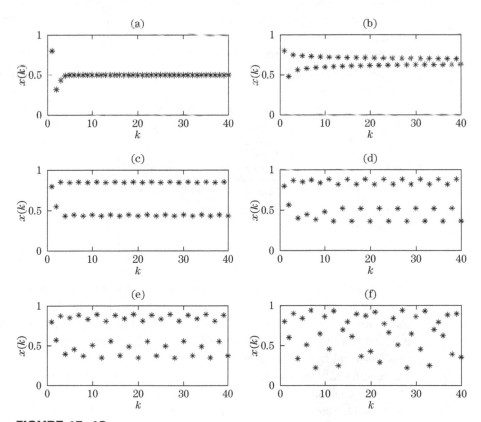

FIGURE 15–18

Numerical solution of the logistic equation for various values of the parameter r.
(a) $r = 2$. (b) $r = 3$. (c) $r = 3.45$. (d) $r = 3.5$. (e) $r = 3.564$. (f) $r = 3.75$.

in Figure 15–18(a) with $r = 2$, that the quantity $x(k)$ appears to approach a constant nonzero value. This value can be computed from Eq. (15.67) by setting $x(k + 1) \rightarrow x(k)$ and solving for the final asymptotic state

$$x(k \rightarrow \infty) = \frac{r - 1}{r} = \frac{1}{2} \tag{15.68}$$

The choice of $r = 3$ as shown in Figure 15–18(b) appears to lead to a single oscillation with an oscillation period of $\Delta k = 2$. Increasing the value of the parameter r to a value of $r = 3.45$ leads to what appears to be two unrelated solutions that are shown in Figure 15–18(c). Both solutions have a distinct oscillation with an oscillation period of $\Delta k = 4$. This separation into two solutions is called a *bifurcation*. A further increase in the value of the parameter r to a value of $r = 3.5$ leads to another bifurcation into four distinct oscillations with an increased value of the oscillations period to $\Delta k = 8$ as shown in Figure 15–18(d). With the value of $r = 3.564$, there is a further bifurcation into eight distinct oscillationswith a period of $\Delta k = 16$ as shown in Figure 15–18(e). The increase of the period by a factor of two with each successive increase in the value of the parameter r is called *period doubling*.

The ratio of the numerical values of the parameter r when the bifurcation occurs satisfies

$$\frac{r_2 - r_1}{r_3 - r_2} = \frac{3.45 - 3}{3.54 - 3.45} = 5$$

and

$$\frac{r_3 - r_2}{r_4 - r_3} = \frac{3.54 - 3.45}{3.564 - 3.54} = 3.75$$

It was noted by Feigenbaum that as the integer $n \rightarrow \infty$, this ratio will approach a constant value

$$\delta = \lim_{n \rightarrow \infty} \frac{r_{n+1} - r_n}{r_{n+2} - r_{n+1}} = 4.669201 \ldots \tag{15.69}$$

The critical values of the parameter r, which cause the separate bifurcations of the response and the period doubling to occur, lead to a numerical value that is known as the Feigenbaum number, which has the same accuracy as the numerical value of π. The final choice of $r = 3.75$ leads to a chaotic solution as shown in Figure 15–18(f).

A summary of the computed values of the quantity x as a function of increasing values of the parameter r is shown in Figure 15–19. This diagram clearly illustrates the bifurcation response and the eventual transition into the chaotic solution.

Similar chaotic phenomena can be observed in certain sets of coupled nonlinear ordinary differential equations that have been used to model various

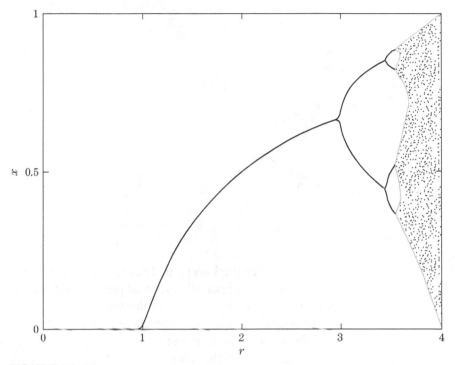

FIGURE 15–19

A summary of the computed values of the quantity x is a function of the parameter r.

physical phenomena found in nature. For example, the first set of coupled nonlinear ordinary differential equations was used to model the weather and it was found that they admitted chaotic solutions. These equations are called the Lorenz equations. With a suitable normalization, they are written as

$$\frac{dX}{dt} = \sigma(Y - X)$$

$$\frac{dY}{dt} = rX - Y - XZ \qquad (15.70)$$

$$\frac{dZ}{dt} = XY - bZ$$

where the three constants have the numerical values of $\sigma = 10, r = 28$, and $b = 8/3$. It is clear that the set of Lorenz equations are nonlinear and numerical solutions will be required.

A three-dimensional perspective graph that illustrates the temporal evolution of the three variables is shown in Figure 15–20. In the ever-expanding chaos literature, the three-dimensional "butterfly" picture is frequently

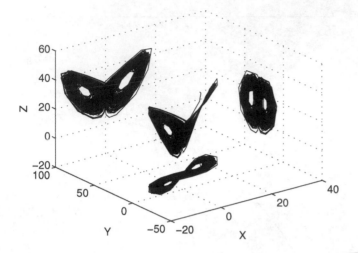

FIGURE 15–20

Solution of the Lorenz set of nonlinear differential equations. In this figure, the solution which looks like a butterfly is plotted as a three-dimensional picture. In addition, the variables of X versus Y, X versus Z, and Y versus Z are shown as two-dimensional plots on the 3 "walls."

encountered for the Lorenz and related sets of equations. The projection of the butterfly on the two-dimensional planes illustrates the temporal evolution of two of the variables that are solutions of the Lorenz equations.

In addition to obtaining the chaotic solution for a particular equation or set of equations, there is interest in synchronizing one system that will be called the *slave*, with another system that will be called the *master*, using the terminology of modern control theory. There are several techniques that can be used to achieve this synchronization and one of them is developed in the following discussion. The idea is to develop a controller with active control theory using the linear equations of a harmonic oscillator that was introduced in Chapter 1. The second-order ordinary differential equation will be subdivided into two first-order ordinary differential equations.

The master system is defined with two first-order ordinary differential equations

$$\frac{dx_{master}}{dt} = v_{master}$$
$$\frac{dv_{master}}{dt} = -m_{master}x_{master} \tag{15.71}$$

Similarly, the slave system is defined with two first-order ordinary differential equations

$$\frac{dx_{slave}}{dt} = v_{slave} + \alpha(t)$$
$$\frac{dv_{slave}}{dt} = -m_{slave}x_{slave} + \beta(t) \tag{15.72}$$

where $\alpha(t)$ and $\beta(t)$ are control parameters that are as yet unknown and are to be determined using the following procedure. Subtract Eq. (15.71) from

Eq. (15.72) and obtain the difference between the two systems,

$$\frac{d(x_{slave} - x_{master})}{dt} = (v_{slave} - v_{master}) + \alpha(t)$$

$$\frac{d(v_{slave} - v_{master})}{dt} = -m_{slave}x_{slave} + m_{master}x_{master} + \beta(t) \quad (15.73)$$

If we choose the control parameters to satisfy the following relations

$$\alpha(t) = 0$$

$$\beta(t) = m_{slave}x_{slave} - m_{master}x_{master} - a_1(x_{slave} - x_{master})$$

$$+ a_2(v_{slave} - v_{master}) \quad (15.74)$$

Eq. (15.73) with this active controller can be written as

$$\begin{pmatrix} \dfrac{d(x_{slave} - x_{master})}{dt} \\ \dfrac{d(v_{slave} - v_{master})}{dt} \end{pmatrix} = \begin{pmatrix} 0 & 1 \\ -a_1 & a_2 \end{pmatrix} \begin{pmatrix} x_{slave} - x_{master} \\ v_{slave} - v_{master} \end{pmatrix} \quad (15.75)$$

The differences between the slave and the master systems approach zero if the roots of the following determinant have negative real parts

$$\begin{vmatrix} \zeta & 1 \\ -a_1 & \zeta + a_2 \end{vmatrix} = \zeta^2 + a_2\zeta + a_1 = 0 \quad (15.76)$$

which is guaranteed if the constants $a_1 > 0$ and $a_2 > 0$.

We note that the slave system that initially had a larger amplitude and higher frequency than the master system quickly follows the master system after the application of the active controller, as shown in Figure 15–21. It

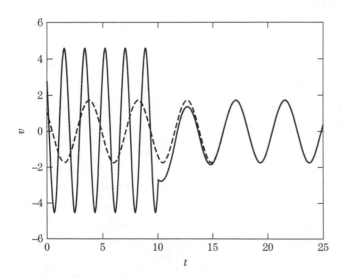

FIGURE 15–21

The synchronization of two linear signals. The controller is applied at the time $t = 10$.

should be mentioned that other controllers could have been developed. Rather than further discuss the topic of synchronization, we will just point out that this is an active area of research using techniques that are borrowed from modern control theory.

 ## Conclusion

Including nonlinear effects in the analysis of both oscillations and wave propagation has introduced several new phenomena that will require additional mathematical skills in order to obtain theoretical results. As far as wave propagation, the soliton and the shock wave are nonlinear waves that are currently receiving considerable attention in the scientific and engineering community. The intuitive approach that we have taken here has been introduced to allow the reader to appreciate the complexity of the soliton. The subject of chaos theory has also been introduced.

 ## Problems

1. Perform an experiment on a swing to verify the frequency dependence upon the values of the initial angle of deflection of the swing.

2. Describe an electrical circuit that would have the nonlinear frequency response described in the mechanical pendulum. Discuss an application for such a circuit. Find the resonant frequency of this particular circuit.

3. Assume that

$$\theta(t) = A \cos \omega t + B \cos(3\omega t)$$

is an approximate solution for Eq. (15.3). If this solution is substituted into the differential equation, show that the frequency of oscillation will depend upon the amplitude A as

$$\omega = \sqrt{\frac{g}{l} \left(1 - \frac{A^2}{16}\right)}$$

4. There is another important equation called the nonlinear Schrödinger equation, or the NLS equation:

$$i\frac{\partial \psi}{\partial t} + \frac{1}{2}\frac{\partial^2 \psi}{\partial x^2} \pm |\psi|^2 \psi = 0$$

It has the property that the amplitude A of the soliton times its half-width W is a constant. Describe the collision of two of these solitons using the intuitive model presented here and Figure 15–11.

5. Contrast the geometrical behavior of a soliton and a linear wave in spherical coordinates.

6. For the NLS soliton with the property described in problem 7, predict the spatial behavior for a spherical soliton propagation.

7. There is another soliton equation called the sine Gordon equation:

$$\frac{\partial^2 \phi}{\partial x^2} - \frac{\partial^2 \phi}{\partial t^2} - \sin \phi = 0$$

The deflection angle is ϕ and this angle rotates about the center. This perturbation will propagate in the x direction. Suggest and perform an experiment using a rubber band and a number of small nails or pins that can be used to study this equation.

8. Determine the critical angle θ_r that will lead to the soliton resonance starting from Eq. (15.56):

$$\theta_r = 2 \tan^{-1} \sqrt{2A_j}$$

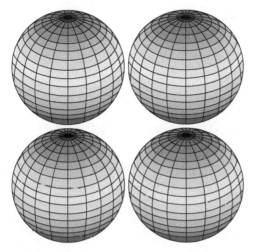

FIGURE 15–22

Problem 9.

9. There has been a laboratory experiment in which a collision of four equi-amplitude KdV solitons that are launched from four spherical surfaces produces a new soliton on the line perpendicular at the center of the four spherical surfaces to the launching antenna. Find the amplitude of the new soliton.

10. Design and demonstrate a controller to synchronize the nonlinear equation that models the oscillation of a swing with the linearized equation.

11. Design and demonstrate a controller to synchronize two different logistic equations that have different values of the parameter r.

12. Synchronize a master system described with the Lorenz equations with the parameters $\sigma = 10$, $r = 28$, and $b = 8/3$ with another slave system described with the identical Lorenz equations with the same parameters. The only difference between the two systems is that the initial conditions are not the same.

Constants and Units

Fundamental Physical Constants

Speed of light	c	defined by 299,792,458 m/sec (exact)
Permittivity	ε_0	$8.85 \times 10^{-12} \simeq \dfrac{1}{36\pi} \times 10^{-9}$ F/m
Permeability	μ_0	$4\pi \times 10^{-7}$ H/m
Elementary charge	e	1.6×10^{-19} C
Electron mass	m_e	9.11×10^{-31} kg
Proton mass	m_p	1.67×10^{-27} kg
Boltzmann's constant	k_B	1.38×10^{-23} J/K
Gas constant	R	8.31 J/mol/K
Absolute zero	0 K	$-273.15°$ C
Planck's constant	h	6.63×10^{-34} J · sec
Avogadro's number	N_0	6.02×10^{23}/mol
Gravitational constant	G	6.67×10^{-11} N · m^2/kg^2
Gravitational acceleration	g	9.81 m/sec^2

Definition of Standards

Meter	the distance travelled by light in vacuum in 1/299792458 seconds
Kilogram	the mass of the international kilogram in Paris, France
Second	9,192,631,770 vibrations of the hyperfine transition $4,3 - 3,0$ of the fundamental state $2S_{1/2}$ in the ^{133}Cs atom
Coulomb	1.0 A sec
Ampere	if two equal straight, infinitely long currents separated by 1 m exert a force of 2×10^{-7} N/m on each other, the current is defined to be 1 A.

Derived Units of Physical Quantities and Notation

Symbol	Quantity	Standard Units
$\mathbf{a}$	acceleration	m/sec^2
E	energy	J
$\mathbf{E}$	electric field	V/m
$\mathbf{F}$	force	N
G	shear elastic modulus	N/m^2
i	unit imaginary number $\sqrt{-1}$	
k	wave number	radians/m
k_B	Boltzmann's constant	J/K
k_s	spring constant	N/m
k_τ	torsional constant	N m
K	elastic modulus of spring	N
M	mass	kg
M_B	bulk modulus of liquids	N/m^2
m	magnification (optics)	
n	number density	1/m^3
N_A	Avogadro's number	number of molecules/mol
P	gas pressure	N/m^2
R	gas constant	J/K/mol
T	temporal period	sec
T	temperature	K
T_s	surface tension	N/m
$\mathbf{v}$	velocity	m/sec
U	potential energy	J
V	voltage	V
Y	Young's modulus	N/m^2
x	coordinate, displacement	m
γ	Ratio of specific heats	
ε_0	permittivity of free space	F/m
ε_r	Relative dielectric constant	
η	Resistivity	Ohm m
μ_0	permeability of free space	H/m
μ_r	Relative permeability constant	

(Continued)

Symbol	Quantity	Standard Units
γ	ratio of specific heats of gas $\gamma = \dfrac{(f+2)}{f}$ f = number of degrees of freedom in molecular motion	
γ_{rel}	relativity factor	
λ	wavelength	m
ν	frequency of oscillation	Hz (= cycles per second)
ν_c	collision frequency	1/sec
ρ_L	linear charge density	C/m
ρ_S	surface charge density	C/m^2
ρ_V	volume charge density	C/m^3
ρ_l	linear mass density	kg/m
ρ_s	surface mass density	kg/m^2
ρ_v	volume mass density	kg/m^3
ω	angular frequency	radians/sec

Trigonometric Identities, Calculus, and Laplace Transforms

B.1 ## Trigonometric Identities

$$\sin(-\theta) = -\sin(\theta) \text{ (odd function)}$$

$$\cos(-\theta) = \cos(\theta) \text{ (even function)}$$

$$\tan(-\theta) = -\tan(\theta) \text{ (odd function)}$$

$$\sin(\alpha \pm \beta) = \sin\alpha\cos\beta \pm \cos\alpha\sin\beta$$

$$\cos(\alpha \pm \beta) = \cos\alpha\cos\beta \mp \sin\alpha\sin\beta$$

$$\tan(\alpha \pm \beta) = \frac{\tan\alpha \pm \tan\beta}{1 \mp \tan\alpha\tan\beta}$$

$$\cot(\alpha \pm \beta) = \frac{\cot\alpha\cot\beta \mp 1}{\cot\beta \pm \cot\alpha}$$

$$\sin\alpha + \sin\beta = 2\sin\left(\frac{\alpha+\beta}{2}\right)\cos\left(\frac{\alpha-\beta}{2}\right)$$

$$\sin\alpha - \sin\beta = 2\cos\left(\frac{\alpha+\beta}{2}\right)\sin\left(\frac{\alpha-\beta}{2}\right)$$

$$\cos\alpha + \cos\beta = 2\cos\left(\frac{\alpha+\beta}{2}\right)\cos\left(\frac{\alpha-\beta}{2}\right)$$

$$\cos\alpha - \cos\beta = -2\sin\left(\frac{\alpha+\beta}{2}\right)\sin\left(\frac{\alpha-\beta}{2}\right)$$

$$\sin\alpha\sin\beta = \frac{1}{2}[\cos(\alpha-\beta) - \cos(\alpha+\beta)]$$

$$\sin\alpha\cos\beta = \frac{1}{2}[\sin(\alpha+\beta) + \sin(\alpha-\beta)]$$

$$\cos\alpha\sin\beta = \frac{1}{2}\left[\sin\left(\alpha+\beta\right)-\sin\left(\alpha-\beta\right)\right]$$

$$\cos\alpha\cos\beta = \frac{1}{2}\left[\cos\left(\alpha+\beta\right)+\cos\left(\alpha-\beta\right)\right]$$

$$\sin 2\alpha = 2\sin\alpha\cos\alpha$$

$$\cos 2\alpha = \cos^2\alpha - \sin^2\alpha = 2\cos^2\alpha - 1 = 1 - 2\sin^2\alpha$$

$$\sin 3\alpha = -4\sin^3\alpha + 3\sin\alpha$$

$$\cos 3\alpha = 4\cos^3\alpha - 3\cos\alpha$$

$$\sin^2\alpha = \frac{1-\cos 2\alpha}{2}$$

$$\lim_{x\to 0}\frac{\sin x}{x} = 1$$

$$\cos^2\alpha = \frac{1+\cos 2\alpha}{2}$$

$$\lim_{x\to 0}\frac{\tan x}{x} = 1$$

$$e^{\pm i\alpha} = \cos\alpha \pm i\sin\alpha$$

B.2   Calculus

An arbitrary constant of integration is not written.

$$\frac{d}{dx}\sin x = \cos x \qquad \int \sin x\,dx = -\cos x$$

$$\frac{d}{dx}\cos x = -\sin x \qquad \int \cos x\,dx = \sin x$$

$$\frac{d}{dx}\tan x = \sec^2 x \qquad \int \tan x\,dx = \ln|\sec x|$$

$$\frac{d}{dx}\cot x = -\operatorname{cosec}^2 x \qquad \int \cot x\,dx = \ln|\sin x|$$

$$\frac{d}{dx}x^n = nx^{n-1} \qquad \int x^n\,dx = \frac{x^{n+1}}{n+1}\ (n\neq -1)$$

$$\frac{d}{dx}\ln x = \frac{1}{x} \qquad \int \frac{1}{x}\,dx = \ln|x|$$

$$\frac{d}{dx}(fg) = \frac{df}{dx}g + f\frac{dg}{dx}$$

$$\int f\frac{dg}{dx}dx = fg - \int \frac{df}{dx}g\,dx \text{ (integration by parts)}$$

$$\frac{d}{dx}e^{ax} = ae^{ax} \qquad\qquad \int e^{ax}dx = \frac{1}{a}e^{ax}$$

$$\int \frac{dx}{1+x^2} = \tan^{-1}x$$

$$\int \frac{dx}{\sqrt{x^2+a^2}} = \ln\left[\sqrt{x^2+a^2} + x\right]$$

$$\int \frac{dx}{\sqrt{a^2-x^2}} = \sin^{-1}\left(\frac{x}{a}\right)$$

$$\int \frac{x}{(x^2+a^2)^{3/2}}dx = -\frac{1}{\sqrt{x^2+a^2}}$$

◆ B.3 ◆   **Power Series**

$$(1+x)^n = 1 + nx + \frac{n(n-1)}{2!}x^2 + \cdots$$

$$e^x = 1 + x + \frac{x^2}{2!} + \frac{x^3}{3!} + \cdots$$

$$\ln(1+x) = x - \frac{x^2}{2} + \frac{x^3}{3} - \frac{x^4}{4} + \cdots$$

$$\sin x = x - \frac{x^3}{3!} + \frac{x^5}{5!} - \cdots$$

$$\cos x = 1 - \frac{x^2}{2!} + \frac{x^4}{4!} - \cdots$$

$$\tan x = x + \frac{x^3}{3} + \frac{2}{15}x^5 + \cdots$$

$$f(x) = f(a) + f'(a)(x-a) + \frac{f''(a)}{2!}(x-a)^2 + \cdots \quad \text{(Taylor series)}$$

$$f(x+\Delta x) \simeq f(x) + \Delta x\frac{df}{dx} \quad (\Delta x \text{ small})$$

Laplace Transforms

$$F(s) = \int_0^\infty e^{-st} f(t)dt$$

$$f(t) = \frac{1}{2\pi i} \int_{Br} e^{st} F(s)ds$$

Br = Bromwich contour integral

$f(t)\,(t > 0)$	$F(s)$
$\dfrac{df}{dt}$	$sF(s) - f(0)$
$\dfrac{d^2 f}{dt^2}$	$s^2 F(s) - [sf(0) + [f'(0)]$
$\displaystyle\int_0^t f(t)dt$	$\dfrac{1}{s}F(s)$
$f(t \pm a)$	$e^{\pm as} F(s)$
$e^{\pm \lambda t} f(t)$	$F(s \mp \lambda)$
$t^n f(t)$	$(-1)^n \dfrac{d^n}{ds^n} F(s)$
$t^{-n} f(t)$	$\underbrace{\int \int \cdots \int}_{n} F(s)(ds)^n$
$\displaystyle\lim_{t\to 0} f(t)$	$\displaystyle\lim_{s\to\infty} sF(s)$
$\displaystyle\lim_{t\to\infty} f(t)$	$\displaystyle\lim_{s\to 0} sF(s)$
$\delta(t)$ (delta function)	1
1 (unit step function)	$\dfrac{1}{s}$
t	$\dfrac{1}{s^2}$
t^n (n = integer)	$\dfrac{n!}{s^{n+1}}$

$t^\alpha (\alpha > -1)$	$\dfrac{\Gamma(\alpha + 1)}{s^{\alpha+1}}$ (Γ = Gamma function)
e^{at}	$\dfrac{1}{s - a}$
te^{at}	$\dfrac{1}{(s - a)^2}$
$\ln t$	$-\dfrac{1}{s}(\ln s - 0.5772 \cdots)$
$\sin \omega t$	$\dfrac{\omega}{s^2 + \omega^2}$
$\cos \omega t$	$\dfrac{s}{s^2 + \omega^2}$
$e^{\alpha t} \sin \omega t$	$\dfrac{\omega}{(s - \alpha)^2 + \omega^2}$
$e^{\alpha t} \cos \omega t$	$\dfrac{s - \alpha}{(s - \alpha)^2 + \omega^2}$
$\dfrac{1}{\sqrt{t}}$	$\sqrt{\dfrac{\pi}{s}}$
$J_0(at)$ (Bessel function)	$\dfrac{1}{\sqrt{s^2 + a^2}}$

For a more complete table, see, for example, "Table of Laplace Transforms" by G. E. Roberts and H. Kaufman (W. B. Saunders Company, Philadelphia, 1966).

References

As waves are ubiquitous in nature, they are the subject of one or more chapters in many books. The following will be useful to the reader who wants an alternative or more advanced treatment of the material covered in this text.

Baldock, G. R., and T. Bridgeman, *Mathematical Theory of Wave Motion*, Ellis Horwood Ltd., Chichester, 1983.

Crawford, F. S. Jr., *Waves*, Berkelcy Physics Course, Vol. 3, Mcgraw-Hill, New York, 1968.

Halliday, D., R. Resnick, and J. Walker, *Fundamentals of Physics*, 8th ed., Wiley, New York, 2007.

Karpman, V. I., *Nonlinear Waves in Dispersive Media*, Pergamon, Oxford, 1975.

Lighthill, J., *Waves in Fluids,* Cambridge University Press, New York, 1978.

Lonngren, K. E., S. V. Savov, and R. J. Jost, *Fundamentals of Electromagnetics with MATLAB*, 2nd ed., SciTech Publishing, Raleigh, 2007.

Morse, P. M., *Vibration and Sound*, 2nd ed., McGraw-Hill, New York, 1948.

Nettel, S., *Wave Physics: Oscillations, Solitons, and Chaos*, 4th ed., Springer Verlag, Berlin, 2009.

Ramo, S., J. R. Whinnery, and T. Van Duzer, *Fields and Waves in Communication Electronics*, 3rd ed., Wiley, New York, 1994.

Remoissenet, M., *Waves Called Solitons*, 3rd ed., Springer Verlag, Berlin, 1999.

Sprott, J. C., *Chaos and Time-Series Analysis*, Oxford University Press, Oxford, UK & New York, 2003.

Whitham, G. B., *Linear and Nonlinear Waves*, Wiley-Interscience, New York, 1973.

Zeldovich, Ya. B., and Yu. P. Raizer, *Physics of Shock Waves and High Temperature Hydrodynamic Phenomena*, Academic, New York, 1966.

Answers to Selected Problems

Chapter 1

5. (a) 6.6 N/m, (b) $-10\cos(2.1t)$ cm, (c) 3.3×10^{-2} J.
6. 0.61 Hz, 0.65 Hz.
8. $\omega = \sqrt{2ag}$. Note that a has dimensions of inverse length, m^{-1}.
11. Yes, 0.28 ms.
12. $B = \gamma q_0/\omega$, $\gamma = R/2L$, $\omega = [(1/LC) - (R^2/4L^2)]^{1/2}$.

Chapter 2

1. (a) 20 m/s, (b) 2 m, π rad/m, (c) 10 Hz, 20π rad/s, (d) positive x, (e) 2 cm.
5. 556 m.
6. 0.52 mm, 0.35 mm.
10. 970 m/s, 910 m/s.
11. 4.8 Hz, 62.8 m.

Chapter 3

8. $\pm(1 + i)/\sqrt{2}$.

Chapter 4

1. 109.5 m/s.
2. (b) $c_w = \sqrt{k_t(\Delta x)^2/I}$.
3. 155 W, 6.3 N s/s $=$ N.
4. kinetic energy $=$ potential energy $= (\sqrt{\pi/2}\rho_l)(\xi_0 c_w)^2/2a$
   momentum $= \sqrt{\pi/2}\rho_l\xi_0^2 c_w/a$.
7. 40.4 m/s.
8. 112 m/s.
11. 22.4 m/s, 3.2 Hz, 7.0 m, 9 mW.
12. 1.3×10^{-11} N/m^2

Chapter 5

1. 3.3×10^3 m/s.
2. 3.8×10^{10} N/m^2.
3. (b) 4.8×10^{-3} J/m, (c) 24 W, (d) 24 W.
5. (a) 1260 m/s, (b) 307 m/s.
7. 880 m.
9. (a) 8.8×10^{-9} m, (b) 9.1×10^{-3} N/m^2, (c) 50 dB.
10. (a) 18 rad/m, (b) 8.1×10^{-7} W/m^2, 59 dB, (c) 2.56×10^{-2} N/m^2.
11. $\xi_{air}/\xi_{water} = 59.4$.

Chapter 6

1. 19.4 N, 8.3 cm.
2. 7.1 Hz.
3. 340 Hz, 680 Hz, ...
4. 4.25 Hz, 12.75 Hz, 21.25 Hz, ...
6. Soft, -0.225 cm, 0.775 cm.
8. $\omega_{l,m,n} = \pi c_s [(l/a)^2 + (m/b)^2 + (n/c)^2]^{1/2}$ where l, m, n are nonzero integers.
9. About 7.5 Hz.
10. (a) 3.95×10^7 & 3.13×10^7 kg/(s m^2), (b) 0.12 & 1.12, (c) 1.3% & 98.7%
13. With $k_1 = \omega/\sqrt{T/\rho_{l1}}$, $k_2 = \omega/\sqrt{T/\rho_{l2}}$, the equation is given by

$$k_2 \tan(k_1 L_1) + k_1 \tan(k_2 L_2) = 0.$$

Chapter 7

1. 0.13 W/m^2, 3.6 km.
3. 1.6×10^{-3} W/m^2, 3.16:1, 23 mi.

Chapter 8

1. (a) 842 Hz, (b) 767 Hz.
2. (a) 843 Hz, (b) 768 Hz.
3. (a) 600.9 Hz, (b) 599.1 Hz.
4. 0.64 m/s.
5. $43°$.
7. 375 m/s.
8. (a) $u/c = 0.16$, (b) receding.

Chapter 9

4. (a) 2.1×10^8 m/s, (b) $115 \, \Omega$.

5. $\varepsilon = 2.04\varepsilon_0$, 0.86 mm.

10. $E_{rms} = 1.94 \times 10^{11}$ V/m.

11. $E_{rms} = 0.76$ V/m.

12. $p = 1.67 \times 10^{-8}$ kg · m/s, $v = 1.67 \times 10^{-5}$ m/s.

25. $\lambda < 3 \times 10^{-10}$ m. X rays have a wavelength shorter than this and can penetrate into copper.

27. For a thickness d, penetration time $\simeq \mu_0 d^2/\eta \simeq 30$ ms. This is the exponentiation time and should be regarded as an order of magnitude estimate.

Chapter 10

1. 4.8×10^{-2} V/m.

3. (a) $3.5 \, \Omega$, (b) 120 A.

5. (a) $a = 1.0 \times 10^{19}$ m/s^2, (b) 6.2×10^{-16} W $= 3.9$ keV/s. This is the initial loss rate. The electron gradually loses energy and the cyclotron radius becomes smaller.

Chapter 11

1. 500 nm $= 5000$ Å.

5. n (film) $= 1.23$, film thickness $= 143$ nm.

8. 24 cm.

9. The peak intensity is quadrupled and the angular spread becomes one half. Check the energy conservation, i.e., explain why the peak intensity is quadrupled even though the amount of light energy passing through the slit is only doubled.

13. (a) 7700/cm.

17. 750 Å, reflection 36%, transmission 64%.

18. $E_\theta(r, \theta) = \dfrac{\sqrt{\mu_0\varepsilon_0}}{2\pi r} \dfrac{I_0}{\sin\theta} \cos\left(\dfrac{\pi}{2}\cos\theta\right)$.

19. $\dfrac{\omega^4 M^2}{12\pi \varepsilon_0 c^5}$.

Chapter 12

1. $25.7°$.

4. A circle around the face center with a radius of 0.89 cm.

7. -20 cm, $m = 2$, virtual, erect.

8. -6.7 cm, $m = 0.67$, virtual, erect,

9. $f = 20$ cm, $i = 30$ cm.

10. No. The dimension along the axis appears as m^2 where m is the magnification in the vertical direction.

11. 50 cm in front of the lens; $m = +1.0$.

12. 11.6 cm to the right of the lens; $m = -0.04$.

14. 15.8 cm to the right of the second surface.

15. $0.125R$ in front of the silvered surface.

17. (a) 30 cm, (b) 16, (c) 19 cm.

18. (a) No, (b) Yes.

22. 2.8×10^8 m/s.

26. incident angle $48.6°$, minimum deviation $37.2°$.

27. 62.8 cm.

28. (a) $i = 12.12.6$ cm, $m = 0.027$, (b) $i = 10.3$ cm, $m = 0.055$. Note a significant change by a factor of 2 in the magnification for a relatively small change in the image location.

29. $f = 33.0$ cm.

30. (c) image at 30 cm to the right of the second lens, $m = -1.5$.

31. effective focal length $= 25.7$ cm, H2 at 34.3 cm to the left of second vertex, H1 at 22.8 cm to the right of the first vertex.

32. glass thickness 2 cm, radius of curvature of the second surface 8.5 cm.

33. Case 1: $x = 23.3$ cm, f (mirror) $= 11.7$ cm, Case 2: $x = 36.7$ cm, f (mirror) $= 18.3$ cm.

34. lens separation $= 54.9$ cm, $m = -23.2$.

35. $m = +8.0$. Move the eyepiece away from the objective by 0.5 mm.

36. $R1 = 6.095$ cm, $R2 = -10.161$ cm.

37. $0.286R$. $f = 1.715R$.

38. $i = -R$ (at the center), $m = +4/3$.

39. Trapezoid. Area 66.1 cm^2.

40. 20.5 cm.

Chapter 13

2. 5.0×10^9/cm^3.

3. (a) 1.0×10^{15} Hz, 2960 Å, (b) 4.1 eV.

4. (a) -13.6 eV, -0.85 eV, (b) 12.8 eV, (c) 970 Å.

5. $2.7 \times 10^{-11} n^2$ (m), $-54.4/n^2$ (eV).

6. 2.7×10^{-12} m.

7. 9.0×10^{22} m/s^2, 4.7×10^{-11} s. This indicates that the classical model of a hydrogen atom consisting of a discrete electron revolving around a proton is unstable against radiation.

Chapter 14

2. $\Delta\omega \simeq 2\pi \times 10^5$ s^{-1} centered at $\omega = 2\pi \times 10^6$ s^{-1}.

3. $f(t) = 0.5 + (4/\pi^2)\left(\cos\omega_0 t + \frac{1}{3^2}\cos 3\omega_0 t + \frac{1}{5^2}\cos 5\omega_0 t + \cdots\right) \omega_0 = \pi \times 10^3$ s^{-1}.

4. $F(\omega) = \sqrt{\pi}ae^{-a^2\omega^2/4}$.

5. $\omega^2/\omega_{pi}^2 = k^2/(k^2 + k_D^2)$.

6. 315 km. Dielectric loss is negligible. In practice, the loss in conductors is more important.

7. $x(t) = (p/\sqrt{Mk})\sin\omega t$, $\omega = \sqrt{k/M}$.

Chapter 15

8. $A_{new} = 16\,A_{old}$.

Index

Printed in the USA
CPSIA information can be obtained
at www.ICGtesting.com
JSHW060426061223
53305JS00001B/1